TRANSPORT AND RELAXATION
IN
RANDOM MATERIALS

TRANSPORT AND RELAXATION IN RANDOM MATERIALS

15-17 October 1985
Maryland, USA

Edited by **J. Klafter**
R. J. Rubin
M. F. Shlesinger

World Scientific

Published by

World Scientific Publishing Co Pte Ltd.
P. O. Box 128, Farrer Road, Singapore 9128.
242, Cherry Street, Philadelphia PA 19106-1906, USA

Library of Congress Cataloging-in-Publication Data

Transport and relaxation in random materials.

  Papers presented at a conference held Oct. 15-17, 1985 at the National Bureau of Standards, Gaithersburg, Md.
  1. Amorphous substances — Congresses. 2. Transport theory — Congresses. 3. Relaxation phenomena — Congresses. 4. Glass-Congresses. 5. Order-disorder models — Congresses. 6. Molecular dynamics — Congresses. I. Klafter, Joseph. II. Rubin, Robert J. III. Shlesinger, Michael F.
QC176.8.A44T73   1986   530.4'1   86-18908
ISBN 9971-50-133-3
ISBN 9971-50-134-1 (pbk.)

Copyright © 1986 by World Scientific Publishing Co Pte Ltd.

All rights reserved. This book, or parts thereof, may not be reproduced in any form or by any means, electronic or mechanical, including photocopying, recording or any information storage and retrieval system now known or to be invented, without written permission from the Publisher.

Printed in Singapore by Kim Hup Lee Printing Co. Pte. Ltd.

INTRODUCTION

It is well known that the properties of ideal crystals can be drastically altered by the introduction of defects. Random amorphous materials in a sense are only comprised of defects, and they offer many new avenues to explore both experimentally and theoretically. The last decade has seen many advances in the physics of random materials. This conference, which was held 15-17 October 1985 at the National Bureau of Standards in Gaithersburg, Maryland, focused on dynamic processes in random materials involving transport, reaction, and relaxation. The purpose of this conference was to bring together researchers in the fields of amorphous semiconductors, polymers, glasses, porous rocks, micelles, and molecular systems to present the similarities and differences of the dynamics in these random media. Emphasis is placed on how sufficient randomness (spatial or energetic) can alter well known physical relationships which apply to homogeneous systems. For example, Debye's law of exponential dielectric relaxation in fluids is replaced by the Williams-Watts law of stretched exponential relaxation in glassy materials. An underlying unity exists in these fields in that many scales (spatial, temporal, energetic, etc.) exist with none of them dominating. This is the paradigm of fractals (in space and time) and their role in dynamic scaling in random materials is a key feature of this book.

CONTENTS

Introduction	v
Program of the Conference	ix
Heat Transport in Rocks, Glasses, and Disordered Crystals R. O. Pohl	1
Nuclear Magnetic Relaxation in Random Porous Materials M. H. Cohen	20
Structure of Small Scale Flow through Porous Media J. F. Willemsen	42
Fractal Surfaces S. Alexander	59
Molecular Diffusion in Porous Structures J. C. Tarczon, A. H. Thompson, W. A. Ellingson, and W. P. Halperin	72
Dynamics in Porous Silica J. M. Drake, W. D. Dozier, and J. Klafter	86
A Fractal Model of Structure, Vibrations, and Spin Relaxation in Paramagnetic Proteins H. J. Stapleton	96
Spin Dynamics in Concentrated Metallic Spin Glasses K. Moorjani, S. M. Bhagat, and M. A. Manheimer	110
Excitation and Recombination of Photocarriers in Trans Polyacetylene Z. Vardeny and E. Ehrenfreund	127
Dispersive Recombination in an Epitaxial Semiconductor M. A. Tamor	142
Changes in Electronic Transport during Structural Relaxation of Glassy Chalcogenides M. A. Abkowitz	152
Models for Dynamically Controlled Relaxation A. Blumen, J. Klafter, and G. Zumofen	163

Fractal Nature of Geometric and Energetic Disorder: Exciton Kinetics in Crystalline Films, Glasses and Membranes *R. Kopelman*	177
Monte-Carlo Simulations: A Means to Gain Insight into Difficult Transport Problems in Random Media *M. Silver, B. Ries, and H. Bassler*	212
Molecular Motion in Glassy Polycarbonate *A. A. Jones, P. T. Inglefield, J. F. O'Gara, and A. K. Roy*	228
Relaxation and Recovery of Glassy Polycarbonate *J. T. Bendler, D. G. LeGrand, and W. V. Olszewski*	240
On the Origin of Non-Exponential Decay Processes in Amorphous Systems with Application to Polymers *E. A. Di Marzio and I. C. Sanchez*	253
Classical and Quantum Ants in a Labyrinth *T. Odagaki*	278
Self-Consistent Mode-Coupling Theory of Excitation Transport in Disordered Media: Application to Anderson Localization and Incoherent Transport *R. F. Loring and S. Mukamel*	292
Carrier Transport in Dynamically Disordered Systems *S. D. Druger*	319
Effective Media Equations for the Conductivity of Macroscopic Mixtures *D. S. McLachlan*	339
Non-Markovian Dynamics of One-Dimensional Trapping Problems *J. Masoliver, B. J. West, and K. Lindenberg*	357
Trapping of Particles and Depolarization of Spins by Random Walks on Lattices *K. W. Kehr, J. K. Anlauf, and R. Czech*	376
On a Generalized Transport Equation for Chromatographic Systems *G. H. Weiss*	394
Diffusion, Anomalous Diffusion, and 1/f-Noise in Chaotic Systems *T. Geisel*	407
Lévy Walks in Chaos and in Turbulence *M. F. Shlesinger and J. Klafter*	408

CONFERENCE ON

TRANSPORT AND RELAXATION PROCESSES IN
RANDOM MATERIALS

October 15-17 1986

Sponsored by:

Joseph Klafter
Corporate Research Science Laboratories
Exxon Research and Engineering Company
Annandale, NJ 08801

Robert J. Rubin
Polymer Science Division
National Bureau of Standards
Gaithersburg, MD 20899

Harvey Scher
SOHIO Research Center
4440 Warrensville Center Road
Cleveland, OH 44128

Michael F. Shlesinger
Physics Divsion (Code 1112)
Office of Naval Research
800 North Quincy St.
Arlington, VA 22217-5000

TUESDAY, OCTOBER 15

A.M.
"Heat Transport in Disordered Crystals, Glasses and Rocks", R.O. Pohl, Cornell University

"A Fractal Model of Structure, Vibrations and Spin Relaxation in Paramagnetic Proteins", H. J. Stapleton, University of Illinois

"Relaxation and Non-Radiative Decay on Fractal Networks", R. Orbach, UCLA

"What are Fractal Surfaces", S. Alexander, Hebrew University

P.M.
"Transport in Dynamically Disordered Lattices", R. Zwanzig, University of Maryland

"Classical and Quantum Ant in a Labryinth", T. Odagaki, Brandeis University

"Models for Dynamically Controlled Relaxation", A. Blumen, Tech. U. Munchen

"Trapping of Particles and Depolarization of Spins by Random Walks on Lattices", K. Kehr, Julich

"Non-Linear Optics and Coherent Transients as a Probe for Disordered Transport", S. Mukamel, University of Rochester

Banquet Lecture: "Occam's Razor and the Mobility of Electrons in Amorphous Materials", Albert Rose, Exxon

WEDNESDAY, OCTOBER 16

A.M. "Dispersive Transport in Amorphous Semiconductors", H. Scher, SOHIO Research Center

"Relaxation and Interlayer Transfer of Photoexcitation in Tetrahedrally Bonded Amorphous Semiconductor Multilayers", T. Tiedje, Exxon

"The Application of Rate Equations to the Interpretation of Hopping Conductivity Data", P. N. Butcher, University of Warwick

"Diffusion, Anonmalous Diffusion and 1/f Noise in Chaotic Systems", T. Geisel, University of Regensberg

P.M. "Transport Processes of Tracer Polymers in Solutions and Melts", H. Yu, University of Wisconsin

"Self-Diffusion in Disordered Systems", P. M. Chaikin, University of Pennsylvania

"A Molecular Level Model for Motion and Relaxation in Glassy Polycarbonate", A. A. Jones, Clark University

"Molecular Motion and Mechanical Relaxation in Polymer Glasses", J. Bendler, General Electric

"Fractal Nature of Geometric and Energetic Disorder: Exciton Kinetics in Semicrystalline Films, Glasses and Membranes", R. Kopelman, University of Michigan

THURSDAY, OCTOBER 17

A.M. "Topology and Morphology of Random, Granular and Porous Materials", M. H. Cohen, Exxon

"Small Scale Structure of Diphasic Flow through Porous Media", J. F. Willemsen, Schlumberger-Doll

"Fractal Pores in Sedimentary Rocks - Implication for Transport Properties", A. H. Thompson, Exxon

"Fractal Asymmetry in Chromatography", G. H. Weiss, National Institutes of Health

P.M. Excitation and Transport of Photocarriers in Conducting Polymers", Z. Vardeny, Technion

"Changes in Electronic Transport During Structural Relaxation of Glassy Chalcogenides", M. A. Abkowitz, Xerox Corporation

"Monte Carlo Simulations: A Means to Gain Insight into Difficult Transport Problems in Random Media", M. Silver, University of North Carolina

"Spin Dynamics in Concentrated Spin-Glasses", K. Moorjani, APL of Johns Hopkins

"On the Origin of Non-Exponential Processes in Amorphous Polymer Systems", E. A. DiMarzio and I. C. Sanchez, NBS

POSTERS

M. Alexanian - (Southern Illinois University), "Glass Transition in the Hard Sphere Fluid"

J. K. Anlauf - (IFF - Kernforschungsanbage Julich), "Some Exact Results in the Theory of Trapping for One-Dimensional Correlated Random Walks"

S. Druger - (Northwestern University), "Carrier Transport in Dynamically Disordered Systems"

J. J. Fontanella - (US Naval Academy), "Effect of High Pressure on Electrical Transport and Relaxation in Amorphous PPO and PPO Complexed with Lithium Salts"

R. Guyer - (University of Massachusetts at Amherst), "The Geometry of Swiss Cheese"

W. Halperin - "Molecular Diffusion in Porous Structures"

S. K. Lyo - (Sandia National Laboratories), "Diffusion and Nonlinear Conduction in a Linear Lattice with Correlated Random Rates: Exact Results and Monte Carlo Studies"

J. Matcha - (University of Massachusetts at Amherst) "Transport in Percolation Systems with a Broad Distribution of Bond Strengths"

D. S. McLachlan - (University of Witwatersrand), "Equations for the Conductivity of Macroscopic Mixtures"

K. S. Mendelson - (Marquette University), "Nuclear Magnetic Relaxation of Water in a Fractal Pore Space"

D. Monroe - (AT&T Bell Laboratories), "Simple Approach to Hopping in Band Tails"

P. Phillips - (MIT), "Tunneling Conduction in Disordered Dissipative Systems"

M. F. Shlesinger (Office of Naval Research) and J. Klafter (Exxon Research and Engineering Company), "The Nature of Temporal Hierarchies Underlying Relaxation in Random Materials"

M. A. Tamor - (Ford Motor Company), "Dispersive Transport in an Epitaxial Semiconductor"

H. J. Yuh - (Xerox Corporation), "Time-Dependent Electrical Transport in Molecularly Doped Organic Polymers"

J. M. Drake – (Exxon Research and Engineering Company), "Characterization of the Fractal Structure of Porous Silicas"

J. P. Stokes, M. D. Robbins and S. Bhattachaya – "1/f" noise in Charge Density Wave Conductors

G. M. Drake, S. K. Sirka, J. S. Huang and J. Klafter – (Exxon Research and Engineering Company), "Characterization of the Fractal Structure of Porous Silicas"

TRANSPORT AND RELAXATION IN RANDOM MATERIALS

HEAT TRANSPORT IN ROCKS, GLASSES, AND DISORDERED CRYSTALS

R. O. Pohl
Laboratory of Atomic and Solid State Physics
Cornell University
Ithaca, NY 14853-2501, USA

ABSTRACT

The thermal conductivities of some disordered crystals and of certain feldspars are very similar to the universal thermal conductivities of structurally disordered, amorphous, solids. The physical origin of the phonon scattering giving rise to this universal behavior is not understood. An exception are the mixed crystals of alkali halides - alkali cyanides. In these crystals, the low temperature (T < 1K) thermal conductivity seems to be determined by resonant scattering of phonons by a very small fraction of the cyanide ions which can undergo quasi-rotational tunneling at these temperatures. The connection is discussed with other disordered crystals, amorphous solids, and feldspars.

1. Introduction

In the temperature range of interest to the geophysicist, i.e. near and above 300 K, the thermal conductivities of all dense rocks and minerals fall within a fairly narrow range; its width is less than one order of magnitude. Nevertheless, for geophysical and geotechnical applications, variations even within such a comparatively narrow range are of considerable importance, and an understanding of the origin of such variations is, therefore, of fundamental as well as of practical interest.

In this paper, we will review the thermal conductivity measurements on rocks and minerals, and some of the important sources of thermal resistance in these solids. We will draw heavily on what has been learned in glasses and in carefully prepared and disordered crystalline solids.

2. Rocks and Minerals

The early measurements by Birch and Clark[1] offer to this date the broadest overview over the thermal conductivity of rocks in the practically important temperature range of 300-600 K. Some of their results are reproduced in Fig. 1. They summarize the variation found in single crystalline and polycrystalline sedimentary rocks (left picture) and igneous rocks (right picture).

Among all the rocks shown, anorthosite, a plagioclase feldspar is especially remarkable. Its conductivity is the lowest measured, and in addition, it increases with increasing temperature, in contrast to all other rocks shown in Fig. 1, which have a negative temperature coefficient.

Birch and Clark noticed the crucial role feldspars played in determining the thermal conductivity of all igneous rocks. They showed that their conductivities were determined by the fraction of feldspars they contained. The lower their feldspar contents (e.g. dunite, bronzitite, hypersthenite), the higher their conductivity. Westerly granite and syenite had a low conductivity, in proportion to their feldspar contents. From the geophysicists point of view, this means that the heat flow in the earth's crust, which consists mainly of igneous rocks, is governed by the feldspars, which make up approximately one half of the igneous rocks.[2] For the solid state physicist, it means that crystal defects, which he would intuitively assume to be of major importance for the thermal resistance in these highly disordered solids, are only of secondary importance. Given the importance of feldspars, the origin of their low conductivity and of their anomalous temperature coefficient, are two important questions. These questions have been studied at Cornell for a number of years[3-8], and will be reviewed below.

In dielectric solids, heat is carried by elastic waves, or phonons in the quantum picture. Their spectrum shifts to lower frequencies as the temperature is lowered. Thus, by measuring the thermal conductivity over a wide temperature range, information can be obtained about the frequency dependence of the scattering, and this,

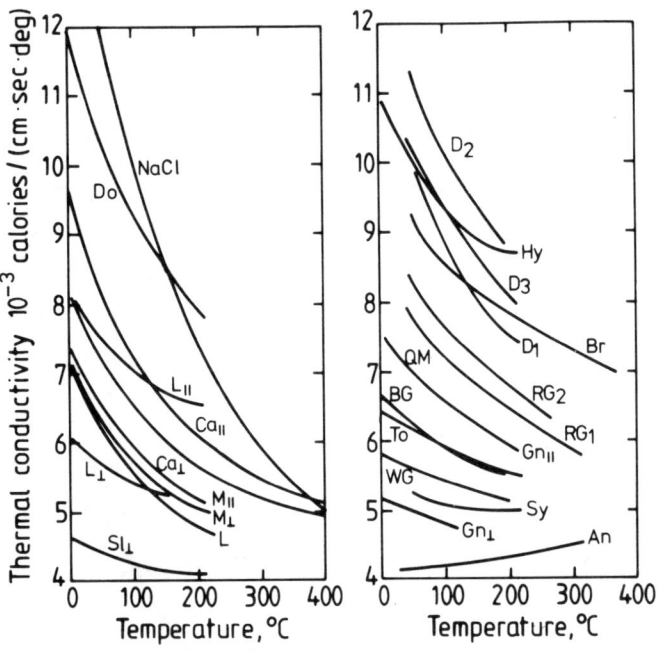

$Sl_\perp$	Slate, Penn., $\perp$ bed-plane.	Sy	Syenite, Ontario.
$L_\perp$	Limestone, Penn., $\perp$ bed-plane.	WG	Westerly Granite
$L_\parallel$	Limestone, Penn., $\parallel$ bed-plane.	To	Tonalite, Calif.
L	Limestone, Solenhofen.	BG	Barre Granite.
$M_\perp$	Marble, Vermont, $\perp$ bed-plane.	QM	Quartz monzonite,
$M_\parallel$	Marble, Vermont, $\parallel$ bed-plane.	RG_1	Rockport Granite 1.
$Ca_\perp$	Calcite, single crystal, $\perp$ optic-axis	RG_2	Rockport Granite 2.
$Ca_\parallel$	Calcite, single crystal, $\parallel$ optic-axis	Br	Bronzitite
Do	Dolomite, Penn.	Hy	Hypersthenite.
NaCl	Halite	D_1	Dunite 1.
An	Anorthosite, Quebec.	D_2	Dunite 2.
$Gn_\perp$	Gneiss, Pelham, $\perp$ bed-plane.	D_3	Dunite 3
$Gn_\parallel$	Gneiss, Pelham, $\parallel$ bed-plane.		

Fig. 1. Thermal conductivity of a variety of rocks and minerals, after Birch and Clark, ref. 1.

in turn, can lead to the identification of the scattering processes or scattering centers.

Fig. 2 summarizes several of these measurements on a number of rocks and minerals, taken from ref. 7. The data are, basically, an

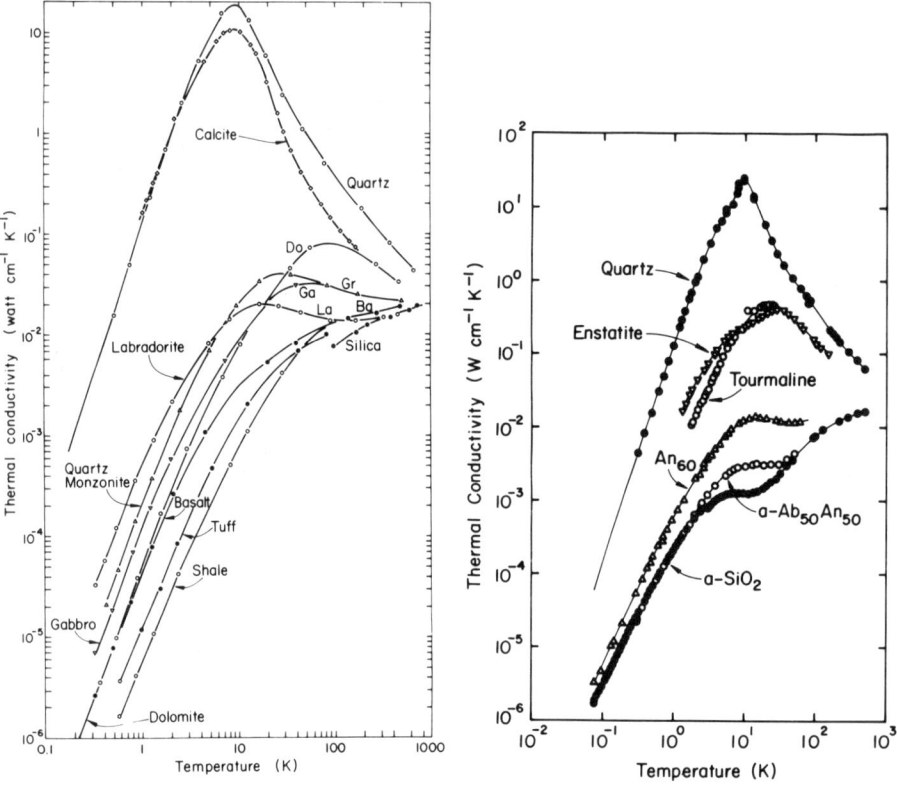

Fig. 2 (left): Thermal conductivity of various rocks, compared with that of nearly perfect single crystals of α-quartz and calcite, and of vitreous silica, Ref. 7. Do: dolomite; La: labradorite, $An_{65.5}$; Gr: quartz monzonite; Ga: gabbro; Ba: basalt.

Fig. 3 (right): The thermal conductivity of a highly perfect single crystal plagioclase feldspar, An_{60}, ref. 10, is very similar to that of an amorphous sample of similar composition, a - $Ab_{50}An_{50}$. For comparison, tourmaline (ref. 11) and enstatite (a pyroxene) (ref. 10) have conductivities of slightly imperfect single crystals.

extension of the measurements shown in Fig. 1 to as low as 0.3 K; some of the high temperature data shown were in fact taken from ref. 1. Yet, by adding the low temperature data, profound differences become apparent, which could not have been guessed by looking at the Fig. 1.

Single crystal quartz and calcite ($CaCO_3$) show the thermal conductivity of nearly perfect crystals. Starting at high temperatures, the conductivity increases with decreasing temperature, as phonon-phonon (Umklapp) scattering processes become less frequent. The conductivity goes through a maximum and then decreases as the third power of the temperature when phonon scattering in the volume of the crystal becomes negligible relative to that of the roughened, and therefore diffusely scattering surfaces (Casimir boundary scattering). In this temperature range, the phonon mean free path approximately equals the sample diameter (~ 5mm in these two cases; the merging of the data at the low temperatures is an accident, though).

Polycrystalline dolomite (a calcium - magnesium carbonate) has a conductivity close to that of calcite only at high temperatures. Below 100 K, the conductivity drops rapidly. The steep drop (approaching a T^3-dependence at the lowest temperatures) suggests scattering by internal surfaces, like grain boundaries. The phonon mean free path calculated from the low temperature data, however, is smaller than the grain size[5], suggesting planar scattering centers within the grains, like twin boundaries, which are also known to scatter phonons with a phonon-wavelength independent mean free path.[9] A similar behavior is seen in the other rocks. Again, phonon scattering by planar defects within the grains, or defects of dimensions large relative to the phonon wavelength, has been inferred from these measurements.[5]

The thermal conductivity of single crystal labradorite, a plagioclase feldspar, shows a rather peculiar behavior in this graph, too. At high temperature, the conductivity is close to that of an amorphous solid, silica (a-SiO_2). As the temperature is lowered, the conductivity cuts through that of all other rocks, and at the lowest temperatures indicates a phonon - wavelength independent path as it could arise from the large ilmenite inclusions observed in this impure sample.[5]

The high temperature thermal conductivity, being equal to that of structurally fully disordered silica, indicates a very high degree of disorder. How can this occur in a crystal? Fig. 3 shows that the thermal conductivity of both solids remains surprisingly similar even at low temperatures, if additional defect scattering is avoided. The sample labelled An_{60} is another single crystal plagioclase feldspar, with the relative composition Albite (Ab, $NaAlSi_3O_8$) to Anorthite (An, $CaAl_2Si_2O_8$) of 40:60 mole %. It contained no visible inclusions. Only in the temperature region around 10 K exceeds its conductivity that of an amorphous (melt-quenched) sample of $Ab_{50}An_{50}$ by a noticeable amount (a factor of ~ 5).[10] Data obtained on two other single crystal silicates, enstatite (a pyroxene)[10] and tourmaline[11] are shown in Fig. 3 for comparison. They show the characteristic behavior of slightly imperfect crystals.

The similarity between the crystalline feldspar and its amorphous counterpart (a-$Ab_{50}An_{50}$) is even more remarkable considering the fact that the thermal conductivity of all structurally amorphous solids is practically identical, as shown in Fig. 4, taken from ref. 12. Independent of their microscopic structure and bonding (covalent, ionic, metallic and polymeric), all glasses have very similar conductivities, which vary as T^2 below 1 K, go through a plateau around 10 K, and then continue to rise (up to their softening points). Even the absolute magnitudes are very close; the curves shown in Fig. 4 show the range of conductivities observed in a large number of studies.

The thermal conductivity of the crystalline plagioclase feldspar lies in the range of all glasses, with the exception of the region of the plateau, where it is approximately ten times higher than that of the average glass. It is interesting to mention that MacDonald and Anderson[13] have recently reported that the real part of the dielectric constant of a labradorite crystal below 1 K also shows the same behavior (a logarithmic variation with T) known for structural glasses.

While a glass-like behavior of the thermal conductivity of crystalline plagioclase feldspars in the middle of the compositional range (~ $Ab_{50}An_{50}$, e.g. (labradorite) is reasonably well established, it is not common to all feldspars, as illustrated in Fig. 5, taken from ref. 7. The conductivities of the albite (Ab_{100}) and the microcline ($KAlSo_3O_8$) crystals are much closer to those of imperfect crystals

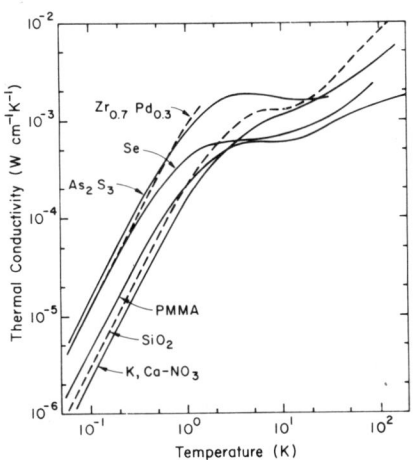 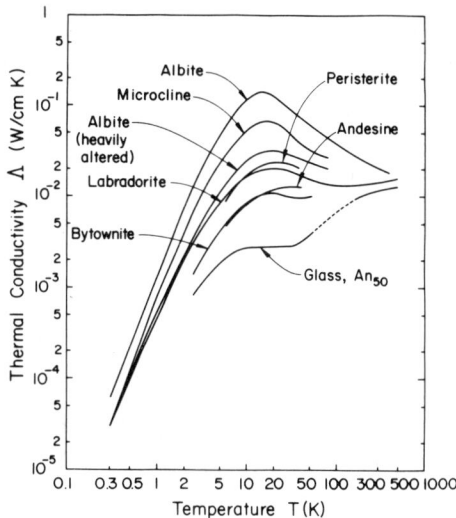

Fig. 4. (left) Thermal conductivity of six different glasses which are characteristic for different bonding types. Ref. 12. Note that the data of the nitrate glass have been extended to 100 K. T. Klitsner, unpublished data.

Fig. 5. (right) Thermal conductivity of different feldspars and a glass of composition $Ab_{50}An_{50}$ (see also Fig. 3). The albite is a low albite, the labradorite is $An_{65.6}$ (same as in Fig. 3). Ref. 7.

than those of glasses. On the basis of the information available at that time, Linvill et al.[7] had suggested that lamellar structures on the scale of tens of Å, as they occur in plagioclase feldspars as a result of intergrowth, could give rise to their low conductivity.

Linvill[10] has since then found low, glass-like, thermal conductivities also in highly perfect single crystal potassium feldspars in which such intergrowths do not occur. We must, therefore, conclude that the origin of the glass-like thermal conductivity observed in certain feldspars is not known. But, what is known about the origin of the thermal conductivty in glasses themselves? This will be reviewed next.

3. Glasses

In a phenomenological theory of the low energy excitations in glasses, proposed by Anderson et al.[14] and by Phillips,[15] the low temperature (T < 1K) thermal conductivity varying as T^2 was explained through resonance interaction with some unspecified tunneling centers which have a uniform density of states (the latter being indicated by a linear specific heat anomaly, also common to all glasses[16]). This model has been succesfully applied to a wide variety of experimental studies, even though the physical nature of these excitations continues to be a mystery.[17] The major puzzle of these excitations is their universality; specifically, how they can lead to the narrow range of the thermal conductivity that has been observed experimentally (Fig. 4). Now, we turn to the temperatures above 1K.

The thermal conductivity in the temperature region of the plateau and above can be described with a phonon scattering rate $\tau^{-1} \propto \omega^4$, i.e. a Rayleigh scattering term (ω is the phonon angular frequency).[16] This is demonstrated in Fig. 6 for the thermal conductivity of a-SiO$_2$. The solid curve drawn through the crystal data is the conductivity Λ computed with the Debye model of thermal conductivity

$$\Lambda = \frac{1}{2\pi^2 v_D} \int_D^{\omega_D} \tau \frac{\hbar^2 \omega^2}{k_B T^2} \frac{e^{\hbar\omega/k_B T}}{(e^{\hbar\omega/k_B T} - 1)^2} d\omega \qquad (1)$$

where $v_D = 4.4 \times 10^5$ cm s^{-1} and $\omega_D = 7.2 \times 10^{13}$ s^{-1} are the Debye velocity and cut-off frequency of quartz, respectively ($\Theta_D = 550$ K), and $\hbar$ and k_B have the usual meaning. The combined phonon relaxation time τ was determined in the usual way as

$$\tau = (\sum_i \tau_i^{-1})^{-1}. \qquad (2)$$

In the crystalline solid, three scattering processes enter into (2), boundary, with the relaxation time τ_{bd}, intrinsic (Umklapp), τ_U, and residual impurity scattering, τ_R. The fit in Fig. 6 was achieved with

$$\tau^{-1} = 9.0 \times 10^5 \text{s}^{-1} + 3.0 \times 10^{-20} \omega^2 T^2 e^{-40K/T} \text{ s K}^{-2}$$
$$+ 6.0 \times 10^{-46} \omega^4 \text{ s}^3. \qquad (3)$$

The second and the third term were freely adjusted, while the first term is the Casimir boundary term for a sample with cross section 5×5 mm^2.

In order to fit the glassy SiO$_2$, the second term in Eq. (3) was removed since Umklapp processes should not occur in solids lacking a periodic structure, and the third term was adjusted for a best fit. However, the Rayleigh scattering rate τ_R^{-1} required was found to be so large that it led for large ω to unreasonably short phonon mean free paths $\ell = v_D \tau$, shorter than the interatomic spacing. This was avoided by choosing instead a Rayleigh term with a cutoff

$$\tau_g = A^{-1} \omega^{-4} + d/v_D, \qquad (4)$$

where d is the Si-Si distance, d = 3.14 Å. The phonon mean free path $\ell_g = v_D \tau_g$ resulting in a good fit to the thermal conductivity of

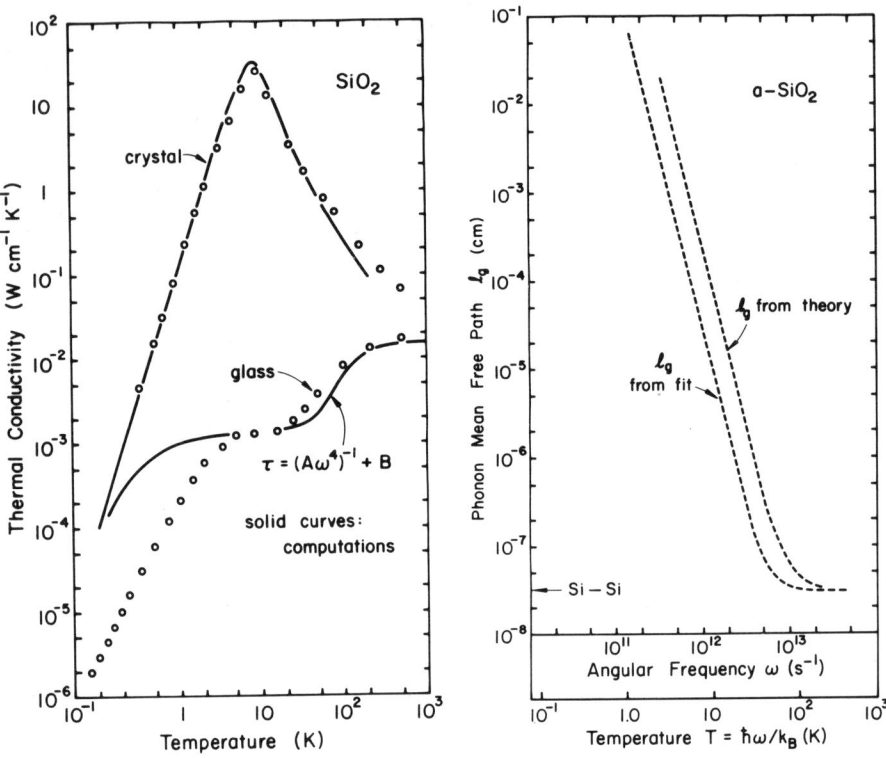

Fig. 6. (left) A Rayleigh scattering term and a high frequency cut-off (eq. 4) give a good description of the high temperature (T > 10 K) thermal conductivity of a-SiO$_2$.

Fig. 7. (right) Phonon mean free path ℓ_g (eq. 4) used to fit the a-SiO$_2$ data (Fig. 6); ℓ_g from theory: Raychaudhuri, ref. 20. The temperature at which phonons of a certain frequency contribute most to the thermal conductivity is roughly one third of the temperature indicated on the lower scale.

a-SiO$_2$ above 10 K (Fig. 6), using $A^{-1} = 0.7 \times 10^{38} s^{-3}$ in eq. (4), is shown in Fig. 7 (below 10 K, the data can be described with the tunneling model[17], but this scattering has been omitted here for simplicity).

Note that for all frequencies $\omega > 10^{13}$ s^{-1}, the phonon mean free path needed in this fit is equal to the Si - Si spacing. It follows that for the majority of the excitations in this glass one can hardly talk about waves, since their mean free paths are shorter than their wavelength. Clearly, the picture of Debye waves has reached its limitations, a fact first pointed out by Kittel;[18] see, however, also the earlier discussion by Birch and Clark.[1]

A natural candidate for Rayleigh scattering in amorphous solids is the structural disorder, although it has often been argued that it does not lead to the required magnitude. A review has been given by Jones et al.[19] However, Raychaudhuri[20] has shown that on the basis of structural and spectroscopic data available for a-SiO$_2$, a scattering strength could be predicted which was very close to that needed to fit the observed thermal conductivity. The term A^{-1} as defined in eq. (4) he derived is equal to 5.6×10^{38} s^{-3}, within a factor of eight from that used to fit the data in Fig. 6. A comparison of the two mean free paths is shown in Fig. 7. Considering the uncertainty involved in determining both quantities, an agreement to within one order of magnitude appears satisfactory.

Nevertheless, the question of the origin of the plateau and the subsequent rise of the thermal conductivity is still a subject of considerable debate. At this conference, Orbach has discussed it in terms of heat transport by fractons, while Graebner and Golding have presented their arguments in favor of phonon localization. Whatever the outcome of this debate as well as that regarding the nature of the tunneling states, we expect that the result will also have to be applicable to the feldspar crystals with glass-like thermal conductivities.

4. Disordered Crystals

We now turn to the question whether glass-like thermal properties can also be found in crystals with relatively simple crystal structure which have been disordered in a controlled and well understood manner. These efforts have been recently reviewed by Anderson.[21] Here, we will discuss only two examples which have recently been studied in considerable detail, and which we believe to be rather well understood.

As introduction, we demonstrate the effect of a large number of monatomic point defects. Figure 8 shows the thermal conductivity of cubic $(KBr)_{1-x}(KI)_x$ for $x = 0$, 1, and 0.47.[22] Pure KBr and pure KI show the characteristic conductivity of highly ordered, crystalline solids. By mixing the Br^- and I^- ions, which have different masses and ionic radii, the conductivity is reduced by about a factor of ten above 1 K, but is still much larger than that of amorphous solids. It also continues to show the negative temperature coefficient characteristic for crystals. Below 1 K, Casimir boundary scattering dominates. The kind of disorder shown in this example leads to point defect scattering, enhanced by resonant mode scattering.[23] No resemblance to glasses is found.

An entirely different situation emerges if molecular impurities, which in solid solution maintain some quasi-rotational degrees of freedom, like OH^-, NO_2^-, and CN^-, are added to alkali halide crystals. Fig. 9 shows the lattice model of $(KBr)_{1-x}(KCN)_x$ as an example. In the dilute limit, $x \ll 1$, phonon resonant scattering by the tunnel-split vibrational ground state[24] leads to a strong depression of the thermal conductivity around 1 K, see Fig. 10, taken from a recent study[25,26]. The strength of the scattering increases in linear proportion with x, the cyanide concentration. However, as x increases above ~ 0.01, the low temperature resonant scattering diminishes, and the thermal conductivity increases with increasing x. Only at higher temperatures, above 10 K, does it continue to decrease. This pattern continues as x approaches 0.25 and 0.50, see

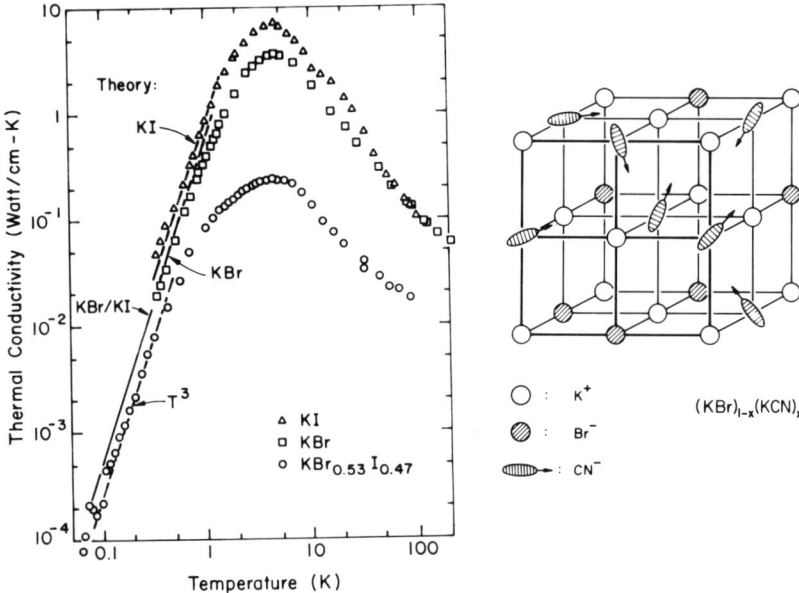

Fig. 8. (left) Thermal conductivities of KBr, KI, and $KBr_{0.53}I_{0.47}$. Also included are Casimir boundary limited T^3 conductivities for each sample calculated from specific heat and sample size. Ref. 22.

Fig. 9. (right) Cubic lattice of $KBr)_{1-x}(KCN)_x$. $x \sim 0.50$.

Fig. 11. For the latter two concentrations, the conductivities show the characteristic behavior found for all glasses; the range of conductivities spanned by all amorphous solids is indicated with the two dashed curves for $a\text{-}As_2S_3$ and $a\text{-}SiO_2$ (taken from Fig. 4). The same conductivity is also observed for $x = 0.70$, see Fig. 12.

For $x = 1$ (pure KCN), the conductivity is again that of a crystal, albeit one with considerable defect scattering. Domain wall

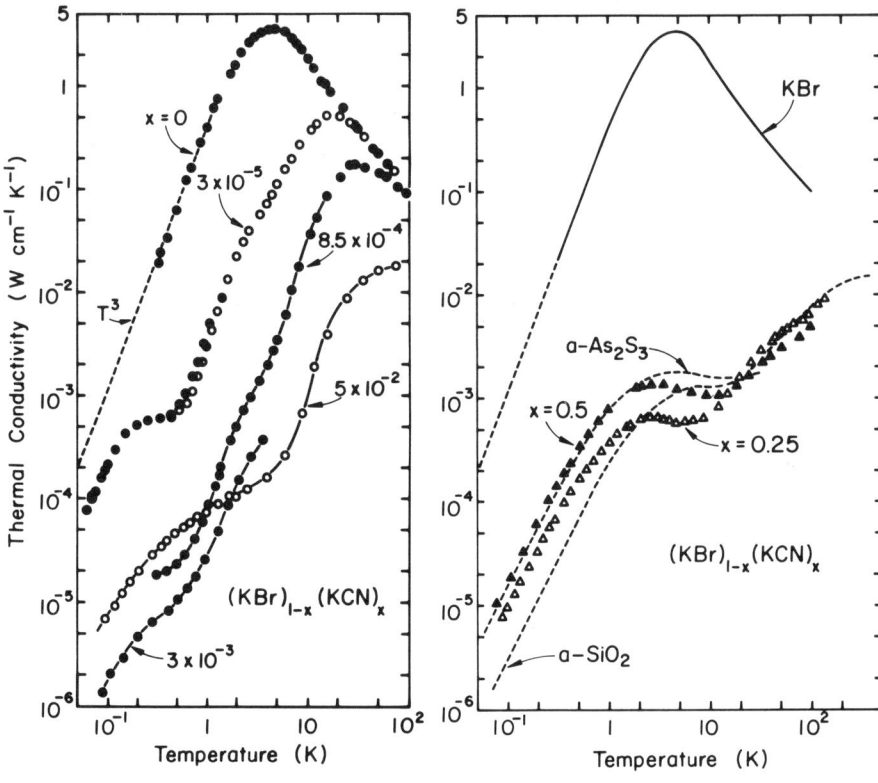

Figs. 10 (left) and 11 (right): Thermal conductivity of single crystal $(KBr)_{1-x}(KCN)_x$. Refs. 25, 26.

scattering in the antiferroelastic KCN has been considered as the cause for the phonon scattering of low temperatures.

We will concentrate on the intermediate composition range, in which the thermal conductivity is clearly like that of all structural glasses. The explanation for the temperature range in which the conductivity varies as T^2, is surprisingly simple; it is supported by measurements of the time dependent specific heat,[26] sound velocity[27] and dielectric constant,[28] all of which show the behavior characteristic for structural glasses.[17] The explanation is as

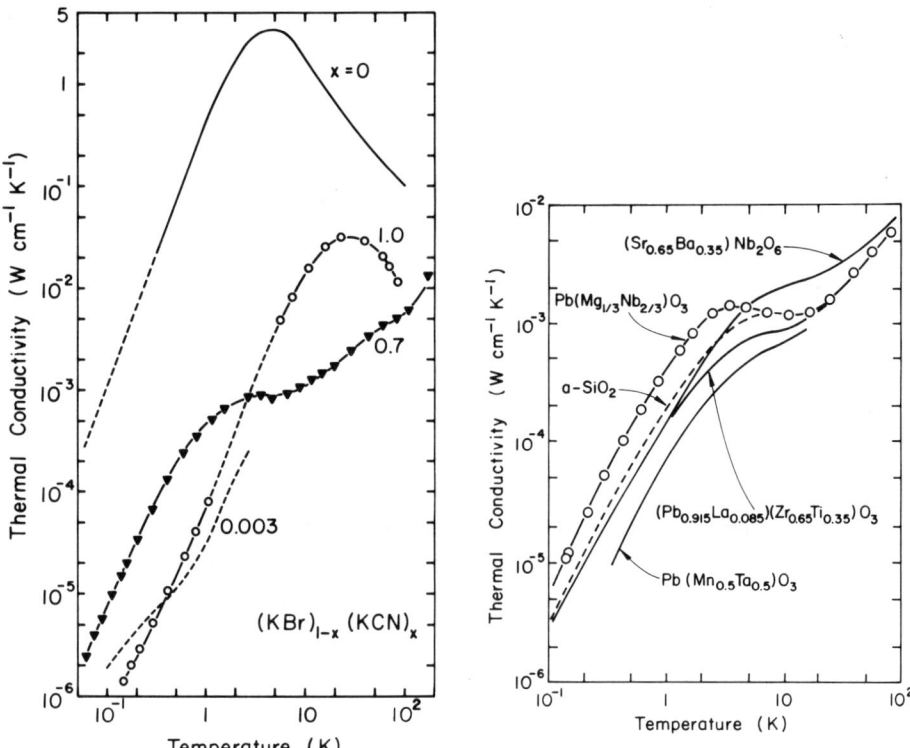

Fig. 12. (left) At $x = 0.70$, the conductivity $(KBr)_{1-x}(KCN)_x$ is still glass-like. Undoped KCN, however has a thermal conductivity of a disordered crystal. Low concentrations of tunneling impurities (example: CN^- at $x = 0.003$, see dashed curve) are unlikely in pure KCN because of the absence of a specific heat anomaly. A possible cause is domain wall scattering. Refs. 25 and 26.

Fig. 13. (right) Four ferroelectrics with glass-like thermal conductivity. Refs. 25 and 31. Single crystals, except for $Pb(Mn_{0.5}Ta_{0.5})O_3$ and $(Pb_{0.915}La_{0.085})(Zr_{0.65}Ti_{0.35})O_3$, which were polycrystalline. a-SiO_2 data, shown for comparison extend all the way to above 100 K, but merge with the $Pb(Mg_{1/3}Nb_{2/3})O_3$ data above 5 K. Similarly, the $(Pb_{0.915}La_{0.085})(Zr_{0.65}Ti_{0.035})O_3$ data (measured to 0.1 K) merge with those of $(Sr_{0.65}Ba_{0.35})Nb_2O_6$ below 1 K. Data points have been omitted for clarity on all but one curve.

follows: Below 1 K, only a very small fraction of the cyanide ions, of the order of 10^{-5}, retain their quasi-rotational tunneling mobility; all others are frozen in because of elastic dipole-dipole interactions. These mobile dipoles have a constant density of states, and scatter phonons with a stress coupling close to that of isolated CN⁻-ions. Sethna and Chow[29] have shown in addition how to determine this mobile fraction from a distribution of potential barrier heights derived from dielectric relaxation measurements above 10 K.[30] Thus, the tunneling model as originally proposed for glasses[14,15] has been applied successfully to the low temperature properties of this crystalline solid. The tunneling entities have been identified for the first time; they are individual cyanide ions, which remain mobile in the random stress field set up by the interacting CN⁻ ion acting as elastic dipoles.

A situation very similar to that described for $(KBr)_{1-x}(KCN)_x$ appears to exist also in certain ferroelectrics[25,31]. Their thermal conductivities are shown in Fig. 13. In these crystalline solids, the order is reduced by the incorporation of a large number of (monatomic) foreign ions, sometimes in conjunction with vacancies. As a result, these crystals do not undergo spontaneous, cooperative ferroelectric phase transitions. Rather, individual unit cells (or perhaps small clusters) will undergo displacive transitions leading to a random polarization, i.e., a glassy polarization phase. Because of the local variations of composition, the transition temperature is spread over a wide range. A more detailed picture of what the tunneling entities may be, how strongly they are coupled to the phonons, and so on, is still missing. Nonetheless, a nearly dense packing of interacting stress dipoles, similar to that found in $(KBr)_{1-x}(KCN)_x$, appears to exist in these ferroelectric cyrstals, too. We mention only in passing that the same kind of dipole configuration can also be envisioned in several other disordered crystals displaying glass-like thermal properties, e.g., Zr-Nb, stabilized ZrO_2, and YB_{66}; we refer to some recent reviews[12,21,32].

5. Summary and Outlook

The study of the glass-like thermal properties of $(KBr)_{1-x}(KCN)_x$ has, for the first time, led to an understanding of the nature of the tunneling centers involved. One may try to extend the picture that has emerged from the study of these mixed crystals to the ferroelectric crystals and to structural glasses. One can envision that glasses also contain a dense packing of interacting stress dipoles, perhaps on account of their random density fluctuations. So far, this picture can only serve as a guide to further experiments. The crucial problem is that of explaining the universality of the glass-like properties. No explanation for this phenomenon has been proposed to date. We must also recognize that the work on disordered crystals has not in any way furthered our understanding of the temperature region above 1 K, in which the thermal conductivity also shows universal behavior.

Finally, we return to the feldspars, the starting point of our discussion. In these solids, alkali and alkaline earth ions are incorporated on interstitial sites of the framework formed by SiO_4 and AlO_4 tetrahedra, in order to compensate for the valence of the Al^{3+} ion. These alkali and alkaline earth ions occupy rather large cages, and might conceivably deform their surroundings to form elastic dipoles interacting with one another. Clearly, the picture must be more complicated, since not all feldspars have the glass-like thermal conductivity. Nevertheless, in looking back at the enhancement of our understanding that has been derived from the study of carefully disordered crystalline solids, we seem justified in our hope that a continuation of these efforts will ultimately also lead to an understanding of amorphous solids and eventually even of the heat flow through the earth's crust.

6. Acknowledgements

The work reviewed in this paper has been largely supported by the National Science Foundation, Grant No. DMR-870-7079. Much of this work has been the subject of the Ph.D. Theses of J. J. De Yoreo and M. L. Linvill. I thank both of them, and also M. Meissner and J. P. Sethna for many stimulating discussions. H. E. Fisher performed the model calculations.

References

1. F. Birch and H. Clark, Am. J. Science 238, 529 (1940).
2. See, e.g., K. C. Condie, Plate Tectonics and Crustal Evolution, Pargamon (1982), Table 4.4, p. 68.
3. R. O. Pohl and J. W. Vandersande, in Alternate Nuclear Waste Forms and Interactions in Geologic Media, CONF-8005107, N. S. Department of Energy, April 1981, L. Boatner, ed., p. 117.
4. M. J. Skvarla, J. W. Vandersande, M. L. Linvill, and R. O. Pohl, in Scientific Basis for Nuclear Waste Disposal, Vol. 3, p. 43, Plenum Press, 1981, J. G. Moore, ed.
5. J. W. Vandersande and R. O. Pohl, Geophys. Res. Lett. 9, 820 (1982).
6. R. O. Pohl and J. W. Vandersande, Mat. Res. Soc. Symp. Proc. Vol. 15 (1983), Elsevier Science Publishing, D. G. Brookins, ed., p. 711.
7. M. L. Linvill, J. W. Vandersande, and R. O. Pohl, Bull. Minéral 107, 521 (1984).
8. M. L. Linvill and R. O. Pohl, in Thermal Conductivity 18, T. Ashworth, ed., Plenum Press, 1985, p. 653.
9. J. W. Vandersande, P. N. Chopra, and R. O. Pohl in Phonon Scattering in Condensed Matter, Solid State Science 51, W. Eisenmenger et al., eds. Springer (Berlin) 1984, p. 182.
10. M. L. Linvill, Ph.D. Thesis, Cornell University, in preparation.
11. W. N. Lawless and R. K. Pandey, Solid State Commun. 52, 833 (1984).
12. R. O. Pohl, J. J. De Yoreo, M. Meissner, and W. Knaak, in Physics of Disordered Materials, D. Adler, H. Fritzsche, and S. R. Ovshinsky, eds., Plenum 1985, p. 529.
13. W. M. MacDonald and A. C. Anderson, Solid State Commun. 53, 343 (1985).
14. P. W. Anderson, B. I. Halperin, and C. M. Varma, Philos. Mag. 25, 1 (1972).
15. W. A. Phillips, J. Low Temp. Phys. 7, 351 (1972).
16. R. C. Zeller and R. O. Pohl, Phys. Rev. B 4, 2029 (1971).
17. Reviewed in Topics in Current Physics, Vol. 24: Amorphous Solids, Low Temperature Properties, W. A. Phillips, ed., Springer (Berlin) 1981.
18. C. Kittel, Phys. Rev. 75, 972 (1949).
19. D. P. Jones, N. Thomas, and W. A. Phillips, Philos. Mag. B 38, 271 (1978).
20. A. K. Raychaudhuri, Ph.D. Thesis, Cornell University, 1980.
21. A. C. Anderson, Phase Transitions 5, 261 (1985).
22. B. D. Nathan, L. F. Lou, and R. H. Tait, Solid State Commun. 19, 615 (1976).

23. R. O. Pohl in *Localized Excitations in Solids*, Plenum Press (1968), R. F. Wallis, ed., p. 434.
24. V. Narayanamurti and R. O. Pohl, Rev. Mod. Phys. **42**, 201 (1970).
25. J. J. De Yoreo, Ph.D. thesis, Cornell University, 1985, unpublished.
26. M. Meissner, W. Knaak, J. P. Sethna, K. S. Chow, J. J. De Yoreo and R. O. Pohl, Phys. Rev. B **32**, 6091 (1985); and J. J. De Yoreo, W. Knaak, M. Meissner, and R. O. Pohl, Phys. Rev. B, to be published.
27. J. F. Berret, P. Doussineau, A. Levelut, M. Meissner, and W. Schön, Phys. Rev. Lett. **55**, 2013 (1985).
28. D. Moy, J. N. Dobbs, and A. C. Anderson, Phys. Rev. B**29**, 2160 (1984); the dielectric data are also discussed in ref. 25.
29. J. P. Sethna and K. S. Chow, Phase Transitions **5**, 315 (1985).
30. N. O. Birge, Y. H. Yeong, S. R. Nagel, S. Battacharya, and S. Susman, Phys. Rev. B **30**, 2306 (1984).
31. J. J. DeYoreo, R. O. Pohl, and Gerald Burns, Phys. Rev. B **32**, 5780 (1985).
32. P. R. H. Türkes, E. T. Swartz, and R. O. Pohl in *Physics of Borides*, Proceedings, International Conference, Albuquerque, NM, July 1985, D. Emin, ed.

NUCLEAR MAGNETIC RELAXATION IN RANDOM POROUS MATERIALS

Morrel H. Cohen
Exxon Research and Engineering Company
Clinton, New Jersey 08801
U.S.A.

ABSTRACT

Relaxation of nuclear magnetism in a liquid within the pore space of a porous material yields interesting geometric information when both relaxation at the liquid-solid interface and nuclear exchange between surface and bulk are rapid relative to bulk relaxation. When the diffusive coupling between adjacent pores is large, the relaxation is exponential with a relaxation rate linearly related to the ratio of the total interfacial area to the pore volume. When it is weak, the relaxation is nonexponential, each pore relaxing independently with a rate linearly related to its surface to volume ratio. The Laplace transform of the relaxation function then gives the probability distribution of the surface to volume ratios of the individual pores. The general theory from which the above conclusions follow is reviewed in the present paper. The theory is then applied to the interpretation of recent measurements of Schmidt, Velasco, and Nur. Evidence for strong coupling, for weak coupling, for the transition between strong and weak coupling, for inhomogeneity above the scale of the grain size, for fractal character, for wetting, and for the persistence of microscopic surface films is drawn from their data, illustrating the utility of nuclear magnetic relaxation studies for the characterization of porous material.

1. Introduction

Consider a multiphase heterogeneous material with at least one homogeneous phase. The particular case we shall have in mind in what follows will be that of a solid porous material containing liquid within its pore space as, e.g. in a sedimentary rock, but the theory we shall present is of much more general applicability. Nuclear magnetic relaxation, e.g. free induction decay, within the liquid can yield information about the pore-space geometry when the relaxation in the bulk is much slower than at the liquid-solid interface, when the mean distance a nucleus would diffuse in the absence of the interface is larger than the mean distance from a point within the pore to the interface, and when the rate of detachment of nuclei from the interface is much larger than the bulk relaxation rate.

These conditions are in fact realizable in actual materials, e.g. for protons in water within the pores of sedimentary rocks. Accordingly, we shall review and extend the theory and apply it to the interpretation of recent data by Schmidt, Velasco and Nur[1] in order to illustrate the utility of nuclear magnetic relaxation studies for the characterization of porous materials.

The subject has a twenty-year history[1-16]. Initially, attention was focussed on oil-field rocks[1-7], but application has also been made to biological systems[8,9] and to oil and water emulsions[10]. Detailed theoretical analysis is more recent, starting initially with the study of single pores by Brownstein and Tarr[11], followed by a detailed examination of the general theory by Cohen and Mendelson[12], a more careful treatment of nuclear desorption by Mendelson[13], further work on single pores by Siddique[14], a study of two coupled spherical pores by Mendelson[15], and an examination of fractal pores by Mendelson[16]. In the present paper, we shall focus on reviewing and extending the theory developed in Refs. 12, 13 and 15 and applying it to the interpretation of the data reported in Ref. 1.

2. The General Theory

We begin with a few definitions. M_b is the bulk nuclear magnetization of the resonant species, and M_s is the surface magnetization of that species. The surface magnetization is in units of magnetic moment per unit area even though it is comprised of the net magentic moment of all nuclear spins associated with the surface, whether they are involved in solvation, physisorption or chemisorption, or are merely in the bulk liquid sufficiently close to electron spins in the surface region to experience an enhanced relaxation rate. $D_{b,s}$ is the bulk or surface diffusion coefficient. In the bulk we are dealing with mass diffusion and not spin diffusion, the latter being orders of magnitude smaller for nuclei in liquids. On the surface, the diffusion process could be mass or spin diffusion. Both, however, are sufficiently slow compared to bulk diffusion in the liquid that D_s will ultimately be set equal to zero. $T_{1b,s}$ are the bulk and surface longitudinal relaxation times. The magnetizations in equilibrium with a magnetic field H are given by

$$M_{b,s}^{\infty} = \chi_{b,s} H, \qquad (1)$$

where $\chi_{b,s}$ is the volume (surface) spin susceptibility. The $\chi_{b,s}$ are proportional to the respective spin densities $N_{b,s}$ so that

$$\frac{\chi_s}{\chi_b} = \frac{N_s}{N_b} \equiv \ell \qquad (2)$$

For the case of protons and a completely hydrated or hydroxylated surface, ℓ is of order Å. Finally, γ is the rate per nucleon of transition from the surface compartment into the bulk compartment.

The Bloch equations governing the time and space dependence of M_b and M_s are

$$\frac{\partial M_b}{\partial t} = D_b \nabla_3^2 M_b - (M_b - M_b^{\infty})/T_{1b} \text{ in } V \qquad (3)$$

$$\frac{\partial M_s}{\partial t} = D_s \nabla_2^2 M_b - (M_s - M_s^{\infty})/T_{1s} \text{ on } S \qquad (4)$$

subject to the boundary condition at the pore-solid interface

$$-D_b \hat{n} \cdot \vec{\nabla}_3 M_b(S) = \gamma [\ell M_b(S) - M_s] \text{ at } S \qquad (5)$$

and the initial conditions

$$M_b(t=0) = M_o \qquad (6)$$

$$M_s(t=0) = M_o \qquad (7)$$

In (3), V is the pore volume, and S is the interface in (4) and (5). In (5), $\hat{n}$ is the outward unit normal of S, M_0 is spatially uniform in (6) and (7), and H, implicit in M_b^∞ in (3) and M_s^∞ in (4) through (1), is both spatially uniform and time independent.

Our problem is to solve (4) and (5) subject to (6), (7) and (8) in sufficient detail to obtain the measurable quantity

$$\langle M \rangle = \langle M_b \rangle_V + \langle M_s \rangle \frac{S}{V} \qquad (8)$$

as a function of time. In (8), $\langle M \rangle$ is the total magnetic moment per unit volume of the sample, $\langle M_b \rangle_V$ is the average of M_b over the pore volume V, and $\langle M_s \rangle$ is the average of M_s over the area of the interface S. From $\langle M \rangle$ we can construct the relaxation function

$$g(t) \equiv \frac{\langle M(t) \rangle - M_\infty}{M_0 - M_\infty} \qquad (9)$$

and express it in terms of the probability distribution P(R) of relaxation rates R,

$$g(t) = \int dR \; P(R) \; e^{-Rt}, \qquad (10)$$

where

$$g(0) = \int dR \; P(R) = 1 \qquad (11)$$

It is straight forward to obtain a formal expression for P(R), via

$$P(R) = \frac{1}{V} \sum_\lambda [\int_V d^3r \; U_\lambda(\vec{r})]^2 \; \delta(R - R_\lambda) + O\left(\frac{\ell S}{V}\right) \qquad (12)$$

In (12), the correction terms of order $\ell S/V$ are small because

$$\ell S/V \ll 1, \qquad (13)$$

as will become clear below, and

$$R_\lambda = w_\lambda + \frac{1}{T_{1s}}, \tag{14}$$

$$\nabla^2 U_\lambda + (w_\lambda/D) U_\lambda = 0 \text{ in } V \tag{15}$$

$$(D/\ell)\, \hat{n} \cdot \vec{\nabla} U_\lambda - (w_\lambda - 1/\tau) U_\lambda = 0 \text{ on } S \tag{16}$$

$$\frac{1}{\tau} \equiv \frac{1}{T_{1s}} - \frac{1}{T_{1b}} \tag{17}$$

In (12)-(17), we have supposed that D_s is negligible and $\gamma\tau \gg 1$.

We see from (10) that $g(t)$ is the Laplace transform of $P(R)$. Thus, if $g(t)$ is measured for a given material, $P(R)$ can be obtained by inverting (10). Our present task is to learn what, if any, geometric information about the pore space is contained in $P(R)$.

3. The Quasiuniform Case

Knowledge of the numerical values of the parameters facilitates the extraction of geometric information from $P(R)$. For protons in pure H_2O at room temperature, D_b is 2.5×10^{-5} cm^2/sec and T_{1b} is 3.27 sec[1]. We have obtained a value of T_{1s} of $\simeq 10^{-4}$ sec from our interpretation of the data of Ref. 1 (see below) for protons at a typical rock-water interface. As discussed above, we expect ℓ to be of order an Angstrom. Typical values of average grain sizes R_g in sedimentary rocks range from 100 Å to 100 μ, with values of order 1 μ quite common. Because the porosity of a sedimentary rock must be less than about 40%, it is clear that the mean shortest distance from a point in the pore space to the interface must be less than the grain size.

T_{1s} and T_{1b} are the bounds of the spectrum of relaxation times, i.e.

$$T_{1s} < T_1 < T_{1b} \tag{18}$$

The mean square distance of diffusion of a proton in water before appreciable relaxation has occurred is $6 D_b T_1$, which satisfies the inequality

$$6 D_b T_1 > 6 D_b T_{1s} \simeq 1.5 \times 10^{-8} \text{ cm}^2 \simeq (1\ \mu)^2 \tag{19}$$

The magnetization will be spatially almost uniform within the radius $\sqrt{6 D_b T_1}$, which, according to (19) can be substantially larger than the mean shortest distance to the interface. Thus, it is possible that diffusion maintains the spatial uniformity of the magnetization within individual pores. On the other hand, it is possible, according to (19) that the magnetization can vary from pore to pore.

Let us suppose for now that the latter does not happen and that the magnetization is practically uniform within the entire pore space. It follows[12] that

$$g(t) = e^{-t/T_1}. \quad (20)$$

$$P(R) = \delta(R - 1/T_1) \quad (21)$$

$$\frac{1}{T_1} = [\frac{1}{T_{1b}} + \frac{\ell S}{V} \frac{1}{T_{1s}}]/[1 + \frac{\ell S}{V}] \quad (22)$$

$$\frac{1}{T_1} \simeq \frac{1}{T_{1b}} + \frac{\ell S}{V} \frac{1}{\tau} \simeq \frac{1}{T_{1b}} + \frac{\ell S}{V} \frac{1}{T_{1s}} \quad (23)$$

Thus, if one finds an exponential $g(t)$, one can be assured that one is in this quasiuniform regime and obtain a value of $1/T_1$ which contains information on the surface to volume ratio S/V of the pore space, Eq. (23).

Having measured $1/T_1$ and $1/T_{1b}$, one can extract via (23), a value for $(\ell S/V)(1/T_{1s})$. If the pore liquid in fact wets the solid, as in water-wetting rocks, there is a way to determine $1/T_{1s}$ experimentally. Let $\mathscr{S}$ be the degree of saturation of the pore space by the liquid, the water saturation in our example. As one reduces the water saturation towards zero, there remains only a surface film essentially equivalent to the interfacial layer of effective width ℓ which is present when $\mathscr{S} = 1$. Thus if one tracks the upper bound of the relaxation spectrum as $\mathscr{S}$ approaches zero, one expects to see a peak emerge at $R = 1/T_{1s}$. Such a peak is in fact observed in Ref. 1, and from it we learn that $1/T_{1s} = 1.4 \times 10^{-4}$ sec for protons at the rock-water interface in berea sandstone, as discussed below.

Having determined $1/T_{1s}$, one now has a value for $\ell S/V$, the ratio of the total number of nuclei in S to that in V. One can go further and obtain S/V only if one knows the value of ℓ. Estimates of ℓ can be obtained from other methods of measuring S/V, e.g. BET adsorption isotherms. Once ℓ is known for a particular mineral, it becomes portable from rock to rock.

For the magnetization to be quasiuniform, the protons must be able to diffuse through many pores before relaxation occurs. In that case, the relevant diffusion coefficient is the macroscopic diffusion coefficient D_M', smaller than the bulk diffusion coefficient D by a factor f_M' which takes into account the fact that the pore space occupies only a fraction of the actual volume of the material and the constrained diffusion paths are tortuous. The condition for uniformity is thus

$$(6 D f_M' T_1)^{1/2} > R_g \qquad (24)$$

For diffusion entirely without relaxation, i.e. simple self diffusion when γ is large, the relevant diffusion coefficient differs some what from the D_M' entering our considerations, it has the form

$$D_M = D f_M, \qquad (25)$$

where f_M is the same quantity which enters the macroscopic conductivity

$$\sigma_m = \sigma_w f_m, \qquad (26)$$

σ_w being the bulk conductivity of the water in the pore space and the rock grains being insulators. f_M differs from f_M' because the boundary condition from which (25) or (26) is derived just involves setting the normal derivative equal to zero instead the more complex condition (4) and (5). However, the difference cannot be large, and f_M provides us with a realistic estimate of f_M'.

For porous rocks,

$$f_M = \phi^m \qquad (27)$$

holds[17] with m roughly 2. Typical values of the porosity are of order 0.1. With the resulting values of f_M, we can estimate upper limits to R_g, R_g^M, for uniformity for various values of T_1.

The condition (24) for overall uniformity of the magnetization is more severe than the condition for local uniformity within a pore, which is

$$(6\, Dt_1)^{1/2} > d, \qquad (28)$$

where d is the mean shortest distance from a point in the pore space to the interface. In Table 1, we list values of R_g^M and d^M, the upper limit to d for local uniformity, for the expected range of T_1 values. Note that for the assumed values of $f_M' = f_M = \phi^m = 10^{-2}$, $d^M = 0.1\, R_g^M$.

Table 1: Values of G_g^M and d^M

T_1 (sec)	R_g^M (μ)	d^M (μ)
10^0	4	0.4
10^{-1}	1	0.1
10^{-2}	0.4	0.04
10^{-3}	0.1	0.01

We conclude from Table 1 that global uniformity within the entire pore space is a realizable possibility in actual sedimentary rocks. We also conclude that (24) can be violated while (28) is still satisfied so that one can have local uniformity within a given pore with variation of the magnetization from pore to pore. The smaller the value of f_M' and therefore of the porosity ϕ, the easier is this case to realize. However, before we can set up a detailed theory of the only locally uniform case, we need to understand more about the topology and the geometry of the pore space.

4. The Topology and Geometry of the Pore Space in Porous Materials

The pore space can be mapped into a graph which is topologically equivalent to it by a process of skeletization[18,19]. In the special case that the solid material is granular, as for a sedimentary rock, the graph is comprised of space-filling polyhedral cells bounded by nonplanar faces. The mapping process yields a precise definition of a grain, and each cell contains one grain. In well consolidated material, each face contains one or more contacts between grains, but in less well consolidated material, material of higher porosity, there can be faces without contacts. The perimeter of each face consists of curvilinear edges. Each edge is an intersection of three or more faces, associated with three or more contacts in well consolidated material, material of relatively low porosity. Each edge passes through a channel in the pore space. The corners of the polyhedra are vertices at which four or more edges intersect to form a vertex. Each vertex lies within a pore chamber connected by four or more channels to its neighboring pore chambers.

The partitioning of the pore space into channels and chambers is arbitrary. The pore space can be divided entirely into chambers partitioned from one another by surfaces passing through narrow throats along the channels. Alternatively, the pore space can be partitioned entirely into a network of channels, abutting upon one another in the vicinity of the vertices. Or, the pore space can be partitioned partly into chambers and partly into channels connecting them. The partitioning can be selected on the basis of convenience for the study of a particular property or behavior of the material. The case of nuclear magnetic relaxation, the convenient partitioning is into chambers or pores only. However, topology need not be enough. The cross section of a channel may show multiple minima in areas along the edge passing through the channel. If a minimum area is small enough, a criterion to be quantified in the next section, it must be chosen as a partition between two pores. Thus a channel can be divided into a string of two-fold coordinated pores. The entire pore space is thus divided into pore chambers around vertices connected by at least four strings of pore chambers to similar pore chambers around vertices. The strings may be decorated with disjoint trees or pore chambers, as may the pore chambers around the vertices. In the following, we shall substitute the term pore for pore chamber.

5. Coupled but Distinct Pores

Consider first an isolated pore bounded by a closed interface. If $(6\ DT_1)^{1/2}$ substantially exceeds any linear dimension of the pore, the magnetization within the pore will be uniform during the time over which appreciable relaxation has taken place. The considerations of Section 3 hold, the magnetization is quasiuniform and

$$g(t) = e^{-t/T_1}, \qquad (29)$$

where

$$\frac{1}{T_1} = \frac{1}{T_{1b}} + \frac{\ell S}{V}\frac{1}{\tau}. \qquad (30)$$

In (3), S and V are the surface and volume of the isolated pore, not of the entire pore space as in Section 3.

Now consider two isolated pores, 1 and 2, in contact over a surface S_{12} which partitions them, as shown in Figure 1. S_{12} is the partitioning surface of minimum area and will not be planar in general, it defines the throat between the pores. Define the volume average magnetization within each pore as

$$M_n(t) = M_i(t) - M^\infty \qquad (31)$$

$$M_i(t) = \frac{1}{V_i} \int_{V_i} d^3r\, M(\vec{r},t), \qquad (32)$$

where $i = 1,2$. If S_{12} were zero, then the longitudinal relaxation times T_i of the individual pores would not be equal if $S_1/V_1 \neq S_2 V_2$ according to (30). In that case the relaxation function for the total magnetization would not be exponential. Suppose now that S_{12} is increased gradually from zero. It is clear that at least initially the nonexponential character remains, and g(t) contains at least some information about the geometric features of the individual pores. If $S_{12} \ll S_1, S_2$, the pores remain distinguishable, but become coupled.

We have proposed[12], and it has been shown rigorously for the case of two spherical pores[15,20], that

$$\frac{dm_1}{dt} = -\frac{m_1}{T_1} + K_{12}(m_2 - m_1) \qquad (33a)$$

$$\frac{dm_2}{dt} = -\frac{m_2}{T_2} + K_{21}(m_1 - m_2), \qquad (33b)$$

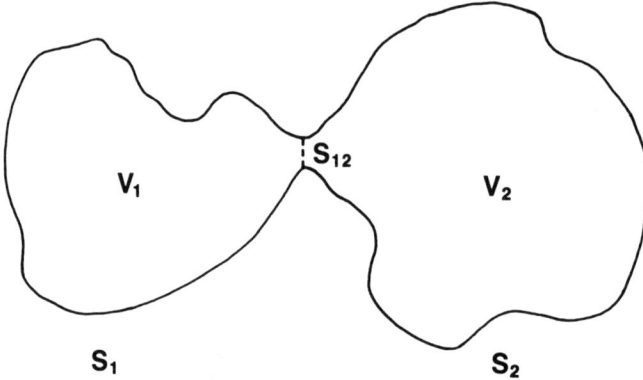

Figure 1. Two Coupled Pores 1 and 2 with Volumes V_1 and V_2, Solid-Pore Interfaces S_1 and S_2, and Partitioning Surface S_{12}

where

$$\frac{1}{T_i} = \frac{1}{T_{1b}} + \frac{\ell\, S_i'}{V_i}\frac{1}{\tau}, \quad i = 1, 2 \tag{33c}$$

$$K_{12} = \alpha\ell\,\frac{S_{12}}{V_1}\frac{1}{\tau} \tag{33d}$$

$$K_{21} = \alpha\ell\,\frac{S_{12}}{V_2}\frac{1}{\tau} \tag{33e}$$

$$S_i' = S_i - (\alpha-1)\,S_{12} \tag{33f}$$

Eqs. (33) hold provided S_1, $S_2 \gg S_{12}$, i.e. the pores remain distinct. This is essentially the same as the condition that $1/T_i \gg K_{12}$ and K_{21}, $i = 1, 2$. α is a number which depends on the details of the pore geometry; its value is 2 for spheres[20].

Eqs. (33a,b) are easy to solve subject to the initial condition $m_1(t = 0) = m_2(t = 0) = m_0$. The results show two very interesting limiting cases. In the weakly coupled case,

$$\sqrt{K_{12}K_{21}} \ll |(\frac{1}{T_1} + K_{12}) - (\frac{1}{T_2} + K_{21})|, \tag{34}$$

each pore relaxes independently, with the relaxation times

$$\frac{1}{T_1'} = \frac{1}{T_{1b}} + \ell\,\frac{(S_1 + S_{12})}{V_1}\frac{1}{\tau} \tag{35a}$$

$$\frac{1}{T_2'} = \frac{1}{T_{1b}} + \ell\,\frac{(S_2 + S_{12})}{V_2}\frac{1}{\tau} \tag{35b}$$

That is, the surface to volume ratio which enters the relaxation rate of an individual pore is the pore-solid interfacial area augmented by the area of the throat which, however, is a small correction. In the strongly coupled case,

$$\sqrt{K_{12}K_{21}} \gg 1 \,(\frac{1}{T_1} + K_{12}) - (\frac{1}{T_2} + K_{21})|, \tag{36}$$

the two pores relax at the same rate and the magnetization remains the same in both pores. In that case, an argument like that in §3 leads to the value

$$\frac{1}{T_1} = \frac{1}{T_{1b}} + \ell \frac{S_1' + S_2'}{V_1 + V_2} \frac{1}{\tau} \qquad (37)$$

for the relaxation time, as in §3. The correct answer should contain the S_i instead of the S_i', indicating a small, first-order error in the analysis.

Eqs. (33) generalize straightforwardly to the case of the coupled network of pores described in §4:

$$\frac{dm_i}{dt} = -\frac{m_i}{T_i} + \sum_j{}' K_{ij}(m_j - m_i) \qquad (38a)$$

$$m_i(t = 0) = m_o \qquad (38b)$$

$$\frac{1}{T_i} = \frac{1}{T_{1b}} + \ell \frac{S_i'}{V_i} \frac{1}{\tau} \qquad (38c)$$

$$S_i' = S_i - \sum_j (\alpha_{ij} - 1) S_{ij} \qquad (38d)$$

$$K_{ij} = \alpha_{ij} \ell \frac{S_{ij}}{V_i} \frac{1}{\tau} \qquad (38e)$$

In Eqs. (38), m_i is the mean magnetization in the ith pore, the summation is over nearest neighbor pores j of i, and S_{ij} is the area of the throat between pores i and j. For the pores to be distinct, each S_{ij} should be $\ll$ each S_i. Once again, there are weak and strong coupling limits. The weak-coupling limit occurs when the mean square value of the matrix elements K_{ij} is small compared to the variance of the $\frac{1}{T_i} + \sum_j K_{ij}$, and in the strong-coupling limit the inequality reverses. In the weak-coupling limit, each pore relaxes with the rate

$$\frac{1}{T_i} = \frac{1}{T_{1b}} + \ell \frac{S_i''}{V_i} \frac{1}{\tau}, \qquad (39a)$$

where

$$S_i'' = S_i + \sum_j S_{ij}. \qquad (39b)$$

Thus, the pore behaves as though it is bounded by a fictitious closed surface; the open surface S_i is closed by the addition of the throat surfaces S_{ij}. The probability distribution of relaxation rates $R \sim \frac{1}{T_{1b}}$ immediately yields the probability distribution of surface to volume ratios S''/V for the individual pores up to the scale factor ℓ/τ. As discussed in §3, τ can be measured, but the problem of determining ℓ remains open though accessible. In the strong-coupling limit, a single relaxation time is obtained

$$\frac{1}{T_1} = \frac{1}{T_{1b}} + \ell \frac{S'}{V} \frac{1}{\tau} \tag{40a}$$

$$S' = \sum_i S'_i \tag{40b}$$

$$V = \sum_i V_i \tag{40c}$$

The correct result, given in §3 contains S instead of S', indicating as before a small first order error in the analysis.

Before going on to the analysis of actual data, we should mention here Mendelson's result[13] for a single spherical pore. He relaxed the condition that $\gamma\tau \gg 1$ and obtained a result more general than (23) or (30),

$$\frac{1}{T_1} = \frac{1}{T_{1b}} + \ell \frac{S}{V} \frac{1}{\tau + 1/\gamma} \tag{41}$$

When $\gamma\tau \gg 1$ holds, (41) reduces to the previous expression. However, when $\gamma\tau \ll 1$ holds, (41) reduces to

$$\frac{1}{T_1} = \frac{1}{T_{1b}} + \frac{\ell S}{V} \frac{1}{\gamma} \tag{42}$$

so that the relaxation rate is determined by the nuclear exchange rate. In general, we must require that

$$T_{1b} \gtrsim \tau + 1/\gamma \tag{43}$$

for measured values of T_1 to contain useful geometric information.

6. Interpretation of the Measurements of Schmidt, Velasco, and Nur[1]

Schmidt, Velasco, and Nur have measured $g(t)$ and obtained $P(R)$ from inversion for five separate cases. The first case was that of pure bulk H_2O at room temperature, for which they obtained a value of 3.27 sec for T_{1b} at room temperature. The remaining four cases are discussed in turn below.

a. Beaver sandstone

The pore space of a carefully dried sample of Beaver sandstone was fully saturated with pure water. We define the water saturation $\mathscr{S}$ as the volume fraction (in percent) of the pore space occupied by water. In the present case, $\mathscr{S} = 100\%$. The probability distribution P(R) showed two sharp peaks. The sharpness of the peaks suggests that the quasiuniform regime holds and the data are interpreted in Table 2 with that in mind.

Table 2: Data for Beaver Sandstone ($\mathscr{S} = 100\%$)

Quantity	Peak #1	Peak #2
T_1	0.8 s	0.27 s
$\frac{1}{T_1} - \frac{1}{T_{1b}} = \frac{\ell S}{V} \frac{1}{\tau}$	0.93 s^{-1}	3.4 s^{-1}
$\frac{V}{S}$ ($\frac{\ell}{\tau} = 10^4$)	1 μ	0.3 μ
$(6 D_{eff} T_1)^{1/2}$	14 μ	8 μ
$(6 D T_1)^{1/2}$	140 μ	80 μ

In Table 2 we first show the values of T_1 corresponding to the R values at the two peaks of P(R). Next, by subtracting the measured value[1] of $1/T_{1b}$ from those of $1/T_1$, we obtain values of ($\ell S/V\tau$) corresponding to each peak. By supposing, as discussed in §3, that $\ell/\tau = 10^{-4}$ cm/sec[12], a value to which we shall return below, we obtain values of V/S of 1 μ and 0.3 μ, respectively. These values of V/S establish a size scale for the pore space. Next using our rough estimate of D_{eff}, we obtain values of the diffusion distance through the pore space of roughly 15 and 30 times larger and of the diffusion distance through bulk water rough 150 and 300 times larger than V/S. It is thus clear that we are in the quasiuniform regime as supposed initially.

Why, then, if we are in the quasiuniform regime, are there two distinct peaks? One, and perhaps the only, internally consistent answer to this question is that the rock itself is heterogeneous on a scale larger than the grain size. There are two local conditions in the rock, one corresponding to each peak in P(R). Each local condition must persist for distances larger than the corresponding diffusion length, 14 µ and 8 µ, respectively. This heterogeneity can manifest itself through variation in ℓ/τ, that is in the nature of the surface layer, or through variation in S/V, that is in the geometry of the pore space.

b. Tight Gas Sand

The pore space of a carefully dried sample of a tight gas sand was fully saturated with pure water. The probability distribution P(R) showed three well separated peaks, the data for which are displayed in Table 3.

Table 3: Tight Gas Sand (ϕ = 100%)

Quantity	Peak #1	Peak #2	Peak #3
T_1	5×10^{-2} s	10^{-2} s	8×10^{-4} s
$\ell S/V\tau$	20 s^{-1}	100 s^{-1}	1200 s^{-1}
V/S	500 Å	100 Å	10 Å
$(6\, D_{eff}\, T_1)^{1/2}$	4 µ	2 µ	0.5 µ
$(6\, D\, T_1)^{1/2}$	40 µ	20 µ	5 µ

One sees immediately from Table 3 that we are well within the quasiuniform regime. As for the Beaver sandstone, one can infer that the tight gas sand studied is heterogeneous on distance scales of at least 0.5 - 4 µ. The value of V/S of 10 Å for peak #3 strongly suggests that the corresponding region of the pore volume is fractal with a surface fractal dimension necessarily the same as the volume fractal dimension. 10 Å is then the lower bound to fractal behavior. Thompson and Katz have observed equality of the two fractal dimensions in some rocks they have studied, and they have estimated the lower bound to fractal behavior to be about 20 Å[21].

  c. Fused Glass Beads (ϕ = 100%)

An artificial sandstone formed by the fusing of glass beads was saturated with pure water. Two peaks both very close together and close to the bulk relaxation rate were found in P(R). Because of the proximity of the peaks and the small shift from $1/T_{1b}$, the data in Table 4 obtained by measurement and analysis of the peaks is of limited accuracy.

Table 4: Fused Glass Beads (ϕ = 100%)

Quantity	Peak #1	Peak #2
T_1	2.75 S	2.70 S
V/S	52 μ	47 μ
$(6\, D_{eff}\, T_1)^{1/2}$	27 μ	26 μ
$(6\, D\, T_1)^{1/2}$	270 μ	260 μ

In the present case, we cannot conclude that we are in the quasiuniform case because the diffusion distance is only half of V/S. Since, however, the diffusion distance in pure water is five times the value of V/S, we are in a regime in which the magnetization is uniform within the pores. Why, then, are we seeing two sharp peaks? The reason is that the sample is formed by fusing a compaction of glass beads of uniform size initially at or near the density of random close packing. As pointed out to me by Mendelson and observed in studies of dense random packing[22], there are two types of pores in such a structure. The first are tetrahedral, and the second are the residue of larger spaces into which an additional atom could not fit. Empirically, the latter have a somewhat larger surface to volume than the former. Thus, the two peaks reported in Ref. 1 are consistent with the geometry expected from fused, dense-random-packed spheres. The surface to volume ratio of the fcc packing of spheres is 8.56/a, where a is the radius of the spheres. Using this as a rough estimate for S/V, we obtain a value of about 400 μ for a from the smaller V/S values in Table 4. As the pore to pore separation is larger than a, this confirms the conclusion that we are not in the quasiuniform case but are dealing with nearly monodisperse pores.

   d. Partially Saturated Berea Sandstone ($\mathscr{S} \leq 100\%$)

Berea sandstone was fully saturated with pure water, and its proton relaxation spectrum taken. Water was then extracted from the sample and the process repeated at successively lower values of the water saturation $\mathscr{S}$. Their results[1] and our proposed interpretation can be summarized as follows:

   1. For $\mathscr{S} > \mathscr{S}_c$, where $\mathscr{S}_c$ is between 70 and 80%, there are three peaks in the relaxation spectrum.

   2. As $\mathscr{S}$ decreases towards $\mathscr{S}_c$, each peak shifts upward in relaxation rate without significant broadening. All three peaks are well within the quasiuniform domain.

   3. There is heterogeneity in the Berea sandstone with three local conditions persisting for distances > 10 μ.

   4. The two higher rate peaks (peaks 2 and 3) shift roughly linearly with S, whereas the lowest peak (peak 1), while shifting at the same rate from $\mathscr{S}$ = 100% to $\mathscr{S}$ = 90%, does not appear to shift further from $\mathscr{S}$ = 90% to $\mathscr{S}$ = 80%.

   5. There may thus be heterogeneity also in the affinity for water.

   6. A very simple interpretation of the upward shift of the relaxation peaks with decreasing $\mathscr{S}$ can be made by supposing that the water wets the rock-water interface. Thus the value of $\mathscr{S}$ would remain constant, but V would decrease linearly with decreasing $\mathscr{S}$. The resulting formula for T_1 would be

$$\frac{1}{T_1} = \frac{1}{T_{1b}} + \frac{\ell S}{V_0}\frac{1}{\mathscr{S}\tau}, \qquad (44)$$

where V_0 is the actual value of the pore volume (corresponding to S = 1). The very rapid variation of peaks 2 and 3 with $\mathscr{S}$, contrasted with the much slower variation of peak 1, suggests that $\mathscr{S}$ is heterogeneous, comment 5. above. Thus the local value of $\mathscr{S}$ must be inserted into (44), and it is clearly changing very rapidly for peaks 2 and 3.

7. Once we have assumed that the pore surface remains wet, the only remaining source of heterogeneity in $\mathcal{S}$ can be heterogeneity in the surface roughness. As S decreases below unity, a water-water vapor interface of high interfacial free energy must develop. This will initially be a small sphere, but, as it grows, its shape must begin to follow that of the pore surface. However, it would cost much too much water surface free energy to follow the pore surface near regions of high positive curvature. Thus, the water film covering the pore surface thins most rapidly near regions of high positive curvature. The rapid variation of peak position with S for peaks 2 and 3 therefore indicates that the associated regions of the pore space are rough. That is, the heterogeneity indicated by the presence of three peaks in the relaxation spectrum may be caused by heterogeneity in the roughness of the pore surface, consistent with the higher R values of peaks 2 and 3 as well as their more rapid movement with $\mathcal{S}$.

8. For $\mathcal{S} < \mathcal{S}_c$, peaks 1 to 3 disappear and the spectrum consists of a broad, monotomically decreasing part between 1000 s^{-1} and 5000 s^{-1} and a sharp peak at about 7000 s^{-1}.

9. The abrupt transition in the spectrum at $\mathcal{S}_c$ is probably the most interesting experimental result of Schmidt, Velasco, and Nur[1]. The macroscopic water film covering the pore surface thins as $\mathcal{S}$ is reduced. However, as pointed out in 6. above, it does not thin uniformly. It thins most rapidly at the regions of highest positive curvature of the pore surface. As $\mathcal{S}$ decreases, thinning can proceed until the bulk component locally disappears leaving only the surface component. These thin patches of microscopic surface film can spread out until the bulk water is disconnected by them at a true percolation threshold. On the other hand, the conditions for the quasiuniform regime will almost certainly cease to hold before the percolation threshold is reached because diffusive coupling between distinct regions of the bulk film associated with smaller positive and arbitrary negative curvature of the surface is drastically reduced. Thus the observed transition is between the strong coupling regime in which only large scale homogeneity is seen (the three peaks for $\mathcal{S} > \mathcal{S}_c$) to the weak coupling regime in which the full probability distribution of S/V is seen for the decoupled regions of the water film (the broad distribution for $\mathcal{S} < \mathcal{S}_c$). The broad distribution is, as one would expect, fractal like and extends down below distance scales of 10 Å.

10. The peak at $R = 7100$ s^{-1}, $\Delta = 70\%$, we attribute to the microscopic surface layer of effective thickness ℓ which grows out from the regions of highest positive curvature and separates portions of the pore surface or space holding bulk water. In this surface layer, which we assume to be the same irrespective of whether bulk water or water vapor lies above it,

$$\ell S/V = 1. \tag{45}$$

Thus the peak position immediately yields a value of 1.4×10^{-4}s for T_{1S}. Analyzing the data of Senturia and Robinson[5] has already given us a value of 1.4 s/µ for T_{1S}/ℓ in a similar material[12]. Thus a value of ≈ 1 Å for ℓ, as assumed in §3, is entirely consistent with the data in the present interpretation.

11. It should be clearly understood that the above interpretations are only proposals based on the present state of the data. Important as the pioneering work of Schmidt, Velasco, and Nur[1] is, much more complete and precise information is required for definitive interpretations.

7. Conclusions

Measuring $g(t)$ and inverting it to obtain $P(R)$ comprises a very powerful structural tool for the characterization of porous materials, particularly when studies of partial saturation of the pore space are utilized. From the data of Ref. 1, evidence for submacroscopic heterogeneity can be obtained with limits on the persistence lengths and estimates of the surface to volume ratios of the pore space in the different regions of the rock. The heterogeneity can be in values of S/V or of ℓ/τ or of both, but the studies of the partial saturation of Berea sandstone show that there is heterogeneity in surface roughness and therefore S/V. The transition from strong to weak coupling has been observed with concomitant evidence for microscopic heterogeneity. There is evidence for fractal structure, and there is evidence for the persistence of a microscopic surface film of water on the pore surface of Berea sandstone as the bulk water is withdrawn. Such a remarkable richness of findings in an initial investigation is quite extraordinary.

REFERENCES

1. E. J. Schmidt, K. K. Velasco and A. Nur, Proceedings of the Symposium on the Chemistry and Physics of Composite Media, edited by M. Tomkiewicz and P. N. Sen, Proceedings Vol. 85-8, the Electrochemical Society, Inc., Pennington, NJ, p. 292.

2. R. J. S. Brown, Petroleum Transactions AIME 207, 262 (1956).

3. R. J. S. Brown, Petroleum Transactions AIME 219, 199 (1960).

4. A. Timur, J. Petroleum Tech., 775 (June 1969).

5. S. D. Senturia and J. D. Robinson, Society of Petroleum Engineers Journal, 237 (Sept. 1970).

6. R. C. Herrick, S. H. Couturie and D. L. Best, paper SPE 8361, 54th Annual Fall Technical Conference, Society of Petroleum Engineers of AIME, Las Vegas, Sept. 23-26, 1979.

7. W. P. Halperin and A. H. Thompson, present volume.

8. J. E. Tanner and E. O. Stejskoal, J. Chem. Phys. 49, 1769 (1968).

9. L. Borghi, F. Savoldi, R. Scelsi and M. Villa, Experimental Neurology 81, 89 (1983).

10. K. J. Packer and C. Rees, J. Coll. Interf. Sci. 40, 206 (1972).

11. K. R. Brownstein and C. E. Tasr, Phys. Rev. A19, 2446 (1979).

12. M. H. Cohen and K. S. Mendelson, J. Sppl. Phys. 53, 1127 (1982).

13. K. S. Mendelson, J. Appl. Phys. 53, 6465 (1982).

14. M. A. B. Siddique, "Study of NMR in a Bounded Region," M. S. Thesis, Marquette University, 1984.

15. K. S. Mendelson, Proceedings of the Symposium on the Chemistry and Physics of Composite Media, edited by M. Tomkiewicz and P. N. Sen, Proceedings Vol. 85-8, the Electrochemical Society, Inc., Pennington, NJ, p. 282.

16. K. S. Mendelson, present volume.

17. G. E. Archie, Trans. AIME 46, 54 (1942).

18. M. H. Cohen and C. Lin, Macroscopic Properties of Disordered Media, edited by R. Burridge, S. Childress, G. Papanicolau, Springer (1982); C. Lin and M. H. Cohen, J. Appl. Phys. 53, 4152 (1982).

19. M. H. Cohen and M. P. Anderson, J. Electrochem. Soc. (1985), in press.

20. M. H. Cohen and K. S. Mendelson, to be published.

21. A. J. Katz and A. H. Thompson, Phys. Rev. Lett. <u>54</u>, 1325 (1985).

22. K. S. Mendelson, private communication.

STRUCTURE OF SMALL SCALE FLOW THROUGH POROUS MEDIA

Jorge F. Willemsen
Schlumberger-Doll Research
Old Quarry Road
Ridgefield, CT 06877

ABSTRACT

Hydrocarbon reservoirs are structures which encompass heterogeneities occuring over roughly ten orders of magnitude in length scales - - from the crystaline roughness on individual grains to the variability of the geological facies. It is generally appreciated that the large scale heterogeneities of a reservoir impose severe problems for the mathematical modelling of fluid transport. However, even the smaller scale heterogeneities affect flow in important ways. For example, low flow rate experiments on water displacing oil through sandstones produce residual oil saturations of the same general order of magnitude as what is observed at the reservoir scale. I discuss both experiments and theoretical models which have been developed in the past few years to attempt to understand how small scale heterogeneities affect flow, and touch briefly upon how this information might be used to move up into larger scale modelling.

INTRODUCTION

My aim in this article is to review recent advances in understanding the mechanisms of the flow and entrapment of two liquids in porous materials, with special emphasis on processes which occur at small spatial scales.

The typical oil-bearing rocks are sandstones, limestones and dolomites, as we might expect from the geological processes involved in the formation of oil. Rocks of these types have several characteristic length scales, with substantial heterogeneity at each scale. Each scale of heterogeneity contributes to making oil recovery difficult and inefficient. For our present purposes, the most fundamental scale is that of the pores and grains.

Grains have "diameters" on the order of 10's of microns. The voids are often classified as pores and throats, where pores are cavities of dimensions that can be as large as grains, and throats are elongated, narrow passages connecting pores, with sizes typically smaller than these. The walls of the pores have substantial microroughness, often accentuated by clay deposits, which can also play a role in flow, but we will not dwell on this topic here. The micro-heterogeneity of interest arises from the broad size distributions of the pore and throat components of the void space. It is natural to set up the calculations on a lattice model geometry, with the bonds interpreted as throats, and the sites interpreted as pores.[1,2]

Before proceeding to the mathematical formulation, however, I want to explain the aim of the calculational scheme. Experimental observations have led to some very interesting questions.[3] First, what are the mechanisms that trap large quantities of oil during displacement? What parameters of the flooding process (e.g. fluid viscosities, capillary pressures, flow rates) significantly affect the results? How does the heterogeneity of the pore structure of the rock affect displacement? What spatial signatures of specific flow processes might be observed and interpreted for planning future recovery strategies? These are the broad aims of the research, but we can address only some of them in what follows.

QUALITATIVE CONSIDERATIONS

The mechanisms through which water displaces oil in a porous medium depend on the relative magnitudes of the forces involved. These, in turn, depend on the properties of the two fluids, the geometry and topology of the pores, and the externally imposed conditions. Due to the great variability of all these parameters of the problem, a large variety of microdisplacement cases can be identified. Rather than confuse matters by trying to cover every case, we concentrate our attention on a few representative cases, being even a bit loose in our characterization where it does no serious harm. In considering "typical" situations the dimensionless parameters of greatest importance are the capillary number, Ca = ratio of viscous, force to surface tension, and the ratio of the viscosities of the two fluids.

When the capillary number is small, say $Ca < 10^{-7}$, the viscous stresses are negligibly small, because the flow is very slow. Under these conditions, the dominant stresses are the capillary pressure and the externally imposed pressure. Only near the moving interfaces can the viscous stresses play a modest role in deciding the pattern of displacement. If we pass to a limit in which the displacement proceeds in only one pore at a time we obtain a situation called quasistatic imbibition.[4,5,6,7,8]

When the capillary number value is moderate or large, say $Ca > 10^{-4}$, the viscous stresses dominate the capillary pressures. Under these conditions, the viscosity ratio becomes a parameter of primary importance. Especially interesting is the case in which the displaced fluid is much more viscous than the displacing fluid.[9,10,11] We now examine these cases in greater detail.

At the beginning of a quasistatic imbibition the oil fills the medium, and the two fluids meet at a set of interfaces that reside in pores near the entrance. The capillary suction is greatest at the narrowest pore at which an interface exists. Due to the capillary pressure, the water is imbibed into the medium, seeking at each step the narrowest pore. In general, even large throats are smaller than small pores and, therefore, once a pore is invaded by water, all the adjoining throats are invaded before another pore is invaded. Once these throats have been invaded, the smallest pore among those in which an interface now exists gets invaded next, and so on. It follows that during quasistatic imbibition the advancement of the water is controlled by the currently smallest accessible pore.

As the water advances, it creates pathways that are highly branched in all directions. There is a stage at which a connected pathway of water spans the entire length of the of the rock sample for the first time, thus achieving "percolation".

In three dimensions, just before percolation the oil is still mostly connected and branches throughout the entire medium; there is no "front" separating the fluids into two separate domains.[12] (In two dimensions, of course, this is topologically not possible.) As imbibition progresses, the water pathways begin to connect with each other and eventually they surround and detach fragments of oil. An important mechanism for the formation of oil ganglia is the pinch-off of the narrow threads connecting small masses of oil with the main body. In simple terms, one assumes that the thread collapses if the water pressure around the thread is larger than the oil pressure within it.

Other mechanism for oil disconnection have been recognized.[13] For instance, thin films of water may creep along the walls of the pores, well ahead of the water microfingers, driven by the capillary suction of the interfaces that are formed among the asperities of the wall microroughness. Under certain conditions, the film of water forms a "collar" that keeps narrowing until the oil thread in it snaps, thus disconnecting a part of the oil. These processes can be modelled by invoking a fractal picture of the surface asperity, but I will not discuss this further here.

When a ganglion is created during slow imbibition, it becomes stranded immediately, because the force exerted by the flowing water on an oil ganglion is much smaller than the resistance offered by the capillary pressure at the downstream interfaces of the ganglion.

As the capillary number increases to moderate values, the microdisplacement phenomena are similar, except that the water advances in many pores simultaneously. The pathways of the advancing water seek, as a rule, the narrowest pores and throats, but in some instances the pressure drop associated with the viscous stresses may influence the choice. This, in turn, may lead to a somewhat different pattern of displacement. The most important qualitative question is whether finite dynamically determined length scales arise, a point I return to later.

The phenomena described above give qualitative answers to some of the major questions posed earlier. Specifically, we see that the mechanisms responsible for the retention of large amounts of oil by the porous medium are the disconnection of the oil into ganglia, and the subsequent entrapment of these ganglia by capillary forces. The randomness of the material is important in this process, inasmuch as the capillary fingering proceeds by choosing among pores of different sizes.

The second scenario which is of considerable interest is that of viscously unstable flow. In this case the dominant phenomenon is viscous fingering. At the macroscopic scale, using continuum equations to describe the flow, this is a subject with a rich literature too vast to explore in this article. But within the last year some very interesting results have emerged regarding viscous fingering at the scale we are focussing on. These results will be presented after introducing a method of calculation suitable for dealing with this regime.

MATHEMATICAL MODELS OF MICRODISPLACEMENT

Based upon ideas scattered throughout engineering literature, two groups developed a scheme tailored to the problem at hand in recent years.[14,15] First, model a porous medium in the ways described earlier, with as much randomness in the lattice as is desired. For simplicity my porous medium will be on a regular network, with no "pores" per se, but only pipes representing throats. These have radii chosen from some probability distribution. This evidently supplies enough randomness for the interesting phenomena to occur.

Label the points where pipes intersect by node numbers "i,j,..". When a single fluid is flowing through the pipe linking i and j, the volume/sec of flowing fluid, J_{ij}, is governed by Poiseuiile's law:

$$J_{ij} = \frac{A_{ij}^2}{\mu L_{ij}} (P_i - P_j) \qquad (1)$$

In this equation, P_i is the pressure at node i; A_{ij} is the area of the pipe linking i and j; L_{ij} is the length of the pipe; and μ is the viscosity of the fluid in the pipe, times a pipe-shape dependent constant.

Since fluid is conserved at every interior node i (those not source or sink nodes), $\sum_i J_{ij}$ = 0. Denote the coefficient of $(P_i - P_j)$ as g_{ij}. Then current conservation implies that the pressures satisfy Kirchhoff's law at each interior node:

$$P_i = \frac{\sum_j g_{ij} P_j}{\sum_j g_{ij}} \qquad (2)$$

Suppose next that there are two fluids in pipe ij at any given time. It is necessary to know how these fluids share the pipe before proceeding. The simplest scenario is one in which, at the individual pore-throat scale, the flow is piston-like, with the fluids occupying distinct regions in the throat, separated by a meniscus. When this configuration is realistic, one employs the "Washburn approximation" for the flow.

First, introduce an additional pressure drop from i to j given by the Laplace-Young expression, $P_\cap$ proportional γ/R_{ij}, where R is the pipe radius, and γ is the coefficient of surface tension times the cosine of the contact angle of the meniscus with the solid. Next, write a "micro-relative permeability" coefficient reflecting the fact the resistance to flow will depend upon the viscosities of both the fluids in the pipe. I will return to this point later. This completes the setup of the problem, but keep in mind that there may be many situations in which the Washburn approximation fails to describe the actual fluid configuration in the pipe. What happens at the pore scale involves "real" fluid mechanics, and much work remains to be done on this subject.

At this stage, there are node equations like Eq.(2) for all internal nodes joined by only a single fluid, with g_{ij} values determined by the local geometry and by the fluid viscosity. In addition, more complicated expressions describe all pipes containing more than one fluid. Finally, external conditions such as constant influx of fluid or constant macroscopic pressure differential must be specified. The crank can then be turned. The calculation is very time consuming for two reasons. First, one does not "solve the circuit" once only - - at each time step, fluids move around, and so create a new circuit which must be re-solved. But more frustratingly, when capillary forces are important, the interface tends to slosh back and forth as capillary and viscous pressures adjust locally to the global configuration.

The above method has been used to calculate flow in the capillary dominated regime, i.e., when capillary pressures in the pipes are very large compared to the average viscous pressure drops due to the flow. These calculations are slow, and we have only little quantitative information relevant to the regime. Nevertheless, Koplik and Chen have verified that the intuition discussed earlier is largely (though not absolutely) correct:[16] the vast majority of moves (displacement of one fluid by another) occur where the capillary pressure is most favorable. This intuition forms the basis for the "Invasion Percolation" model we will discuss in detail momentarily. IP is thus a shortcut for simulating flow in the capillary dominated regime.

Very recently, Chen and Wilkinson have studied flow in the limit of vanishingly small viscosity for the displacing fluid, and vanishingly small coefficient of surface tension, both experimentally and theoretically. The experiments are performed using two-dimensional networks etched into glass, and yield beautiful flow patterns of highly irregular viscous fingers. Before describing these in more detail, we shall finish the calculational prescription for the pipes containing an interface.

Suppose that at time t fluid I has advanced into a pipe a distance x from its entry node. By analogy with Ohm's law for electricity, $V = IR$, we view the hydraulic resistance to flow as consisting of two elements in series. Thus, if the electrical resistor is of length L, of which x has R_1, while (L-x) has R_2, the resistance of the whole is $R = R_1 \frac{x}{L} + R_2(1-\frac{x}{L})$. Carrying through the analogy to Darcy's law (with zero capillary pressure) we have

$$-\frac{\Delta P}{L} = \frac{J}{A^2}\left[\mu_I \frac{x}{L} + \mu_{II}\left(1 - \frac{x}{L}\right)\right] \qquad (3)$$

Next, because J_{ij} is a volume/time, one has directly

$$J_{ij} = A_{ij}\frac{\Delta x_{ij}}{\Delta t} \tag{4}$$

That is: at time t, the pressure drop across the node is known. From this, one calculates the amount by which the interface advances through the pipe in the interval Δt. This alters the "relative permeability" of the pipe, and one must compute the pressures for $t+\Delta t$ afresh. This determines the next advance of the interface, etc. The procedure is mathematically sound in the limit of infinitesimal Δt, and is approximate for any computationally convenient increment.

Almost all of the above prescription is general. The simplifying elements in Chen and Wilkinson's work are setting the viscosity of the invading fluid equal to zero, and neglecting capillary forces altogether.

We have, then, two limiting regimes of considerable interest: invasion percolation in the limit of capillary dominance, and irregular viscous fingering in the limit of infinite viscosity ratio. In principle, the model is capable of handling the crossover regimes, though it runs extraordinarily slowly in these regimes. We now discuss each limiting region in turn.

CAPILLARY DOMINANCE

Invasion percolation is the name given to a special dynamic process through which a cluster grows into a sample through selection of paths of least resistance. The physical basis for the model is that when capillary forces totally dominate viscous forces, a displacing fluid tends to penetrate a sample saturated with another fluid at points where the capillary forces are instantaneously most favorable. This occurs at bonds or sites which are optimal for the process under consideration.

For the purpose of performing computer simulations of the above process, resistances to invasion are assigned randomly to the sites or bonds of a regular lattice, and are held fixed throughout the process. The invader grows by selectively moving to (through) the best uninvaded site (bond) as determined by the resistance distribution. The displaced fluid is treated as incompressible: once it has been surrounded the invader cannot penetrate it further.

Much has been learned about IP through computer simulations, but I must refer to the literature for the many details. In brief, the invading cluster is has a self-similar geometrical structure with a well-defined fractal dimension. This dimension is different from that of classical static percolation in $d=2$, but equal in $d=3$ and above. In addition, the "trapped" clusters of defending fluid have a power-law size distribution, again reflecting geometrical self-similarity.

Inasmuch as the IP growth cluster is fractal, the sample size sets the only natural length scale. This observation can be used to delineate the domain of applicability of the IP picture. For example, when buoyancy is taken into account, Wilkinson[17] and de Gennes[18] have independently estimated how a second length scale enters, effectively cutting off the size distribution of stranded oil blobs. The buoyancy estimates have been extrapolated to the case that viscous forces are reinstated into the dynamics, and although it is my opinion that these estimates are on less firm ground than those for buoyancy, the important thing is that IP supplies a point of departure for studying how typical sizes arise under realistic flow conditions.

Fortunately, IP is not simply a theoretician's abstraction having nothing to do with the real world. Lenormand and Zarcone[19] have performed experiments using etched two dimensional random networks and observed IP growth at very low capillary numbers, with a measured cluster dimension in excellent agreement with the computer-generated predictions. Work remains to be done in real materials in three dimensions. In particular it is not trivial to do quantitative work although some suggestions have been made regarding tomographic techniques to test the fractal nature of the growth cluster.[20]

VISCOUS FINGERING AND DLA

As mentioned earlier, everyone expects viscous fingering to occur under unfavorable mobility conditions. What has been pleasantly surprising is that small scale viscous fingering occurs with the following novel characteristics:

1. The viscous fingers form a random fractal cluster which seems to be the same as DLA;[21]
2. The flow pattern becomes dendritic rather than fractal as the medium becomes regular rather than highly random;
3. A small amount of surface tension between the fluids does not stabilize the configuration at laboratory size scales (order of a few hundred pores).

As is well known, recent research has led to the speculation that DLA might be relevant to almost any growth phenomena involving solutions of the Laplace equation, and in which multiple time scales exist. Now, for the problem of flow in porous media, material conservation implies a Laplacian situation at the smallest relevant scales of the porous medium (when the Washburn approximation is valid), but this is not quite the same thing. So it is important to examine more carefully how the flow situation might be related to DLA, inasmuch as they both lead to clusters with the same fractal dimension.

To explore this connection, first write a master equation for diffusion which takes into account bond dependent hopping probabilities as follows:

$$n(i,t+\Delta t) = n(i,t) + \sum_{\{j\}} \sigma(j \rightarrow i) n_j(t) - \sum_{\{j\}} \sigma(i \rightarrow j) n_i(t) \qquad (5)$$

Then at steady state, Kirchhoff's law will be satisfied by the concentrations n_i, and in fact this Kirchhoff condition can be identified directly with Eq.( ) through the substitutions

$$P_i = n_i / \sum_{\{j\}} g_{ij}, \ \sigma(i \rightarrow j) = g_{ij} / \sum_{\{j\}} g_{ij} \ . \qquad (6)$$

We see, then, that random walks may be used to solve for the pressures in the flow problem, provided that a steady state is achieved. Having done this, one may advance the interface according to the rules described earlier, and continue the calculation by performing more random walks.

There is clearly nothing "wrong" with this - one may choose whatever method is appealing to solve the Laplace equation. But the advancement of the interface described above is not at all like the accretion of particles onto a DLA cluster. Saying both processes have something to do with diffusion does not convincingly demonstrate their equivalence.

Heuristically, the key to understanding the connection between the two processes resides in Eq.(4). According to this equation, the inviscid fluid advance is governed by both the pressure drop and by the time step Δt. But there is no reason to identify this time step with the diffusional time step (one computer step). Suppose on the contrary that the flow time step corresponds to the random walk relaxation time (time to achieve steady state.) In that case the pressure drops will be approximated accurately. In addition, we can replace the deterministic growth process by a stochastic one, assigning a probability for a site to fill proportional to the pressure drop and to the pipe dimensions.

It would seem, then, that the flow will resemble DLA if DLA is not as keenly dependent on diffusion providing the limiting time for the aggregation process as it seemed at first. But that has actually been established in the literature.[22] In the large, the structure of the aggregate is the same whether adhesion occurs on the first encounter, or occurs only after many.

To summarize these remarks: Assume that the microflood computational algorithm adequately describes the physics of the fluid flow, as IP and now Chen and Wilkinson's results suggest. In the viscously unstable regime, a stochastic computation can be formulated utilizing random walks, which should approach the deterministic result if the aggregation probabilities are small and appropriately weighted. This should give results similar to DLA to the extent that DLA is robust against changes in the aggregation probabilities.

However, one problem persists. It is that DLA clusters are fractal on lattices with equal hopping probabilities, yet the microflood calculation shows growth is dendritic rather than fractal on such lattices. Precisely how the randomness produces the pressure field appropriate for a DLA growth configuration is not presently understood. The interesting idea has been raised that the lattice randomness serves only to trigger self-sustaining instabilities known to occur in continuum systems, but a deeper mathematical investigation seems warranted.[23]

Furthermore, the introduction of an infinitesimal viscosity in the growing fluid may introduce a cutoff length scale beyond which the flow will not be a DLA-like fractal. Exploiting further the analogy between Ohm's law and Darcy's law, we expect the pressure drop across a fractal to scale with the radial extent R of the fractal as

$$\Delta P \propto R^{2-d+d_T} \tag{7}$$

Here d_T is a transport exponent which I will not dwell on here, but which depends on the properties of the fractal.

On the other hand the pressure drop for regular flow would have been dominated by $\mu_0 R^{2-d}$. There is less work done if the fluid flows through the fractal provided

$$R \leq (\mu_0/\mu_W)^{1/d_T} \tag{8}$$

This is an extremely crude estimate, which can certainly be refined. Qualitatively, however, I believe this reflects the mechanism which introduces a cutoff in the extent of fractal behaviour.

SUMMARY AND CONCLUSIONS

1. Adopting a network model for the medium and a local model for the flow within individual pores and throats allows computation of the overall flow.
2. The calculations are in agreement with experimental observations on etched two dimensional networks in the extreme cases of capillary dominance and of infinitely unfavorable mobility ratio;
3. The calculations need improvement, both computationally and conceptually, in cases other than the extremes listed above;
4. More three dimensional experimention to reveal local quantities such as details of residual saturations are needed;
5. No one has yet figured out a good way to scale the calculational scheme up to provide a continuum dynamical formulation which preserves some of the exotic features observed at the smaller scales, although some work in the continuum may provide valuable clues.

Let me conclude by emphasizing my hope that the mathematical modelling described here poses interesting challenges to the random transport community, computationally, but more importantly conceptually: traditionally we would like to compute field-valued quantities such as local pressures and flow rates, yet there are perhaps even more important quantities to be considered such as trapped region sizes, distributions, and spatial configurations, on which we have presently no ready handle. We are challenged not only with discovering computational methods for solving specific equations with random coefficients, but with devising a way of thinking about physically observable quantities which are not intrinsically variables in the equations.

REFERENCES

1a. Fatt, I., 1956a, Trans. AIME 207, 160.
1b. Fatt, I., 1956b, Trans. AIME 207, 160.
1c. Fatt, I., 1956c, Trans. AIME 207, 164.
2. Josselin de Jong, G., 1958, Am. Geophys. Union, 39, 67.
3a. Chatzis, I. and Morrow, N. R., 1981, SPE 10114.
3b. Chatzis, I., Morrow, N. R., and Lim, H. T., 1983, Soc. Pet. Eng. J., 23, 311.
4. Chandler, R., Koplik, J., Lerman, K., and Willemsen, J., 1982, J. Fluid Mech., 119, 249.
5. de Gennes, P. G. and Guyon, E., 1978, U. de Mechanique 17, 403.
6. Lenormand, R. and Bories, S., 1980, C. R. Acad. Sci. Paris 291, 279.
7. Larson, R. G., Davis, H. T., and Scirvan, L. E., 1981, Chem. Eng. Sci. 36, 75.
8. Larson, R. G., Scriven, L. E., Davis, H. T., 1977, Nature, 268, 409.
9. Paterson, L., 1984, Phys. Rev. Lett. 52, 1621.
10. Lenormand, R., 1985, Proc. 167 Meeting of the Electrochemical Society, Toronto.
11. Chen, J. D. and Wilkinson, D., 1985, to appear in Phys. Rev. Lett..
12. Wilkinson, D. and Willemsen, J. F., 1983, J. Phys. A16, 3365.
13. Lenormand, R. and Zarcone, C., 1984, SPE 13264.
14. Koplik, J. and Lasseter, T. J., 1982, SPE 11014.
15. Dias, M. M. and Payatakes, A. C., 1985, to appear in J. Fluid Mec..
16. Chen, J. D. and Koplik, J., 1985, J. Colloid. Interface Sci..
17. Wilkinson, D., 1984, Phys. Rev. A 30, 520.
18. de Gennes, P. G., 1984, private communication.
19. Lenormand, R. and Zarcone, C., 1985, Phys. Rev. Lett. 31, 432.
20. Willemsen, J. F., 1985, Phys. Rev. 31, 432.
21. Witten, T. A. and Sander, L. M., 1983, Phys. Rev. B 27, 5686.
22. Meakin, P., 1983, Phys. Rev. A 27, 1495.
23. Scher, H., private communication.

FRACTAL SURFACES

S. Alexander[+]
Exxon Research and Engineering Company
Corporate Research - Science Laboratories
Annandale, New Jersey 08801
and
Department of Physics
University of California, Los Angeles
Los Angeles, California 90024

ABSTRACT

Most experiments measure the mass dimension of fractal surfaces. It is shown that most existing models for rough surfaces result in self-affine fractals which are strongly anisotropic in the normal direction and have a nonfractal area. Self-affine surfaces will usually look flat in covering experiments and for surface scattering. The distinction between self-similar and self-affine surfaces and the properties of the latter are discussed in detail.

One of the most useful applications of fractal concepts has been in the description of surfaces. Smooth surfaces are rare in nature. Rough surfaces are certainly the rule and their roughness can be important. Assigning a fractal dimension to the surface of a proper, d-dimensional object is an important step in characterizing and describing it--and has implications. Mandelbrot[1] has used the scale dependence of the length of shore lines to determine the fractal dimension of the surface of the earth and Mandelbrot and Voss[2] have used this to generate realistic synthetic landscapes. Avnir Farin and

---

[+] Permanent address: The Racah Institute of Physics, The Hebrew University, Jerusalem, Israel

Pfeifer[3] have shown that the surfaces of many naturally occurring materials are fractal. Passoja[4] has shown that fresh fractures tend to be fractal. Theoretical models for surface formation also sometimes predict fractal surfaces. The best understood is probably the solid on solid roughening transition.

Rough fractal surfaces can be of two types--they can be isotropic and self-similar--examples would be shore lines (in 2-d) or the surface of a Koch Pyramid[5]. They can also be anisotropic, with the vertical playing a special role, and self-affine[6,7]. Examples are the surface of a Mandelbrot-Voss[2] earth or that resulting from a solid-on-solid roughening transition.

The problem I want to discuss is mainly experimental. I want to show that self-affine surfaces will look flat when one tries to measure their fractal dimension by the usual techniques available in condensed matter physics. There is also a conceptual problem because it seems to be difficult to generate simple random surface models which are self similar.

The mathematics of self-affine fractals is discussed in Mandelbrot's book[6] and in more detail in Voss's[7] Yelo lectures. I believe the application to solid surfaces is new and, in any case, not generally appreciated. I was directly motivated by the recent, and ongoing, controversy concerning the modification of Porod scattering when the surfaces are fractal[8-10].

How does one measure the fractal dimension of a surface (D)?

Avnir et al.[3] measure D by covering a surface of external dimension L with rigid molecules of linear dimension ℓ. The measured apparent area is then

$$S(\ell) \propto \ell^{d-1} \cdot N_\ell \propto \ell^{d-1} \left(\frac{L}{\ell}\right)^D \tag{1}$$

where

$$N_\ell \propto \left(\frac{L}{\ell}\right)^D \tag{2}$$

is the number of mutually exclusive molecules (or discs) needed to cover the surface. This is clearly a variant of the Richardson "caliper" method for shore lines[1] (d=2). For a surface in space one has d=3.

A closely related straightforward argument has led Bale and Schmidt[8] to the conclusion that one should observe a scattering intensity

$$I(q) \propto q^{D-6} \qquad (3)$$

for weak scattering from fractal surfaces. Equation 3 replaces ordinary Porod scattering (D=2, $I(q) \propto q^{-4}$). This has given rise to considerable controversy. Po-Zen Wong[9], in a paper which has still not been accepted for publication, obtained a very different result. Using an explicit, and very reasonable, model for surface scattering he calculates I(q) explicitly. He finds ordinary Porod scattering (q^{-4}) independent of the fractal dimension D.

Equations 1 and 3 are undoubtedly correct for self-similar surfaces. I shall show that both fail for self-affine fractal surfaces. Such surfaces will look flat when investigated by either of these techniques.

There are of course other ways of measuring (or defining) D which would give the correct fractal result. This is the reason one does want to consider the surfaces as fractal. For applications to materials they are, however, usually much less accessible experimentally.

Consider a surface described by a single valued function,

$$h = h(x,y) \qquad (4)$$

over the plane (or a line h = h(x)). We assume a height-height correlation function,

$$\langle h(\vec{r})h(\vec{r} + \vec{L})\rangle = b^2 (L/a)^{2H} \qquad (5)$$

where a and b are some, in principle arbitrary, elementary length scales. Surfaces for which Equation 5 holds are fractal with a Haussdorf dimension

$$D = d-H \qquad (6)$$

where d is the dimension of space[2].

It seems obvious that one wants to describe such a surface as fractal. It has roughness at all length scales and a wide range of scale invariant behavior described by Equation 5. For example, the fluctuations associated with Brownian motion or with "fractal time" anomalous transport can always be represented by curves of this type. The correlation function (Equation 5) does however imply a strong anisotropy. The vertical (h) direction is special.

We note first that if $h(\vec{r})$ is to represent the boundary of a (d=2) island or the surface of a compact 3-d object one must have,

$$H < 1 \qquad (7)$$

H > 1 would mean that the vertical fluctuations get larger than the dimension of the object (L) for large L--which is impossible.

We also note that the construction of such a fractal is strongly anisotropic. The vertical (or radial) direction is singled out. In particular (for H < 1) the surface becomes arbitrarily thin for large L,

$$\frac{\delta(L)}{L} \propto \left(\frac{(L/a)^H}{(L/a)}\right) \xrightarrow[L/a \to \infty]{} 0 \qquad (8)$$

In contrast self-similar fractal surfaces, such as a Koch curve or pyramid, always have a finite surface thickness ($\frac{\delta(L)}{L} \to$ const).

Consider now what happens when we try to measure the length of a self-affine curve with yardsticks of length ℓ. The measured length would be,

$$\mathcal{L}(\ell) = \ell \cdot N(\ell,L) = \ell \cdot \frac{L}{\langle \ell_{||} \rangle} \qquad (9)$$

where L is the end to end distance between the ends of the curve and $\langle \ell_{||} \rangle$ is the average horizontal projection of the yardstick. The surface is rough so there must be a vertical projection.

$$\ell_\perp^2 \propto b^2 \left(\frac{\ell_{||}}{a}\right)^{2H} \tag{10}$$

and therefore,

$$\tan\theta \approx b/a\,(\ell_{||}/a)^{H-1}; \quad \ell_{||}^2 + \ell_\perp^2 = \ell^2 \tag{11}$$

The choice of a and b is in essence arbitrary but Equations 10 and 11 define a length scale. Let $\ell_{||}(\ell) = \ell_\perp(\ell)$ then from Equation 10:

$$\ell_{||}(\ell_c) = \ell_\perp(\ell_c) = a \cdot \left(\frac{b}{a}\right)^{\frac{1}{1-H}} \approx \ell_c \tag{12}$$

For small $\ell\ (\ll \ell_c)$ one will have,

$$\ell \approx \ell_\perp \gg \ell_{||} \tag{13}$$

The measuring rods are essentially, vertical and N_ℓ will be much larger than L/ℓ. On the other hand for large ℓ - the rods are flat,

$$\ell_{||} = \ell[1-0(\tfrac{a}{\ell})^{1-H}] \approx \ell \gg \ell \tag{14}$$

Thus from Equation 9,

$$\pounds(\ell) = \ell \cdot \frac{L}{\ell_{||}} \tilde{=} L(1 + 0(\tfrac{a}{L})^{1-H}) \tag{15}$$

and the fractal character only shows up as a small correction.

The extension to surfaces is immediate.

Now in principle (i.e. formally) one can of course have both types of ℓ. For physical surfaces the small $\ell(\ll \ell_c)$ range is however not accessible.

One notes first that ℓ_c is, by construction, a lower bound on the grain size. One cannot have a grain with this sort of boundary and $L < \ell_c$. It is very hard to imagine how a physical process could produce such a minimum size in the grain distribution and at the same time retain the scale independence implied by Equation 5 across this scale, for larger grains. Reasonable physical processes give $b \stackrel{<}{\sim} a$ near the lower cutoff--e.g. at an atomic scale. All measurements to which the fractal description is relevant are then done in the flat large ℓ/ℓ_c regime of Equation 15. There may be exceptions to this general claim but we are not aware of any.

It is instructive to look at this problem of measuring the "length" of a self-affine curve or the area of a self-affine surface, from a slightly different point of view by measuring the mass of the fractal. Consider first the usual self-similar case. We cut the connected fractal into (L/ϵ) $(d-1)$ dimensional slices of thickness ϵ. The intersection of a fractal of dimension D with a line (or plane) is a fractal set of dimension $(D-1)$. We can now calculate the total mass (i.e. length or area) of the fractal by adding up the contributions from all slices. In particular if we have chosen a small enough ϵ,

$$\epsilon < a \tag{16}$$

where a is the inner cutoff, each slice contains a mass,

$$\delta M \approx L^{D-1} \cdot \epsilon \tag{17}$$

Thus,

$$M(L) \propto L/\epsilon \cdot (L)^{D-1} \epsilon = L^D \tag{18}$$

Thus D is the mass dimension.

We can generalize this slightly by allowing large $\epsilon (>>a)$. One has,

$$\delta M \sim \left(\frac{L}{\epsilon}\right)^{D-1} \epsilon^D \tag{19}$$

so that the result (Equation 18) is indeed independent of ε.

The result for a self-affine fractal is different and one finds that the mass dimension is always d-1 independent of H.

The intersection of a fractal described by Equation 5 with a horizontal plane (or line) is a fractal set of dimension,

$$\hat{D} = d-1-H \qquad (20)$$

This set is isotropic in the plane and self similar. An example are the Mandelbrot coast lines. Another example is the Brown line on line curve[1]--i.e. the plot of a Brownian motion along the y(h) axis versus time (x). One has $H = \frac{1}{2}$. The intersections with the x axis (y=0) are the returns to the origin. They constitute a fractal set (in time) with

$$\hat{D} = \frac{1}{2} = d-1-H \qquad (21)$$

We can now measure the mass by slicing the fractal. One has

$$\delta M \sim L^{\hat{D}} \cdot \varepsilon \qquad \varepsilon < a \qquad (22)$$

where $\hat{D}$ is given by Equation 20. The number of slices is, from Equation 5,

$$N(\varepsilon) = b/\varepsilon \cdot \left(\frac{L}{a}\right)^H \qquad (23)$$

and combining Equations 20, 22 and 23

$$M(L) = \delta M \cdot N(\varepsilon) \propto L^H/\varepsilon \cdot L^{d-1-H} \propto L^{d-1} \qquad (24)$$

The fact that one is dealing with a fractal, which is manifest in the height-height correlation function (Equation 5), and in the sections (Equation 22) drops out completely for the mass. The mass dimension is d-1 as for a flat surface.

As for the self-similar fractal, one obtains the same result if one consistently allows ε to be larger than the cutoff (a).

Thus a self-affine fractal has a well defined area. This result is crucial in the discussion of interphase surface energies for solid-on-solid models[10].

An interesting correlary of these arguments concerns the mathematical theory of hiking on the surface of a Mandelbrot-Voss planet. We want the geodesics, i.e. the the shortest paths between two points on the surface.

Let A and B be two points on the surface. The geometric distance between A and B is L. Consider the intersections of planes through A and B with the surface. Let Θ be the angle between the plane and the vertical direction.

A horizontal section $(\Theta = \frac{\pi}{2})$ is a self-similar fractal of dimension (D-1) with a fractal length (Equation 1).

A vertical section $(\Theta = 0)$ is self-affine so that the distance £(e) as measured by the hiker is simply proportional to L and independent of the size of his steps (ℓ) (Equation 15).

For intermediate values of Θ there is a gradual crossover between these two extremes.

For very small $(\frac{\pi}{2} - \Theta)$:

$$\cot \Theta < b/a \ (a/L)^{(1-H)} \qquad (25)$$

The path is still a self-similar (slightly anisotropic) fractal on the relevant scale (L).

Smaller values of Θ (i.e. more vertical planes) define a length scale.

$$L_{\Theta/a} \approx (\frac{b}{a}\tan\Theta)^{\frac{1}{1-H}} \qquad (26)$$

The fractal is self similar on scales small compared to L_Θ and self affine on larger scales.

On scales large compared to L_Θ the excursions in the cutting plane, perpendicular to the A-B axis, are determined by the vertical fluctuations of the surface. Thus, the intersection is, for large scales, a self affine fractal with a correlation function

$$\langle h(x)h(x+X)\rangle \frac{b^2}{\cos^2\Theta}(X/a)^{2H} \qquad (27)$$

$$X > L_\Theta$$

For the self-affine region one would have,

$$\ell_c \approx L_\Theta \qquad (28)$$

where ℓ_c is defined in Equation 12 and L_Θ in Equation (26). The small ℓ (ℓ/ℓ_c) regime of the self-affine fractal is however missing because the fractal is self similar for these lengths.

For hiking the implication is that one should always follow the vertical section or at least one for which

$$L_\Theta < \ell_s \qquad (29)$$

ℓ_s being the step length of the hiker. More horizontal trails are very long.

$$\mathcal{L}(L,\ell_s) \approx \ell_s \cdot (L_\Theta/\ell_s)^{D'} \cdot \frac{L}{L_\Theta} \qquad (30)$$

We believe the discussion of the intermediate type fractals which are self similar at short length scales and self affine on large scales is new[11].

We now consider surface scattering. In the weak scattering limit the scattering is determined by the Fourier transform of the contrast correlation function,

$$G(R) = \langle \delta(n(\vec{r})-n_1)\ \delta(n(\vec{r}+\vec{R})-n_2)\rangle \qquad (31)$$

where n_1 and n_2 are the refractive indices of the two materials separated by the surface.

The averaging in Equation 31 is both over $\vec{r}$ and over the orientaions of $\vec{R}$ permitted by the surface for given $\vec{r}$. The two ends of $\vec{R}$ have to be on different sides of the surface. We repeat the calculation of Po Zen[9] in slightly different and abbreviated form. One can write the integral as:

$$G(R) = \int dxdy \int_{h(x,y)-A(R)}^{h(x,y)} dz \int d\Omega_z \qquad (32)$$

where $h(x,y)$ is the position of the surface and the angular limits of the integration depend on the position z with respect to the surface. The lower limit $A(R)$ is such that all z for which the angular integral does not vanish are included.

For a flat surface one would have ($z' = h(x,y)-z$)

$$A(R) = R \int_z d\Omega = 2\pi \left(\frac{R-z'}{R}\right) \qquad (33)$$

to obtain the porod result[9].

$$G(R) \propto SR;\ G(q) \propto Sq^{-4} \qquad (34)$$

For a rough surface there are, in general corrections to $A(R)$ and to the angular integration in Equation 33, because of the local slope and curvature of the surface on the scale R. The dominant correction is the change required in $A(R)$ when the surface, at x,y, is not parallel to the x,y plane:

$$A(R) \approx \frac{R}{\langle|\cos\theta|\rangle} \qquad (35)$$

where $\langle\cos\theta\rangle$ is a suitably averaged value. Clearly the fact that the distance from the inclined surface is smaller than the $\tilde{z}$ also shows up in the angular integral. In addition there are of course corrections due to the local curvature of the surface (on scales $\approx R$) but they only contribute to higher order.

Corrections of the type 35 are dominant for self-similar fractal surfaces for which the surface is essentially nowhere locally parallel to the x,y plane. For this reason it is much simpler to replace the horizontal integration in Equation 32 ($\int dxdy$) by an integral over a smoothed surface locally parallel to the surface with z along the normal to the surface. This immediately leads to the Bale and Schmidt[8] result:

$$G(R) \propto S(R) \cdot R \propto L^D R^{d-D} \qquad (36)$$

replacing S in Equation 34 by S(R) (Equation 1). The Fourier transform is then[8],

$$G(q) \propto L^d q^{-(2d-D)} \qquad (37)$$

as in Equation 3.

Self-affine fractal surfaces are always locally flat. From Equation 11 one has (Equation 11).

$$\langle |\tan\theta| \rangle \approx \frac{b}{a}(R/a)^{H-1} \qquad (38)$$

so that one can retain the horizontal integration of Equation 32 as long as

$$R/\ell_c > 1 \qquad (39)$$

Corrections to the flat surface result (Equation 34) are then of order $(a/R)^{1-H}$.

This is the main result of Po Zen Wong[10]. We have already noted that the small R regime, where Equation 39 would not be valid, is not physically, accessible. Also, contrary to claims in ref. 10, one cannot have H > 1.

Finally, we note another implication of this discussion. The roughening transition and related problems[10] are usually discussed in terms of a solid on solid model of the surface which is strongly anisotropic. The result is a self-affine fractal rough surface with a height-height correlation function of the form of Equation 5. The surface energy remains proportional to the area $(S \propto L^2)$ and the surface tension does not diverge. This is a manifestation of the fact that the rough surface is self affine and therefore has a nonfractal area (mass dimension = d-1). The same type of fluctuations could, for a model with less anisotropy, generate a self similar fractal surface with divergences in the surface tension.

The aim of the above discussion was to point out the importance of the distinction between self-affine and self-similar fractal surfaces. Both types of surfaces are fractal and rough on all length scales (down to a physical inner cutoff). They both have well defined fractal Haussdorff dimensions. They differ however when one uses other definitions of the fractal dimension and corresponding experimental techniques which measure them. Self affine surfaces have a nonfractal mass dimension (area) and appear flat for covering and surface scattering experiments. They have a fractal height height correlation function and are expected to result from physical processes which are very anisotropic so that the normal to initial flat surface plays a special role.

Acknowledgment

I am grateful to T. Witten and to G. Grest who brought Po Zen Wong's work to my attention, for discussions and to R. C. Voss and B. Mandelbrot for patiently explaining to me the more general self-affine context of the fractal surfaces discussed above. This research was supported in part by NSF grant 84-12898 and by the Fund for Basic Research of the Israel Academy of Science.

References

1. Mandelbrot, B. B. - The Fractal Geometry of Nature, Freeman 1982 (FGN) Chp. 5, Chp. 26.
2. FGN, Chp. 28, R. F. Voss (preprint). B. Mandelbrot (preprint).
3. Avnir, D., and Pfeifer, P., Nouv. J. Chim. 7, 71 (1983); Pfeifer, P., Avnir, D. and Farin, D., Surf. Scie. 126, 569 (1983); Pfeifer, P. and Avnir, D., J. Chem. Phys. 79, 3558 (1983); Avnir, D., Farin, D. and Pfeifer, P., J. Chem. Phys. 79, 3566 (1983); Nature (London) 308, 261 (1984); J. Colloid and Interface Sci., 103, 112 (1985).
4. Mandelbrot, B. B., Passoja, D. and Panellay, A. (to be published). D. Passoya (private communication, 1983).

5. FGN, p. 13a.
6. FGN, p. 237.
7. Voss, R. F., Proceedings of NATO school, Geilo (1985).
8. Bale, H. D. and Schmidt, P. W., Phys. Rev. Lett. $\underline{53}$, 596 (1984).
9. Po Zen, Wong, 1984 (Preprints).
10. Witten, T., 1985 (Preprints).
11. I am grateful to F. Lavryaz for pointing out to me the importance of looking at the way inclined planes emerge from a self-affine surface.

MOLECULAR DIFFUSION IN POROUS STRUCTURES

J.C. Tarczon[*,t], A.H. Thompson[+], W.A. Ellingson[t], and W.P. Halperin[*]

[*] Dept. of Physics and Astronomy, Northwestern University, Evanston, Illinois, 60201
[t] Materials Science and Technology Division, Argonne National Laboratory, Argonne, Illinois, 60439
[+] Exxon Production Research Co., Houston, Texas, 72001

ABSTRACT

NMR spin-lattice relaxation and diffusion measurements have been performed on protonic fluids infused into several different sandstones, porous vycor glasses, and green ceramic compacts. The spin-lattice relaxation measurements have directly shown that there is a very wide distribution of pore sizes in the sandstones, in contrast to the other materials. The mean-square displacement of diffusing molecules has been measured in the sandstones with pulsed-magnetic-field-gradient techniques.

1. Introduction

The application of NMR techniques to characterize porous structures has been reported previously[1] and there has been effort recently to develop an appropriate analytical approach[2] with particular emphasis on sandstones. In this paper we report measurements on a number of porous materials including compacts of green-state ceramics, vycor glasses, and sandstones. All of the materials have an interconnected pore space. The technique requires filling of the pore space with a protonic fluid such as water or glycerol. NMR measurements were performed on the filler fluid and give information about the physical structure of the porous material. There are two types of experiments that we discuss here:

    a)   Spin-lattice relaxation measurements, in principle, allow one to obtain information concerning the physical structure of the porous material.

    b)   Pulsed-magnetic-field-gradient diffusion measurements allow one to obtain dynamical information concerning the random walk of the filler fluid molecules in the interconnected pore space.

2.   Preparation

The porous samples are filled by capillary wetting or vacuum impregnation, with a protonic fluid. The bulk diffusivity of the fluid will determine the range of length scales for mean-square displacement of the fluid molecules for a given measurement time.

We have used, water ($D = 2.3 \times 10^{-5}$ cm^2/sec, at 25°C), hexane ($D = 4.2 \times 10^{-5}$ cm^2/sec, at 25°C), decanol ($D = 7.5 \times 10^{-7}$ cm^2/sec, at 25°C), and glycerol ($D = 1.5 \times 10^{-9}$ cm^2/sec, at 25°C). We have impregnated alumina compacts, vycor glass samples with average pore dimensions of 100 Å and 550 Å, three different sandstones that have been previously studied by Katz and Thompson[3] using electron microscopy (SEM). Two of these were reported to be fractal over more than three orders of length scale (from approximately 10 Å to more than 10 µ).

3. NMR Spin-lattice Relaxation Measurements

The essential concept is that there is always an enhanced relaxation rate of the liquid in the pore space at the liquid-solid interface, T_{1s}^{-1}, compared to that of the bulk liquid, T_{1b}^{-1}. The surface relaxation can be attributed to magnetic impurities or hindered molecular motion.

Analysis of the relaxation of the nuclear magnetization as a function of time following an excitation of the nuclear resonance can give structural information concerning the distributed pore space. Several approaches will be discussed in the context of possible fractal pore space distributions.

The theory of Cohen and Mendelson[2] can be summarized in the following way. For a unimodal pore space distribution of size r, which we define here as the inverse of the surface-to-volume ratio, $r = (S/V)^{-1}$, one can write that,

$$[M(t) - M_\infty] / [M(0) - M_\infty] \;=\; \exp(-t/T_1), \tag{1}$$

where,

$$1/T_1 \;=\; 1/T_{1b} + (S/V)(L/T_{1s}) \;=\; 1/T_{1b} + a/r, \quad T_{1s} \ll T_{1b} \tag{2}$$

The bulk liquid relaxation rate of the nuclear magnetization, M(t), is denoted by T_{1b}^{-1} and a surface thickness parameter L is used to constrain the range of the surface relaxation which is characterized by the rate T_{1s}^{-1}. This range can be taken approximately to be the thickness of a molecular monolayer, 3 Å.

For a distributed pore space in which the pores are only weakly coupled, the axation is given by a sum of exponentials:

$$[M(t) - M_\infty] / [M(0) - M_\infty] = V^{-1} \sum_j V_j \exp(-t/T_{1j}). \quad (3)$$

$$T_{1j} = 1/T_{1b} + (L/T_{1s})(S/V)_j = 1/T_{1b} + a/r_j, \quad T_{1s} \ll T_{1b} \quad (4)$$

The magnetization recovery in this case is given by the Laplace transform of the pore size distribution.

We can apply this result to the specific case of a fractal pore size distribution characterized by a fractal dimension d_F, in the size interval, $r_1 < r < r_2$. The distribution of pore volumes, $\rho(r)$, of size r can be represented in the fractal range by $\rho(r) \propto r^{-\alpha}$, where $\alpha = d_F - 2$.

$$\rho(r) = \begin{cases} A\,\delta(r - r_s) & r_s < r_1 \\ B\,\delta(r - r_L) & r_2 < r_L \\ C\,r^{-\alpha} & r_1 < r < r_2 \end{cases} \quad (5)$$

We may take $A \approx 0$. (The volume of the small pores of size r_s, is negligible.)

Substituting $\rho(r)\,dr$ for V_j in Eq. 4 and converting the sum to an integral we can find that the relaxation function has the following limiting forms,

$$[M(t) - M_\infty] / [M(0) - M_\infty] = \exp(-t/T_{1b})\,[\,1 + at\,(r_2/r_1)^{\alpha-1}(\alpha-1)/(\alpha r_1) + \ldots\,]$$

$$\text{for the limit } t \to 0 \quad (6)$$

$$[M(t) - M_\infty] / [M(0) - M_\infty] = \exp(-t/T_{1b} - ta/r_L)$$

$$\text{for the limit } t \to \infty \quad (7)$$

This behavior is compared with the Coconino and St Peters sandstone data in Fig. 1 and 2. The long-time limit, Eq. 7, is an exponential relaxation in the time and is consistent with our observations. The short-time limit expression is not consistent with the data.

Guyer[4] has developed a different model for spin-lattice relaxation in a fractal pore space based on the Menger sponge. This is appropriate for the case when there is very slow molecular diffusion compared to the time for surface relaxation, T_{1s}. This leads to the establishing of a non-uniform nuclear magnetization within each pore following an rf excitation, in contrast to the approach of Cohen and Mendelson where this magnetization is supposed to be uniform within a pore. For Guyer's model,

$$[M(t) - M_\infty] / M(0) - M_\infty] = \exp(-t/T_{1b})[1 - (t/T_\phi)^{(1-\alpha)/2}]$$

$$\text{where } r_1 < (\langle x^2 \rangle)^{1/2} < r_2. \tag{8}$$

In the above expression, the time $T_\phi = \phi d^2 / D_b$, where ϕ is the porosity, D_b the bulk diffusion constant, and d the sample size. The root-mean-square displacement of the molecule, $\langle x^2 \rangle^{1/2}$ should lie in the fractal region for the time frame under consideration in order that Eq. 8 apply.

For the Coconino sandstone $d_F = 2.78$, according to Katz and Thompson[3], and using this model we find that the recovery of the magnetization should be a linear function of $t^{0.11}$ in the short-time limit. This behavior is not observed in our data. However the predicted[4] long-time limit is exponential, consistent with the observations.

Although the results for the spin-lattice relaxation at short times could not be

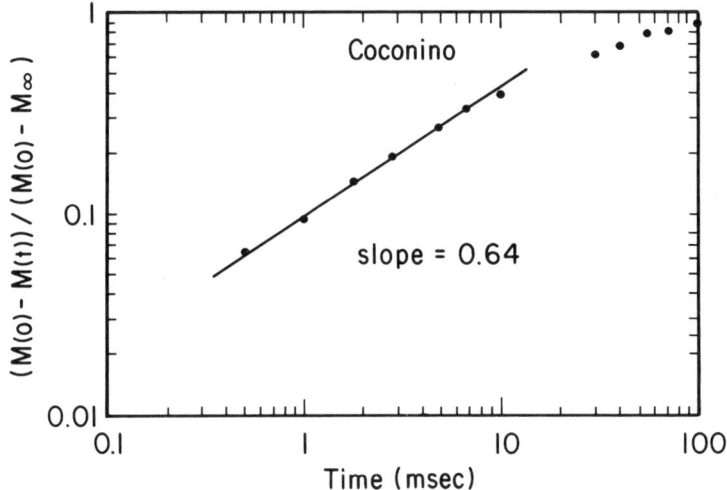

Figure 1. The magnetization relaxation $[M(0) - M(t)] / [M(0) - M_\infty]$ of water (spin-lattice relaxation) infused in Coconino sandstone is shown plotted in log-log form as a function of the time. This is displayed in a format suitable for comparison with Eq. 6 or 8. A linear behavior is observed although the slope is not consistent with that predicted for the models discussed in the text (Ref. 2, slope = 1; Ref. 4, slope = 0.11)

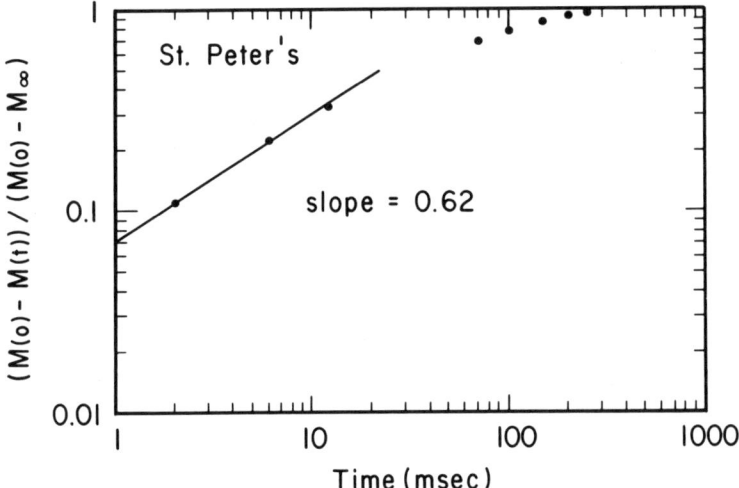

Figure 2. The magnetization relaxation $[M(0) - M(t)] / [M(0) - M_\infty]$ of water (spin-lattice relaxation) infused in St. Peters sandstone is shown plotted in a log-log form as a function of the time. This is displayed in a format suitable for comparison with Eq. 6 or 8. A linear behavior is observed although the slope is not consistent with that predicted for the models discussed in the text (Ref. 2, slope = 1; Ref. 4, slope = 0.11)

understood in terms of the existing models we have been able to show that they are *scale invariant* in time by comparing the data for Coconino and St. Peter's sandstones. The time scaling parameter was found to be consistent with Guyer's[4] parameter $T\phi$ which is proportional to the porosity. This comparison is shown in Fig. 3. The time scale for relaxation of the magnetization of water in St. Peters sandstone has been multiplied by the factor of 0.59. From experimental data[3] the ratio of T_ϕ for Coconino to that for St. Peters sandstone lies in the experimentally defined range of 0.39 and 0.52 consistent with the above factor. This scaling result suggests that the pore space distribution in these sandstones is similar and rather broadly dispersed; however, we emphasize that this comparison is as yet, only phenomenologically motivated.

Measurements of the spin-lattice relaxation of water infused into the pore space of vycor glasses of several different pore sizes, and of a very pure silicate sandstone, Arkansas sandstone, have been performed. The relaxation was predominantly exponential in these cases with the only small deviations observed appearing in the short-time limit for the Arkansas sandstone. These results suggest that the pore space in these materials is not widely dispersed. The results obtained with water in the pore spaces are plotted in Fig. 4 showing consistency with the theory of Cohen and Mendelson[2] and Eq. 2. where a linear relation between inverse pore size, r^{-1}, and the spin-lattice relaxation rate T_1^{-1} is expected. Similar results have been obtained with green compacts of ceramic powders including Al_2O_3, and $MgAl_2O_4$.

4.   Pulsed Magnetic Field Gradient Diffusion Measurements (PFG-NMR)

The technique of PFG-NMR can be used rather generally to determine the diffusion constant as has been summarized recently[5]. This technique gives an

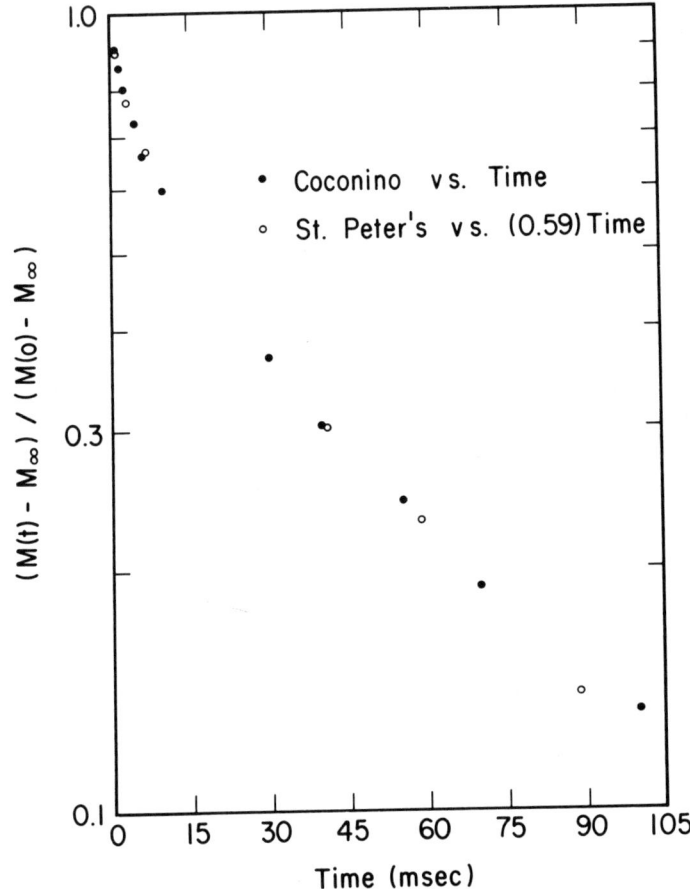

Figure 3. A semi-log plot of the short-time dependence of the spin-lattice relaxation of water in Coconino and St. Peters sandstones is shown where scaling of the results is achieved by normalizing the time with the porosity as was described in the text.

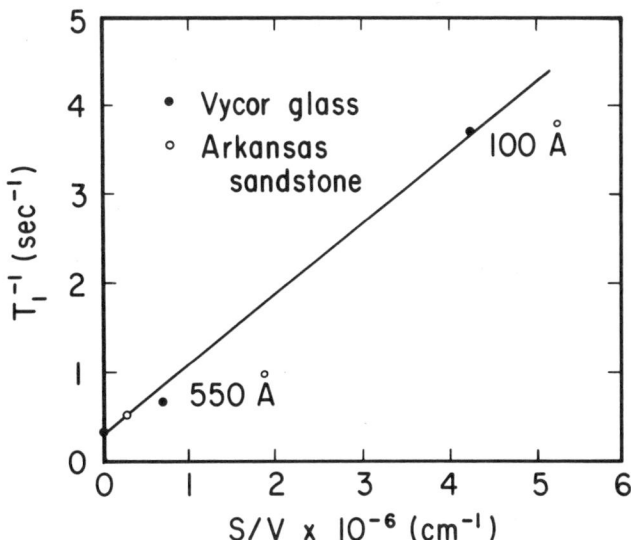

Figure 4. The spin-lattice relaxation time of water imbibed in porous silicates is displayed including two vycor glass samples and a magnetically clean sandstone. For these samples the recovery of the magnetization was found to be a simple exponential relaxation in time characteristic of what is expected for anarrow pore size distribution.

unambiguous measure of the mean-square molecular displacement, $<x^2>$, in terms of the measured relative nuclear magnetization, M/M_0, obtained from a spin-echo experiment after allowing diffusion for a period of time Δ,

$$\ln(M/M_0) = -\gamma^2 G^2 \delta^2 <x^2> \qquad (9)$$

The applied gradient fields are applied for periods of time δ and of magnitude G. This expression is more commonly written in terms of the diffusion coefficient D_b in the following way,

$$\ln(M/M_0) = -\gamma^2 G^2 \delta^2 D_b (\Delta - \delta/3) \qquad (10)$$

Adjustment of either Δ or D_b by choice of appropriate filler fluids allows the exploration of the dependence of the root-mean-square displacement, $(<x^2>)^{1/2}$, on diffusion time over an experimentally accessible range of about three orders of magnitude. In a fractal space we expect that the relationship between the mean-square displacement and the diffusion time to be given by,

$$<x^2> \propto \Delta^{2/d_w}, \qquad (11)$$

where d_w is the random walk dimension of that structure.

In practice, there are limitations to this approach if T_{1s} is very short as is frequently the case in sandstones owing to the presence of electronically active relaxation sites on the sandstone surface. In this situation[6] the fluid in the small pores is relaxed predominantly by surface electron-nuclear dipolar interactions in the form summarized by Eq. 2. On the other hand, during the PFG-NMR experiment, the larger pores allow larger spatial excursions of the diffusing fluid molecules and this leads to more effective relaxation of the nuclear magnetization as described by Eq. 9 as compared with the surface relaxation mechanism. It is not simple to deconvolve these effects unless some assumptions about the pore geometry are made *a priori*. .

We have found evidence of non-classical diffusion in St. Peters sandstone as is suggested by Eq. 11. This is schematically represented in Fig. 5, where we show that the magnetization is independent of diffusion time, Δ, for $\Delta > 5$ msec. The root-mean-square displacement corresponds to pores of size $\approx 4\ \mu$. We have found that this value depends on the gradient field as has been recently suggested[6].

In the porous ceramics and vycor glasses we have only observed classical diffusion behavior where $<x^2> \propto \Delta$. The corresponding measured self diffusion coefficients are substantially reduced from that of the bulk by approximately a factor of two indicating that there is a marked effect of the restricted geometry of these pore spaces. A more detailed quantitative study of these effect is in progress.

5. Conclusions

a) A scaling behavior for the non-exponential time evolution of the magnetization relaxation of water in sandstones has been discovered. This appears to hold during the first part of the recovery and strongly suggests a very wide pore distributions. At long times the magnetization recovery of water in sandstones is exponential in time. The scaling time appears to be proportional to the porosity. In porous green-state ceramic bodies and vycor glass materials, only exponential relaxation is observed suggesting that these pore spaces are more homogeneous.

b) Diffusion measurements of filler fluids in porous structures shows classical

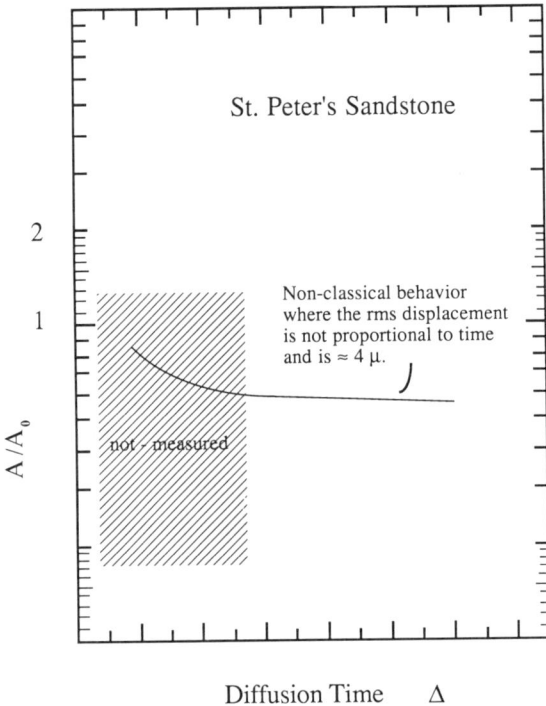

Figure 5. The pulsed-field-gradient diffusion measurements of water in St. Peters sandstone indicate that non-classical diffusion is observed. The molecular root-mean-square displacement appears to be limited to approximately 4 μ.

behavior in green ceramic compacts and in vycor glass, and extremely non-classical behavior in the St. Peter's sandstone.

6. Acknowledgements

This work was performed in the NMR facility of the Materials Research Center at Northwestern University supported by the NSF-MRL program grant no. DMR82-16972. Partial support for this work has been provided through the Department of Energy. We thank Ken Mendelson for stimulating conversations and for pointing out an error in the original manuscript and Bob Guyer for helpful discussions.

7. References
   1. J.D. Loren, and J.D. Robinson, Soc. Pet. Eng. J. (Sept. 1970), 268.
   2. M. H. Cohen, and K. S. Mendelson, J. Appl. Phys. 53, 1127 (1982).
   3. A. J. Katz and A. H. Thompson, Phys. Rev. Lett. 54, 1325 (1985).
   4. R. A. Guyer, private communication.
   5. J. C. Tarczon and W. P. Halperin, Phys. Rev. B 32, 2798 (1985).
   6. M. Lipsicas, J.R. Banavar, and J. Willemsen, (preprint)

DYNAMICS IN POROUS SILICA

J. M. Drake, W. D. Dozier, J. Klafter

Exxon Research and Engineering Company
Route 22 East - Clinton Township
Annandale, New Jersey 08801

Abstract

We have measured the self-diffusion coefficient (D_S) of a dye molecule in solution in porous vycor glass by the technique of forced Rayleigh scattering. We find that the value of D_S for motion in the pore space is about two orders of magnitude less than for free diffusion in bulk solution. By derivatizing the active surface sites we have separated the roles of the chemical and physical features of the pore space on diffusion. We believe this to be the first measurement of self-diffusion in solution in porous vycor which has been proposed to have a fractal pore structure.

The transport properties of molecules in solution in porous media are of great technological importance, especially with regard to catalysis and tertiary oil recovery. Recent studies on the structure of several rocks have shown the possibility that pore spaces are fractals.[1] However superimposed on this geometry there is also chemical heterogeneity.[2,3] Both are probably important for the problem of diffusion. Here we study molecular diffusion in a porous medium trying to distinguish between the geometrical and the chemical contributions. We present what we believe to be the first measurements of the self diffusion coefficient (D_S) of a probe molecule in solution in a porous silica.

The porous silica we studied was vycor glass which is made by phase separation of boron in borosilicate glass. The phase separated boron forms a connected path which produces the pore space when removed by acid leaching.[4] Recent SAXS measurements have shown that the structure of this pore space is reminiscent of spinodal decomposition, with a broad peak in the structure factor corresponding to a length of about 300 Å.[5] Energy transfer measurements which probe small length scales ($\leq$ 100 Å), suggest that the pore structure of vycor glass is fractal with a fractal dimension ($\bar{d}$) of 1.74 $\pm$ 0.05.[6] On length scales over which the pore space is fractal one expects to observe anomalous diffusion which is given by the fractal and spectral ($\tilde{d}$) dimensions[7-10]. Our original motivation was to measure the spectral dimension of vycor and characterize the self-diffusion of a molecule in solution within the pore space.

The vycor was 7930 glass from Corning. Two types of disks were used a 6 mm thick an 25 mm in diameter disk with a scalloped surface and a 1 mm thick and 25 mm diameter disk with a flat but not polished surface. We first measured the pore size distribution using a 500 point nitrogen desorption technique.[11] The distribution is sharply peaked at 30 Å radius, with essentially all the features falling between 20 and 55 Å radius. The self-diffusion coefficient was measured by forced Rayleigh scattering (FRS). The optical configuration for this technique is described in reference (12). Forced Rayleigh scattering is an effective tool for measuring the self-diffusion coefficient. The diffusing molecule is a photochromic dye. In this case, azobenzene was used, which is stable in solution as the trans isomer but can be photoisomerized to the cis isomer. The cis and trans isomers have different optical properties at various visible wavelengths. A laser beam is used to induce the isomerization. The beam is split and recombined in the sample, giving a pattern of interference fringes. This sets up a spatially periodic density of cis-azobenzenes. A second laser can then be Bragg scattered from what is effectively a diffraction grating. Since the first laser beam is only flashed on the sample a short time, the pattern begins to fade as the dye molecules diffuse and the density of cis-azobenzene becomes homogeneous.[12] The interference fringes were created with the 488 nm argon ion laser line ("writing wavelength") and the scattering wavelength was 628 nm from He-Ne ("reading wavelength"). The vycor glass was index-matched at the "writing" wavelength using alcohol/toluene mixtures which gave an index of 1.4684. The two solutions which proved most useful were 4.3 M methanol in toluene and 3.3 M 1-propanol in toluene. The concentration of azobenzene was always 8 millimolar and the measurements were made at $\sim 20°C$.

There are two major properties of porous materials which effect the measurements of self-diffusion. They are related to the extent to which pores are dead ended and tortuous (geometry) and to the degree the diffusing molecules are adsorbed on surface sites (chemistry). Separating the contributions of physical and chemical traps to the diffusion of azobenzene in vycor was a problem which we had to address. Our initial approach was to block the surface adsorption sites by adsorbing alcohol molecules. The order in which the alcohol, toluene and azobenzene are added to the vycor determined the degree of spatial homogeneity which could be achieved in distributing azobenzene in the vycor. Our prescription was to imbibe in methanol over three days, then remove the excess methanol by evaporation and re-expose the vycor to an alcohol/toluene mixture. After about two weeks the sample was nearly index-matched. The excess alcohol/toluene was removed and replaced with a solution of alcohol/toluene with azobenzene. After an additional week there was enough FRS signal to conduct the diffusion measurement.

D_S is determined from the decay rates of the Bragg scattering from fringes of several spacings.[12] For an exponential decay with rate τ^{-1}:

$$D_S q^2 = \frac{1}{\tau} - \frac{1}{\tau_r}, \qquad (1)$$

where q is the wavevector of the fringes (2π/fringe period) and $1/\tau_r$ is the thermal relaxation rate of cis-azobenzene to trans-azobenzene. The result for azobenzene in 4.3 M methanol in toluene are shown in Figure 2. Note the collection of points at $1/\tau \cong$ 1Hz. Relaxations of about this rate were found independent of fringe spacing, showing it to be an intrinisc relaxation of the dye molecules adsorbed to the pore surface rather than arising from diffusion. The decay of the Bragg scattering was also decidedly non-exponential for these cases. By moving the sample, q-dependent decays could be found. These signals displayed a good exponential decay which obeys Eq. (1). If we identify the q-independent, non-exponential signals as those coming from azobenzene chemisorbed to the surface and the rest as coming from free dye diffusing within the pore space, we obtain D_S = 4.2 ± 0.3 X 10^{-8} cm^2/sec.

Although we were able to measure D_S for these samples the extent of spatial inhomogeneity indicated that the methanol did not effectively block all the active sites of the vycor glass. To achieve a more complete chemical passivation of the surface sites we derivatized the silanol groups by reacting them with 1-propanol. This esterification reaction was carried out in neat 1-propanol at its boiling point for 24 hours. The reaction that took place was[13]

$$Si - OH + CH_3CH_2CH_2OH \xrightarrow[98^0C]{\Delta}$$

$$Si - OCH_2CH_2CH_3 + H_2O,$$

which yields a more hydrophobic surface that is less active towards adsorption of azobenzene. Using this derivatized vycor the procedure for preparing samples was the same except no pretreatment with methanol was used. The important result obtained is that it was <u>spatially homogeneous</u> as measured by FRS. These samples showed <u>no</u> non-exponential, q-independent signals. The q-dependent relaxation was the same over the entire sample clearly suggesting that the diffusion of azobenzene was not restricted by chemisorption to the pore surface. For the azobenzene in 1-propanol/toluene solution in 1-propanol derivatized vycor we find $D_S = 2.4 \pm 0.4 \times 10^{-8}$ cm^2/sec. The value of D_S for methanol/toluene in the derivatized vycor, is $3.6 \pm 0.6 \times 10^{-8}$ cm^2/sec. The methanol/toluene value suggests that the derivatization had no effect on the diffusion coefficient, since D_S before and after derivatization are essentially the same. It should also be noted that the difference between D_S in the two solvent systems is completely explained by the differences in the viscosities of the solvents. We also measured the free bulk diffusion D_0 in the two solvents by the same FRS method and found the values to be $2.4 \pm 0.5 \times 10^{-6}$ cm^2/sec and $1.4 \pm 0.3 \times 10^{-6}$ cm^2/sec in methanol/toluene and 1-propanol/toluene, respectively. <u>D_S is two orders of magnitude lower then free bulk diffusion.</u>

Similar behavior has been observed by Karger et. al.[14] who used NMR to study self diffusion in porous glasses. However it is not clear from the work whether the experiments were measured self diffusion in a liquid saturated pore or surface diffusion. But their results are strongly affected by the chemically active surface sites, an important issue which we have eliminated from our results.

We now present two possible models to interpret the slow dynamics observed by the FRS in porous vycor. We first follow the parallel-pore model[14] proposed to describe a monodispersed pore size distribution in porous catalysts. In this case the diffusion is related to D_0 in a phenomenological way through the porosity Φ and the tortuosity factor δ[15]:

$$D_S = D_0 \ (\Phi/\delta) \tag{2}$$

In order to determine the tortuosity factor we also measured the porosity of the vycor using two techniques: nitrogen adsorption and spontaneous liquid imbibition. The nitrogen adsorption method yielded a porosity Φ of 0.40 ± 0.04 for pores above 10 Å radius and indicated a micropore porosity of 0.14 ± 0.05 for pores with a radius between 5-10 Å. The second technique was based on the spontaneous imbibition of a liquid, where the volume of solution imbibed was compared to the volume of the vycor disk. This kind of measurement is prone to interference by the surface sites and by the relationship between the size of the imbibing molecules and the pore structure.[16] Therefore, porosity measurements were made using a solution of the same composition as that used to index-match the pore space for the light scattering experiments. The porosity measured by imbibition was 0.28 ± 0.03 for both 4.3 M methanol in toluene and 3.3 M 1-propanol in toluene. This apparent difference in the porosity using the two techniques may in fact be due to the difference in the molecular yardsticks used to probe the pore space.

For porosity Φ = 0.28 and from Equation 2 we obtain the tortuosity factor $\delta \cong 18$. This δ is much higher then the value 5.6 found previously from the diffusion of hydrogen.[15] From the measurements of D_0 we calculated the hydrodynamic radius of the cis-azobenzene to be about 4.1 ± 0.7 Å in the propanol/toluene and 6.5 ± 1 Å in methanol/toluene. These radii are not much larger than a hydrogen molecule radius with respect to the available pore space, since the pores have a narrow size distribution. It could be that since under steady-state flow conditions (the conditions under which the gas diffusion measurements are done) "dead end" pores do not contribute to the measurement, so one should renormalize Φ to account for only that fraction of the pore space that contributes to the diffusion. The FRS experiment is done under conditions of macroscopic equilibrium. If one assumes that δ is the same in both experiments, then the result suggests that 2/3 of the pore-space is not contributing to the diffusion. If that were the case, the FRS signal would lose 1/3 of its intensity in a relatively short time and the rest would decay at either an extremely slow rate or at a rate of $1/\tau_r$, whichever is faster. This is not what is observed. If instead we take Φ to be the measured value then the difference in δ is understood as an additional effect of "tortuosity" since the lost time spent in dead end pores should affect the diffusion constant in the FRS experiment. The concept of tortuosity bears a clear intuitive meaning, which probably gives the correct trend. It is however not directly measurable and not well defined.[15,17]

A second possibility is to model the geometrical disorder by a fractal, self-similar structure. Thus, we assume, based on previous suggestions,[6] that the pore structure of porous vycor has a fractal character on length scales $\ell_1 < \ell < \ell_2$ where ℓ_1 and ℓ_2 are the lower and upper cut-offs of the fractal range respectively. We calculate the time τ it takes a molecule to diffuse over lengths of the order of the fringe spacing L. It is clear that, since the range of the fringe spacing is between 1 and 100 μm,

$L \gg \ell_2$, namely we measure diffusion over lengths larger than those expected to exhibit anomalous diffusion.[7-10] On scales such that $L \gg \ell_2$ classical diffusion sets in and the scattering intensity should decay exponentially with time, as we experimentally observed. From a cross-over argument for a random walker on a structure which is fractal for $\ell_1 < \ell < \ell_2$ we obtain

$$\tau = \tau_0 \left(\frac{\ell_2}{\ell_1}\right)^{2\bar{d}/\tilde{d}} \left(\frac{L}{\ell_2}\right)^2 \qquad L \gg \ell_2 \qquad (3)$$

where τ_0 is the elementary step time of the walker and $\tilde{d}$ is the spectral dimension. Using the relationship between D_S and τ, Eq. (1) where τ_r is very large, we rewrite Equation 3 in terms of diffusion constants

$$D = D_0 \left(\frac{\ell_1}{\ell_2}\right)^{2(\bar{d}/\tilde{d}-1)} \qquad (4)$$

where $\ell_1^2 \sim D_0 \tau_0$.

Let us assume, for instance, that the fractal and spectral dimensions have the three dimensional percolation values $\bar{d} \sim 2.5$ and $\tilde{d} \sim 1.3$. Then for the ratio $D/D_0 = 1.75 \times 10^{-2}$ measured in methanol/toluene solutions we obtain through Eq. (4) the ratio $\ell_1/\ell_2 \sim 0.1$. This means that if the lower cut-off is taken as the average pore size, $\ell_1 \sim 30$ Å, the upper cut-off is $\ell_2 \sim 300$ Å. The porosity of a fractal pore geometry is given by[1]:

$$\Phi \sim \left(\frac{\ell_1}{\ell_2}\right)^{3-\bar{d}} . \qquad (5)$$

For the assumed values of ℓ_1/ℓ_2 and $\bar{d}$ we calculate a porosity $\Phi \sim 0.31$ which is close to the porosity measured by imbibition. As mentioned above, different molecules may probe different portions of the pore space. Even et. al measured $\bar{d} \sim 1.74$ which may be attributed[15] to the larger molecules used in their experiment. We would like to emphasize that the values we chose to model the system are quite heuristic. One can assume other parameters $\bar{d}$ and $\tilde{d}$ which are also consistent with the observed diffusion and porosity. More independent measurements of the various parameters and molecular size effects are needed.

In summary, although in most porous surfaces chemical heterogeneity is superimposed on the geometrical disorder, we have managed to separate these two contributions by blocking the chemical traps by derivatization. We measured the pore size distribution of the vycor and its porosity and related them to the self-diffusion of the azobenzene molecule through a conventional phenomenological approach which relies on the concept of a tortuosity factor and through a more detailed fractal model which involves the basic parameters $\bar{d}$ and $\tilde{d}$ and the range of the fractal behavior.

We have demonstrated the applicability of the FRS technique as a method to study dynamics in transparent porous media. More details and results on different porous silicas with controlled pore size distributions will be published elsewhere.

References

1. A. J. Katz and A. H. Thompson, Phys. Rev. Lett. $\underline{54}$, 1325 (1985).
2. J. M. Drake and J. Klafter, J. of Lumin. $\underline{31}$ & $\underline{32}$, 642 (1984).
3. P. G. deGennes, J. Fluid Mech. 136, 189, (1983)
4. C. C. Ballard, E. C. Broge, R. K. Iler, D. S. St. John, and J. R. McWhorter, J. Phys. Chem. $\underline{65}$, 20 (1961).
5. S. K. Sinha (private communication).

6. U. Even, K. Rademann, J. Jortner, N. Manor and R. Reisfeld, Phys. Rev. Lett. 52, 2164 (1984).
7. Y. Gefen, A. Aharony and S. Alexander, Phys. Rev. Lett. 50, 77 (1983).
8. S. Alexander and R. Orbach, J. Physique Lett. 43, L625 (1982).
9. R. Rammal and G. Toulouse, J. Physique Lett. 44, L13 (1983).
10. For experimental studies of $\tilde{d}$ see: P. W. Klymko and R. Kopelman, J. Phys. Chem. 87, 4565 (1983); P. Evesque, J. Phys. (Paris) 44, 1217 (1983).
11. Measurements were done using a Omnisorp (TM) - 360 analyzer with W. M. J. Pieters of Omnicron Technology Corp.
12. F. Rondelez, H. Hervet and W. Urbach, Chem. Phys. Lett. 53, 138 (1978).
13. T. H. Elmer, M. E. Nordberg, G. B. Carrier and E. J. Korda, J. Am. Cheramic Soc. 53(4) 171 (1970).
14. J. Kärger, J. Lenzher, H. Pffifer, H. Schwabe, W. Heyer, F. Janowski, F. Walf and S. P. Zdanov, J. Amer. Cer. Soc. 66, 69 (1983).
15. J. M. Smith, Chemical Engineering Kinetics (McGraw-Hill, N.Y. 1981).
16. W. H. Wade, Soc. Petro. Eng. J., April 1974, p. 139.
17. F. A. L. Dullien, Porous Media Fluid Transport and Pore Structure (Academic Press, New York, 1979).

A Fractal Model of Structure, Vibrations, and Spin Relaxation in Paramagnetic Proteins

Harvey J. Stapleton
Department of Physics
University of Illinois at Urbana-Champaign
1110 W. Green St., Urbana, IL 61801

The intent of this talk is to present an account of the experimental results that have led to the development of a fractal theory of electron spin relaxation in paramagnetic biopolymers. The model was first proposed orally at the 1980 March APS meeting. A Phys. Rev. Lett. appeared in October[1]. The further development of an applicable theory was picked up by Alexander and Orbach[2], Rammal & Toulouse[3], and more recently by Alexander, Entin-Wohlman, and Orbach[4]. At this point it is still uncertain if the current theory of relaxation by a two fracton process can explain the observed anomalies in the temperature dependence of spin-lattice relaxation in paramagnetic proteins. I hope that the basis for this rather inconclusive statement will become clearer by the end of the talk.

It is appropriate for this audience to say a few additional words about protein structure before discussing the anomalous behavior observed in them. Proteins are linear, unbranched, occasionally cross-linked polymers of a large number of amino acid residues. The polymerization process is shown schematically in Fig. 1.

The carbon atom to which one of twenty possible residues is attached is called the alpha carbon, C_α. The strongly bonded amide planes and their associated dihedral angles, Φ and Ψ are shown in Fig. 2. Because the peptide bond is planar, the flexibility of the polypeptide backbone is primarily through its two dihedral angles per residue, but even this motion is hindered by the bulkiness of the polymer. The number of amino acid residues varies from protein to protein, but 100-300 is a resonable range for an individual chain. According to the data in Fig. 1, below, this implies molecular weights in the 10,000- 30,000 range. A typical dimension for a globular protein is 40 Å.

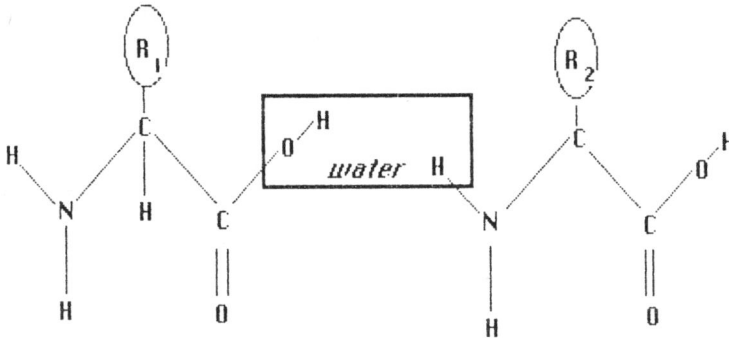

Fig. 1. Polymerization of amino acid monomers. Side chains $R_1....R_{20}$. M. W. of residues 57-186 (wt. av. approx 110).

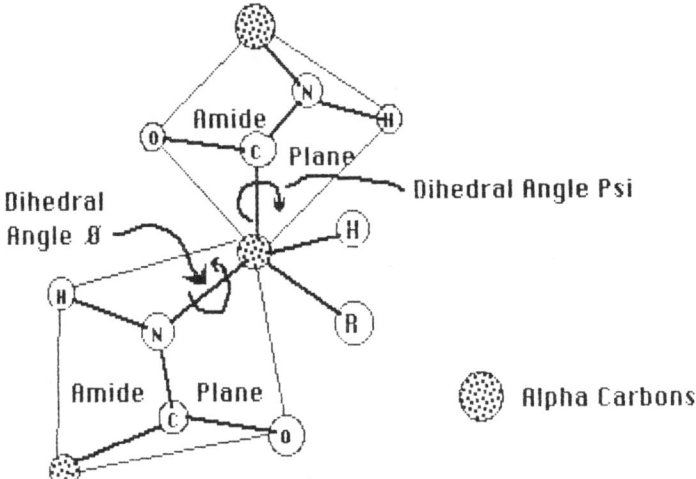

Fig. 2. Schematic diagram showing the strongly bonded amide plane and dihedral angles associated with each amino acid residue in the polypeptide chain.

The amino acid sequence of the protein is called its primary structure. These side chains can be nonpolar, charged polar, or uncharged polar. Each protein folds into a unique structure. The secondary structure consists of regular arrangements of the backbone into helices, sheets, turns, or random coils. The tertiary structure of the protein refers to the folding of the secondary structures in three dimensional space. In proteins containing transition metal ions, the metal is located at a specific site on the chain. This is generally the active site for some important biochemical reaction. To a large extent, the remaining structure of the protein is arranged to provide a proper environment for the reactions at the active site.

The Raman spin relaxation process is a two phonon mechanism in which a phonon of energy $h\nu_1$ is destroyed and one of energy $h\nu_2$ is created, with the energy difference, $h(\nu_1 - \nu_2)$, being equal to the Zeeman energy of the flipped spin, $g\beta H$. If the phonon density of states, $\rho(\nu)$, varies with frequency as a $\nu^{\bar{d}-1}$ power law, then the temperature dependence of the Raman rate is a T^n power law with $n=2\bar{d}+3$ for paramagnetic ions with an odd number of electrons (Kramers' ions). For non-Kramers' ions, $n=2\bar{d}+1$. This simple power law dependence is valid at temperatures well below the phonon Debye temperature. At higher temperatures, the temperature variation of the Raman rate drops, eventually reaching the classical limit of T^2, appropriate to a two phonon process. In ordinary materials $\bar{d}=3$, the dimension of the space. On fractal structures, $\bar{d}$ can be non-integer and is termed the fracton or spectral dimension. I will make a distinction in this talk between the spectal dimension, $\bar{d}$, and m, the spectral exponent, an experimentally obtained parameter defined as $(n-3)/2$. There are three important questions. What are the fractal dimensions of protein structures ? What is the value of the spectral dimension for a structure of known fractal dimension, D ? What is the relationship between $\bar{d}$, D, and n, the Raman rate power law ? Before discussing those issues further, a summary of the experimental data will be presented.

The first indication that Raman spin-lattice relaxation rates in proteins are anomalous appeared in 1973. Mailer and Taylor[5] reported relaxation data on single crystals of ferricytochrome C, an iron containing protein with 103 residues. Figure 4 of their paper indicated that the T dependence of the Raman relaxation rate was T^7 at temperatures as high as 20 K. There were two things wrong about this: (1) a T^7 dependence is not predicted for Fe^{3+}, a Kramers' ion; and (2) the line through their data actually follows a T^6 dependence. It was not until 1980 that this latter feature of the data was known to us.

In 1976 Herrick and I published relaxation data on another iron-containing protein, cytochrome P-450 from <u>Pseudomonas putida</u> [6]. We were aware at the time that the data fit a $T^{6.3}$ dependence very well, but we had no knowledge of fractal dimensions and interpreted our data in terms of a anomalous T^7 dependence with a sufficiently low

Debye temperature (75 K) to make the effective temperature exponent 6.3. This can only be made to work over a limited range of temperatures, however, since the Raman rate is no longer a simple power law under these conditions.

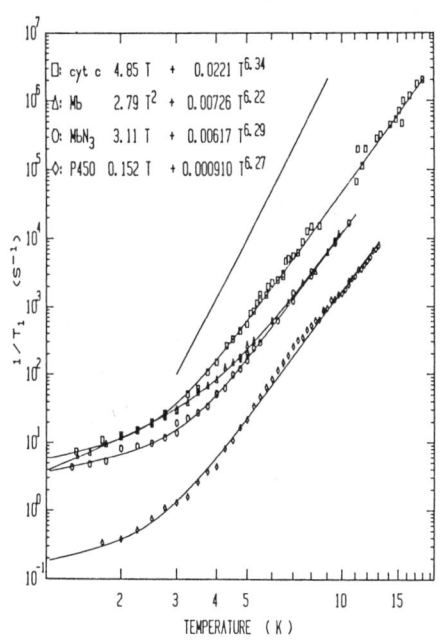

Figure 3 shows relaxation data[1] from four low spin hemoproteins: ferricytochrome C (cyt c), myoglobin.OH (Mb), myoglobin azide (MbN$_3$), and a reanalysis of Herrick's cytochrome P-450 data. Each data set is separately fit to the sum of a direct relaxation process (varying with temperature as T or T^2) and a simple T^n power law. The best fitting values of n are indicated. Standard errors are typically 1%. The straight line in the figure represents a T^9 dependence. For Mb and MbN$_3$, only every third datum is displayed between 5 and 11 K to improve clarity. The values of n are all in the neighborhood of 6.33. It is worth noting at this point that from an n value of 6.33, we infer an m value of 5/3, the fractal dimension of a self-avoiding random walk in three dimensions.

Gayda et al.[7] published relaxation data on a 2 iron – 2 sulfur ferredoxin in 1979. Figure 4 is our reanalysis of their data assuming a simple power law dependence for the Raman rate. The data fit an n value of 5.67 ± 0.11, or equivalently, m=1.34 ± 0.06.

We have also published relaxation data[8] on a similar iron-sulfur protein, putidaredoxin from P. putida. Here solvent conditions were found to exert an influence on n. These data are shown in Fig. 5. The solvents differed in their ionic strengths. The larger m value of 1.34 ± 0.03 correlates well with Gayda's data. In the same paper we reported relaxation data on another hemoprotein, cytochrome C551. Here again, solvent effects were observed. The values of m were approximately 0.88 ± 0.14 in a low ionic strength solvent (I=0.1), and 1.43 ± 0.09 in a high ionic strength solvent (I=1.1).

In the same paper we reported relaxation data on another hemoprotein, cytochrome C551. Here again, solvent effects were observed. The values of m were approximately 0.88 ± 0.14 in a low ionic strength solvent (I=0.1), and 1.43 ± 0.09 in a high ionic strength solvent (I=1.1).

Additional relaxation data will be deferred until after a discussion of an algorithym for computing the fractal dimension of a protein, which as we have seen is neither self-similar nor random, nor strictly speaking even a polymer since the side chains differ among monomer elements. We consider the alpha carbon coordinates from x-ray structural data as specifying the position of each monomer of the biopolymer. For an M residue protein, we have M-1 elemental segments, r_i, from which to compute various end-to-end lengths. The average separation, $<r>$, between alpha carbons in proteins is approximately 3.8 Å.

Fig. 6 is a fractal structure which illustrates properties relevant to a polypeptide. Fractals with geometric self-similarity are comprised of two elements: an initiator, which in this example is a straight line over the interval [0,R]; and a generator, which is the largest N-sided structure spanning the interval [0,R]. The most general relation characterizing a fractal curve is given by the uppermost expression in the figure. If all lengths r_i describing the generator are equal, which is a good approximation for the distances between alpha carbons in proteins ($r \approx 3.8$ Å), then the expression simplifies to the more common form in the second line. A fractal dimension is calculated directly from the properties of the initiator and generator (N=8, r=R/4). The ratio R/r also defines a scaling factor s of the generator that determines the manner of constructing the fractal in an infinite number of stages. Note that each stage of a complete construction in the figure should extend over the entire interval of the initiator and that a fifth stage would be virtually imperceptible. The D of a structure, however, is independent of the number of constructions and the increasing contour length. Estimating D for an arbitrary polypeptide without an explicit form for the initiator or generator, is based on identifying statistically the most common repeating motifs within the structure.

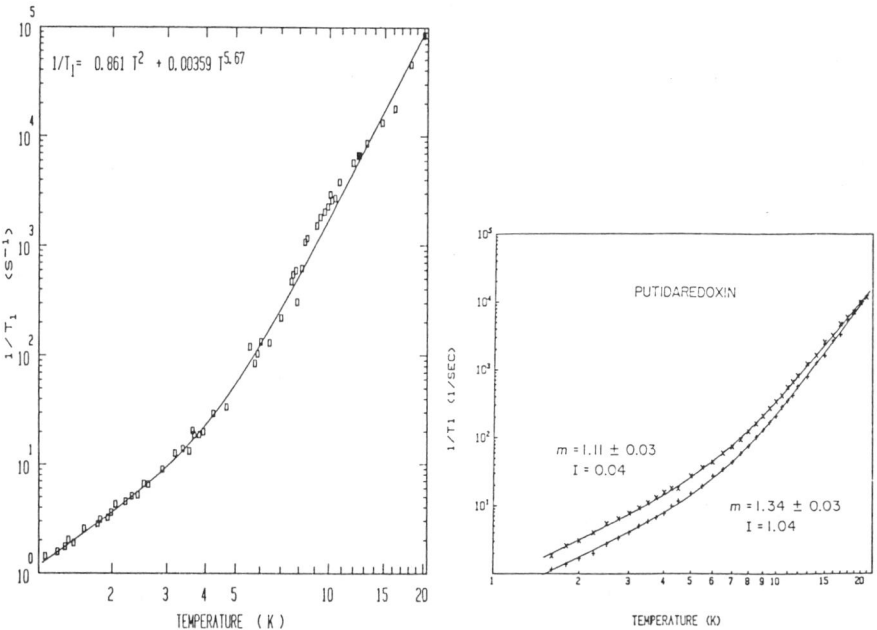

Fig. 4 Reanalysis of relaxation data on a 2Fe-2S protein from Ref. 7.

Fig. 5 Relaxation data on another iron-sulfur protein, putidaredoxin from P. putida, indicating a solvent dependence.

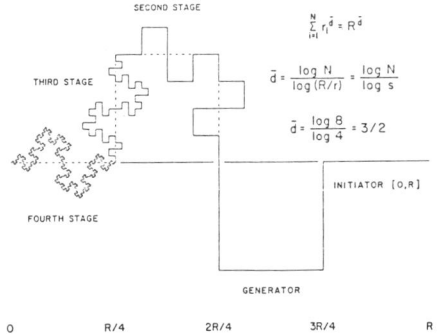

Fig. 6 A fractal structure illustrating properties relevant to a polypeptide chain. In this figure the fractal dimension, D, is denoted by $\bar{d}$.

We want to approximate any self-similarity of biopolymer chains as both statistical (i.e. random) and geometric. We will compute a suitably averaged root-mean square (RMS), end-to-end length, R(N), for N bonds of average length $\langle r \rangle$ and plot Log {R(N)/$\langle r \rangle$} vs. Log N to obtain D from the inverse slope, i.e.

$$R(N)/\langle r \rangle = N^{1/D}. \tag{1}$$

To distinguish this "chain" fractal dimension from others, which might scale all the mass within a sphere of radius R, we will add a subscript c to the symbol, i.e. D_c.

Values of R(N) will be calculated as a weighted mean between two limits, $R_S(N)$ and $R_G(N)$, where the subscripts S and G refer to statistical and geometric self-similarity. For a chain of M residues,

$$R_S^2(N) = \{1/(M-N)\} \sum_{i=1}^{M-N} R^2_{i, i+N}, \tag{2}$$

where $R_{i, i+N}$ is the through space distance between residues i and i+N separated by N sequential alpha carbon segments. An algorithm for identifying geometric self-similarity in arbitrary chain structures is dependent upon the mean square lengths $R_G^2(N)$ from a consecutive sequence of segments, each comprised of N adjacent alpha carbon segments, i.e.

$$R_G^2(N,j) = (1/B) \sum_{i=1}^{B} R^2_{j+(i-1)N, j+iN}, \tag{3}$$

where j is an index of an element of the chain (j=1, 2, 3, ...), and B is limited solely by the total number of elements. For N=8, j=1, the right hand side of Eq. (1) would contain the terms $R^2_{1,9}$, $R^2_{9,17}$, $R^2_{17,25}$, etc. For each N, $R_G(N)$ is determined by the one value of j that yields a minimum in D_c. This algorithym has been tested on the fractal of Fig. 6 and found to produce the proper fractal dimension of 3/2 regardless of the starting point within the fractal.

We have computed the chain fractal dimension of 70 proteins by the technique discussed above. The results for six proteins of particular interest are shown graphically in Fig. 7. The computed values of D_c are: Bovine Pancreatic Trypsin Inhibitor (BPTI), 1.34(4); Myoglobin, 1.67(2); Cytochrome P450, 1.63(4); Cytochrome C, 1.61(6); Cytochrome C551, 1.60(1); and Ferredoxin, 1.37(7). As shown in Table I, there is a remarkable correlation between these values of D_c and the largest m values observed in corresponding relaxation data.

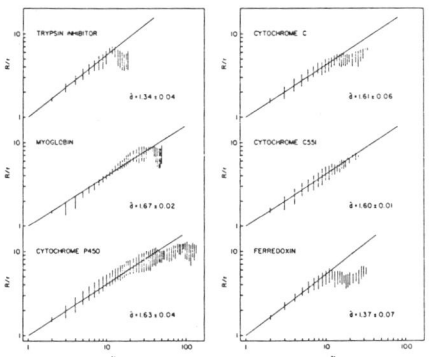

Fig. 7 Computed chain fractal dimensions, D_c, of six selected proteins. The slope of the straight line determines D_c, which in this figure is denoted by $\bar{d}$.

Table I. Correlations between the largest m values observed in the relaxation data of various proteins and the chain fractal dimension computed from x-ray data on the same or very similar protein structures.

Protein	Source	m_{max}	D_c
HEMEPROTEINS			
myoglobin.OH	sperm whale	1.61(5)	1.67(2)
myoglobin azide	sperm whale	1.65(4)	1.67(2)
cytochrome P450	P. putida	1.64(3)	1.63(4)
cytochrome C	horse	1.67(3)	
	albacore		1.61(4)
cytochrome C551	P. aeruginosa	1.43(9)	1.60(1)
IRON SULFUR PROTEINS			
ferredoxin	S. maxima	1.34(6)	
	S. platensis		1.37(7)
putidaredoxin	P. putida	1.34(3)	
	S. platensis		1.37(7)

The solvent effects noted in the data from putidaredoxin and cytochrome C551 led us to a systematic study of solvent effects using myoglobin azide[9]. Eleven experiments were performed under different solvent conditions. Low values of m ($\approx$ 1.2) were obtained in frozen solutions at high (1 M) and low (mM) saltconcentrations, and at high (21 mM) and low (mM) protein concentrations. Values of m in reasonable agreement with the computed chain fractal dimension were obtained on a sample concentrated under partial vacuum and on samples dissolved in a 50% water - 50% glycerol solvent. In all instances, slower direct process relaxation rates were associated with higher m values. Figure 8 shows a feature present to a lesser degree in all the data exhibiting a slow direct relaxation rate. The data of Fig. 8 refer to a dilute MbN_3 sample (1.5 mM protein in 50% glycerol and water). The effective temperature exponent fitting the data rises sharply at approximately 4 K, then drops to a value in agreement with the computed chain fractal dimension for temperatures above 6 K. Treating the interaction as if these vibrations were simply phonons with a spectral density of 1.67, the integrand of the integral over vibrational mode frequencies peaks at a frequency of 660 GHz for a temperature of 6 K. This corresponds roughly to a wavelength of 38 Å if a sound velocity of 2.5×10^5 cm/sec is assumed. This characteristic length is in approximate agreement with myoglobin's molecular size (44 x 44 x 25 Å). Vibrations of wavelength greater than the molecular size become dominant in the Raman integral at temperatures below 6 K. The density of states for these excitations is not expected to reflect the protein structure, but rather that of the solvent.

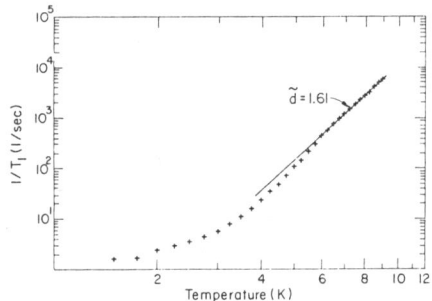

Fig. 8 Relaxation data on 1.5 mM myoglobin azide in a solvent of 50% glcerol and 50% water. Note the abrupt change in the slope of the data at an approximate temperature of 6 K. Above this temperature a spectral exponent of 1.61 is indicated.

A model calculation has been made[9] utilizing a two component vibrational density of states with a transition frequency at 538 GHz. Below that frequency a spectral dimension of 3 was assumed, and above it, a value of 1.61 was used. In addition there can be a discontinuity in the mode density at the transition frequency. Figure 9 shows the results of such a calculation, for a 4-fold increase in the density (or coupling) of vibrational states as the transition frequency is crossed in an increasing direction. With this model two experimentally observed features are reproduced. The Raman relaxation data of Fig. 8 are nicely fit, but more surprising, the stronger the relative strength of the three dimensional solvent modes, the lower the m value for the Raman rate above the transition temperature. With a strong contribution from the lower frequency solvent modes, the spectral exponent m can drop below D_c. This is consistent with our experimental observation that the chain fractal dimension is always observed in the Raman rate with the direct process relaxation rate is weakest.

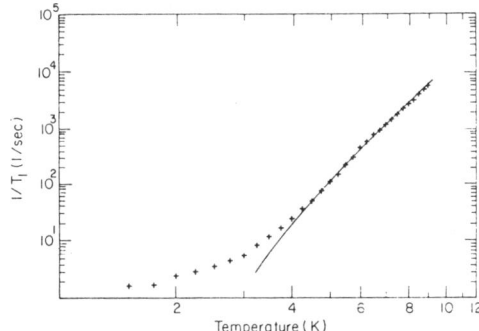

Fig. 9 A computed Raman relaxation rate for a 4-fold increase in the density of states at 538 GHz and a cross-over from a spectral dimension of 3 to 1.61. The calculated rate is forced to fit the datum at T=5.5 K. The data are from Fig. 8.

Gō et al.[10] have computed the low frequency normal modes of Bovine pancreatic trypsin inhibitor (BPTI) using a method appropriate to low frequency modes. There are only 58 residues in BPTI and that implies 116 dihedral angles along the backbone. The authors also included 125 side chain angles. An incomplete histogram of the normal mode frequency distribution they reported is shown in Fig. 10. It corresponds to the lowest 201 modes out of a total of 241. The higher frequency modes were found to be quite localized, while vibrations below 3600 GHz (120 cm^{-1}) were predominantly delocalized, i.e. they correspond to concerted movements of the entire polypeptide backbone (including side chains). Since the lowest 117 vibrational modes are below 75 cm^{-1}, an estimate of 2250 GHz for ν_{max} is not unreasonable.

Superimposed on the histogram of Fig. 10 are normalized densities of states predicted using spectral dimensions of $\bar{\bar{d}} = 1.34$ (see Fig. 7) and $\bar{\bar{d}} = 3$. An unweighted, non-linear, least-squares analysis of the data in the frequency range 0 - ν_{max} yields a best-fit normalized spectral dimension of 1.24.

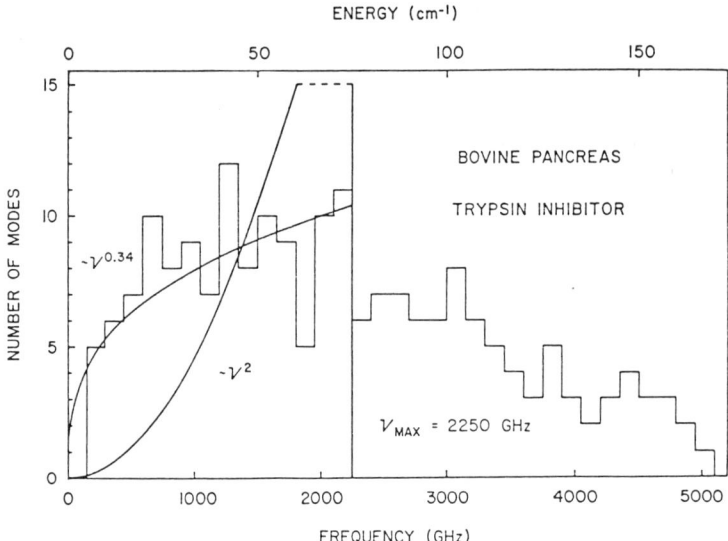

Fig. 10 Histogram of the lowest 201 vibrational modes in bovine pancreatic trypsin inhibitor. Data are reproduced from Ref. 10.

The relationship between the spectral and fractal dimensions of a linear chain is:

$$1 \leq \bar{\bar{d}} = D/(1+\theta/2) \leq D, \tag{4}$$

where D is the fractal dimension of the structure, not the chain fractal dimension, and θ is the exponent that gives the range dependence of the diffusion constant[2]. The important point here is the connectivity of the structure.

Alexander et al.[4] have computed the temperature dependence of the Raman spin relaxation rate for a two fracton process. For Kramers' ions their theory predicts a temperature exponent

$$n = 2\bar{\bar{d}}(1 + 2d_\varphi/D) - 1, \tag{5}$$

where d_φ is an exponent that characterizes the localization of the fracton wavefunction. It is expected to lie between unity and the chemical length dimension. Eq. (5) can be rewritten as

$$n = \{2D + 4d_\varphi\}/\{1 + \theta/2\} - 1. \qquad (6)$$

We can estimate how self-consistent these equations are in their application to our biopolymers by setting n = 6.3 (from experimental data), D = 2 (from x-ray data), and allowing d_φ to vary between 1 and D = 2. The results for $\bar{\bar{d}}$ and θ are shown in Fig. 11.

Note that the spectal dimension varies between 1.217 and 1.825, so that some sort of off-backbone connectivity is required, even with the additional localization parameter, d_φ, of this theory[4].

By contrast, we have a simpler, empirical relationship

$$n = 2D_c + 3 \qquad (7)$$

which requires: (1) that there is no localization of the fracton vibrations, (2) complete connectivity within biopolymer chain, and (3) the concept of a strong backbone that can be characterized by a chain fractal dimension (or Flory constant).

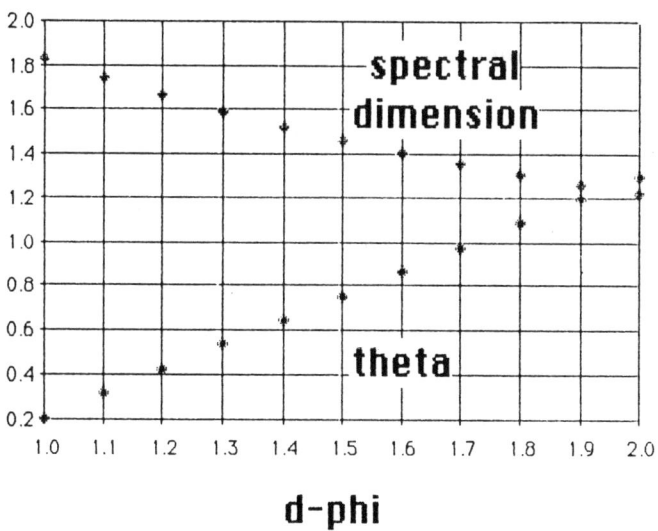

Fig. 11. Values of the spectral dimension and θ (the range exponent of the diffusion constant) for values of the localization parameter, d_φ, in the interval between 1 and 2, that would result in a $T^{6.3}$ Raman relaxation mechanism.

Returning now to additional relaxation data, Scholes et al.[11], have reported relaxation data on a large (M.W. = 200,000) copper and iron containing protein (cytochrome C oxidase). The Raman rates of both the low spin iron and the copper exhibited exhibited fractal behavior with spectral exponents of 1.77 ± 0.08 and 1.79 ± 0.07, respectively. It is not known if one of the paramagnetic ions is relaxing the other, or if both are sampling the same fractal density of vibrational states.

We are also in the process of measuring a copper-containing protein, azurin. Our measurements have just begun so we have no results to report at this time. The computed chain fractal dimension of this protein is 1.35 ± 0.05.

We have also made several measurements on the coenzyme of vitamin B_{12}. This is a small (18 Å), non-polymer organic complex containing 100 H atoms, 109 non-H atoms that include a paramagnetic colbalt ion. Because of some experimental problems, we are not yet ready to report our results, but we have made a theoretical computation of the spectral dimension, $\bar{d}$, using the technique of Rammal and Toulouse[3]. We have simulated 1295 random walks on the 109 non-H sites of the complex, assuming equal jump probabilities and determined the number of distinct sites visited, S_N, as a function of N, the number of steps. These are related by

$$S_N \propto N^{\bar{d}/2}. \tag{8}$$

Our computation yields a spectral dimension of 1.388 ± 0.0004. We need to repeat the computation using the H sites as well.

We have attached an NO radical directly to the iron atom in myoglobin, producing nitrosyl ferrous myoglobin, and measured the spin-lattice relaxation rate of the NO radical between temperatures of 4 and 20 K. This system exhibited only a direct relaxation rate that varied linearly with temperature as $1/T_1 = 376$ T. In the future we plan to attach ruthenium ions to the surface of various nonmagnetic proteins and measure their relaxation rates. Other planned experiments involve the partial digestion of proteins such as cytochrome and then measurements of relaxation rates on the remaining fragment of the protein that contains the low spin iron.

This research was supported in part by the U. S. Public Health Service under NIH Grant GM-24488.

[1] H. J. Stapleton, J. P. Allen, C. P. Flynn, D. G. Stinson, and S. R. Kurtz, Phys. Rev. Lett. 45, 1456 (1980).

[2] S. Alexander & R. Orbach, J. Physique LETTRES 43, L625 (1982).

[3] R. Rammal & G. Toulouse, J. Physique LETTRES 44, L13 (1983).

[4] S. Alexander, O. Entin-Wohlman, and R. Orbach, J. Physique LETTRES 46, L555 (1985).

[5] C. Mailer and C. P. S. Taylor, Biochim. Biophys. Acta. 322, 195 (1973).

[6] R. C. Herrick and H. J. Stapleton, J. Chem. Phys. 65, 4778 (1976).

[7] J. P. Gayda, P. Bertrand, A. Deville, C. More, G. Roger, J. F. Gibson, and R. Cammack, Biochem. Biophys. Acta. 581, 15 (1979).

[8] G. C. Wagner, J. T. Colvin, J. P. Allen, and H. J. Stapleton, J. Amer. Chem. Soc. 107, 5589 (1985).

[9] J. T. Colvin and H. J. Stapleton, J. Chem. Phys. 82, 4699 (1985).

[10] N. Gō, T. Noguti, and T. Nishikawa, Proc. Natl. Acad. Sci. USA 80, 3696 (1983).

[11] C. P. Scholes, R. Janakiraman, H. Taylor, and T. E. King, Biophys. J. 45, 1027 (1984).

SPIN DYNAMICS IN CONCENTRATED METALLIC SPIN GLASSES*

K. Moorjani
Milton S. Eisenhower Research Center
The Johns Hopkins University - Applied Physics Laboratory
Laurel, MD 20707

and

S. M. Bhagat
Department of Physics and Astronomy
University of Maryland, College Park, MD 20742

and

M. A. Manheimer
Laboratory for Physical Sciences, College Park, MD 20740

The spin resonance measurements in concentrated metallic spin glasses, over a wide range of temperature and frequency, reveal a variety of dynamical phenomena which collectively establish a lack of time-scale for dynamics of spins in random systems.

---

* The work at JHU/APL is supported by the Department of the Navy, Naval Sea Systems Command, under contract N00024-85-C-5301.

1. Introduction

The dynamical behavior of spins in a broad class of spin glasses has been copiously investigated theoretically and experimentally. The materials of interest include metals, semiconductors and insulators; dilute as well as concentrated alloys, with spins situated either on a crystalline lattice, or on an amorphous structure. A host of experimental techniques that include ac susceptibility, magnetic resonance, positively charged muon spectroscopy and neutron scattering, have been employed to probe the spin dynamics. Much of the data has been reviewed in Chapter VII of reference 1.

The subject of the present paper concerns spin dynamics in concentrated metallic spin glasses where direct exchange interactions, as opposed to indirect RKKY interactions, are expected to dominate. The prototype amorphous alloys that we have investigated include, vapor-quenched Fe_xB_{1-x} and Mn_xB_{1-x}, and liquid-quenched $(Fe_xNi_{1-x})_{75}(P-B-Al)_{25}$ and $(Fe_xNi_{1-x})_y(P-B)_{1-y}$. The choice of mostly Fe-based alloys is due to the sensitivity of the sign of Fe-Fe exchange interaction to the Fe-Fe interatomic separation, which allows both ferro- and antiferro- magnetic interactions to co-exist in a certain compositional range. The amorphous atomic structure of the alloys implies that the frustration and the concomitant spin glass behavior originate not only from the competing exchange interactions, but also from the misfitting anti-ferromagnetic bonds on odd-membered structural rings.

2. Magnetic Phase Diagram

The magnetic phase diagram of the alloys under consideration, obtained from dc measurements, is similar to that shown in Fig. 1 for a-Fe_xB_{1-x}[2]. As the system is diluted magnetically, conventional paramagnetic-ferromagnetic transition is observed at a Curie temperature T_c which decreases with decreasing composition, x, of the magnetic species. The decrease in T_c becomes more rapid as one approaches the multi-critical point, x_m. For $x > x_m$, in a narrow range of composition, another transition from a ferromagnetic state to a frozen state is observed at T_f, with the transition line for all alloys, except a-Fe_xB_{1-x}, having a negative slope. For a-Fe_xB_{1-x}, T_f is a non-monotic function of x (Fig. 1). All such alloys are commonly referred

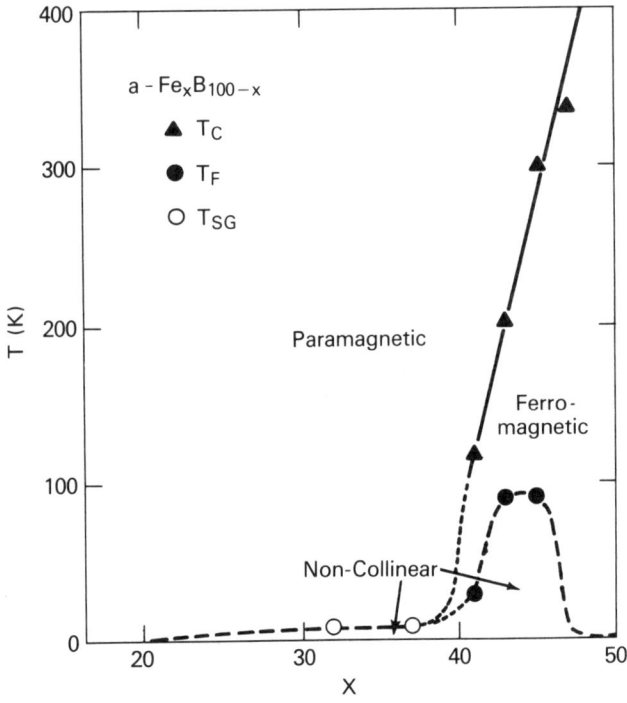

Fig. 1 Magnetic phase diagram of a-Fe_xB_{1-x} alloys.

to as reentrant alloys. For $x < x_m$, no ferromagnetism is observed, but a transition to a frozen state occurs directly from the paramagnetic phase, at a temperature T_{sg}. These alloys are normally referred to as spin glass alloys.

Besides spin dynamics in reentrant and spin glass alloys, the interest also lies in alloys within the concentration regime just above where reentrant behavior disappears. The latter regime is of importance since these alloys are not conventional ferromagnets, but instead exhibit random characteristics in the sense that magnetic moments are not aligned in a specific direction. They thus exhibit non-collinear magnetic structures without the characteristics of a frozen state. The effect of such a behavior on spin dynamics is significant as will be seen below.

3. Experimental Technique

The prime technique in our studies is that of spin resonance, which of course employs an external magnetic field. It is well-established that the spin glass transition, as seen in the appearance of a

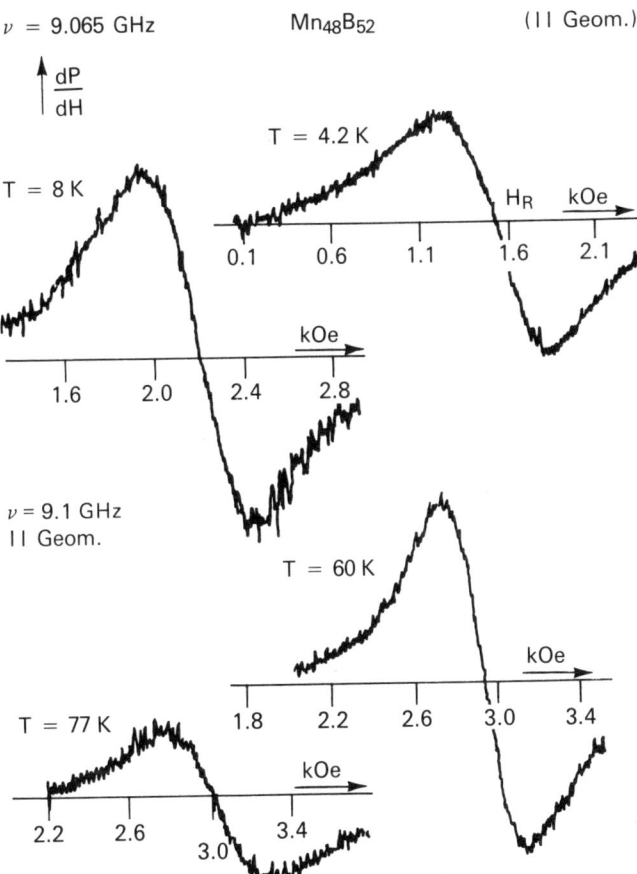

Fig. 2 Derivative curves for resonance absorption at various temperatures.

cusp in temperature dependence of dc mangetic susceptibility, is extremely sensitive to external fields. Nevertheless, as will be clearly demonstrated, the dynamical effects arise at temperatures far above the transition temperatures T_t (T_f or T_{sg}), and in this range the spin resonance technique actually anticipates the transition to the frozen states. The interest, therefore, is not just in the neighborhood of T_t; the random placement of spins has consequences far above T_t, and what appears as a paramagnetic or a ferromagnetic state at $T > T_t$ in static measurements, is not a conventional paramagnetic or ferromagnetic state as probed by dynamical measurements.

Before presenting the main results, we comment on the resonance line shape which undergoes no significant changes or distortions as one goes through the transition temperature. As seen in Fig. 2, the line shape in a-$Mn_{48}B_{52}$ at 9GHz in $||$ geometry remains essentially unchanged as the temperature is varied from 77K ($\approx 4\ T_{sg}$) to 4.2K ($\approx 0.2\ T_{sg}$). This is also true at lower frequencies, and even for reentrant alloys, where the higher internal field leads to more symmetric lines whose shapes are well reproduced by the Gilbert-Landau phenomenological equation. Since the line shapes remain unchanged, the discussion of the linewidth, Γ, is equivalent to that of the relaxation frequency and hence we will only discuss the linewidth data.

4. Paramagnets and Ferromagnets

For a conventional paramagnet, Γ depends linearly on temperature, $\Gamma = -a+bT$ (Bloch-Hasegawa behavior), and is approximately independent of the microwave frequency, ν. In a single crystal ferromagnetic alloy[3], however, Γ is independent of temperature and proportional to ν. The Landau-Lifshitz-Gilbert equation of motion for the magnetization $\vec{M}$,

$$\dot{\vec{M}} = -\gamma[\vec{M} \times (\vec{H}_{eff} + \vec{H}_{int})] + \frac{\lambda}{\gamma M_s^2} \vec{M} \times \dot{\vec{M}} \qquad (1),$$

adequately describes the resonance in presence of an applied field $\vec{H}_{eff}$ that includes the static demagnetization field. In Eq. (1), M_s represents the saturation magnetization, λ is the frequency-independent damping parameter, and γ denotes the gyromagnetic ratio. Since for polycrystalline ferromagnets, in absence of anisotropy, Γ is known to become temperature-independent for $T \leqslant 0.8\ T_c$, we define the ferromagnetic regime for the alloys under consideration as covering the range of temperature for which Γ is temperature-independent and discuss its frequency dependence. This contribution to the linewidth will be denoted by Γ_0. The absence of exchange-conductivity contribution to Γ_0, which would arise if the exchange terms were included in Eq. (1), is justified due to the typical resistivity values ($\sim 1\ \mu\Omega m$) for the alloys of interest.

5. Ramdom Magnets

In contrast to the above results on paramagnets and ferromagnets, the temperature and frequency dependences of Γ for reentrant and spin glass alloys are remarkably different. For example, in the reentrant a-$Fe_xNi_{80-x}P_{14}B_6$ alloys[4] (Fig. 3), while the ferromagnetic alloy ($x = 40$) does indeed show a temperature-independent linewidth, the linewidth rises at low temperatures as x is decreased ($x = 19$, 17), and for alloys with $x \leqslant 15$, exhibits a low-temperature peak besides becoming temperature-dependent even at high temperatures. The various contributions are clearly separable as seen for the spin glass alloy[5] a-$Mn_{48}B_{52}$ in Fig. 4. The Bloch-Hasegawa behavior at high temperatures is followed by a temperature-independent component Γ_0 with a subsequent rise in Γ, that starts at temperatures well above T_{sg}, followed by a peak located at $T < T_{sg}$. The observed frequency dependence of Γ is also shown. Note that lines through the points (Figs. 3 and 4) are not just to guide the eye, but represent a particular functional dependence to be discussed later.

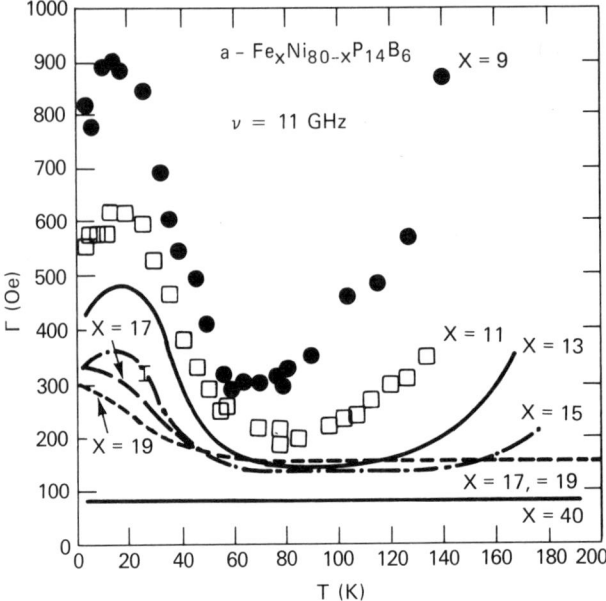

Fig. 3 Temperature dependence of resonance linewidth in a-$Fe_xNi_{80-x}P_{14}B_{16}$ alloys.

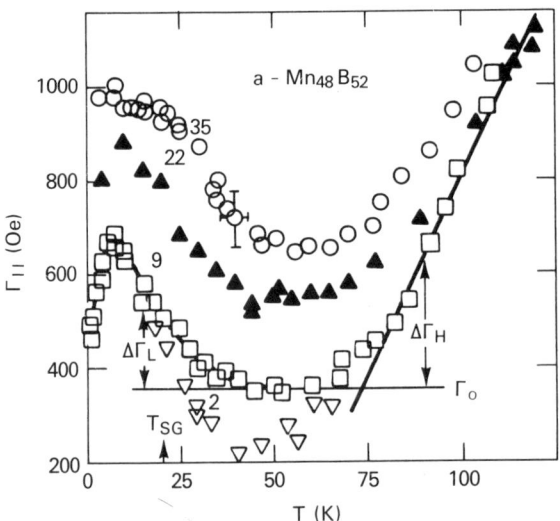

Fig. 4 Variation of resonance linewidth with Temperature and frequency in a-$Mn_{48}B_{52}$.

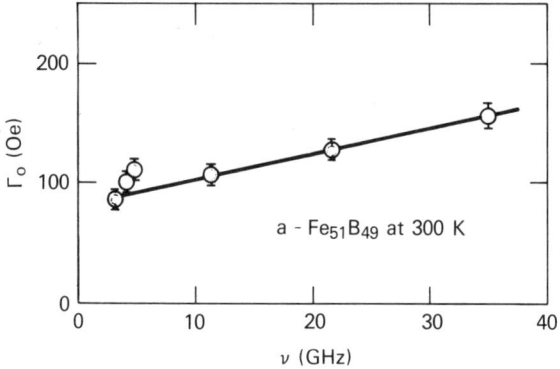

Fig. 5 Frequency dependence of Γ_o in a-$Fe_{51}B_{49}$.

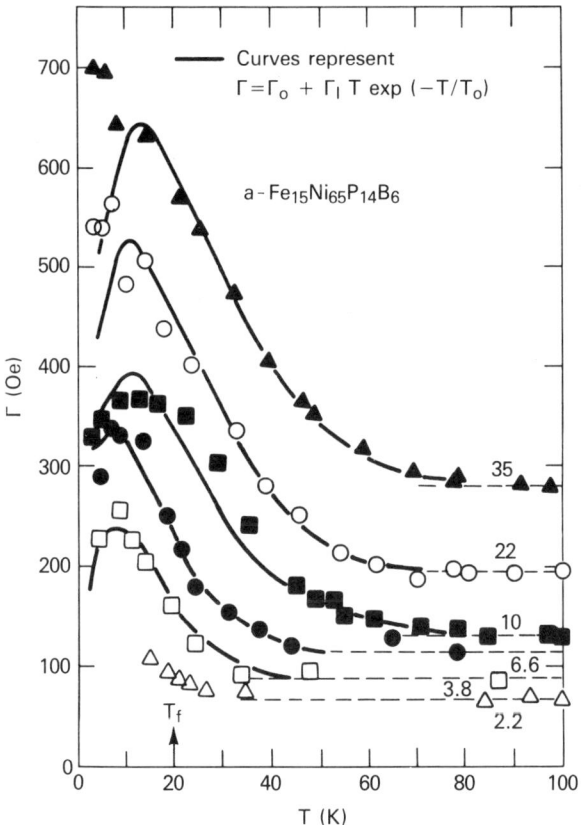

Fig. 6 Variation of resonance linewidth with temperature and frequency in a-$Fe_{15}Ni_{65}P_{14}B_6$.

For ferromagnetic alloys, a-$Fe_{40}Ni_{40}P_{14}B_6$ (Fig. 3) and a-$Fe_{51}B_{49}$ (Fig. 5), Γ_0 is independent of temperature and proportional to frequency just as in crystalline ferromagnets. However, in contrast to crystalline ferromagnets, a finite linewidth results even at $\nu = 0$ (Fig. 5). This is also observed in reentrant[4] (a-$Fe_{13}Ni_{67}P_{14}B_6$, a-$Fe_{19}Ni_{56}P_{16}B_6Al_3$, a-$Fe_{30}Ni_{45}P_{16}B_6$ Al_3), and spin glass[5] (a-$Mn_{48}B_{52}$, a-$Fe_7Ni_{73}P_{14}B_6$) alloys along with the increased Γ at high temperatures.

The full temperature and frequency dependence of the linewidths is shown in Fig. 6 (also see Fig. 3) for typical reentrant alloys and in Fig. 7 (also see Fig. 4) for the spin glass alloys. In these figures

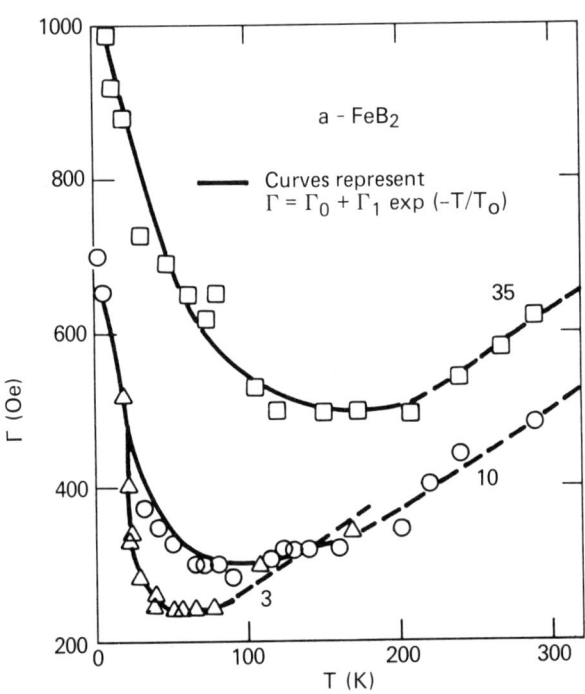

Fig. 7 Temperature and frequency dependence of resonance linewidth in a-FeB_2.

the lines represent the expression

$$\Gamma = \Gamma_0 + \Gamma_1 T^n \exp(-T/T_0) \qquad (2)$$

with n = 1 in Figs. 3, 4 and 6 and n = o in Fig. 7. Except at the lowest temperatures in the highest frequency data (ν = 35 GHz in Fig. 6), Eq. (2) reproduces the observed behavior extremely well and argues against any critical exponent behavior [i.e. $\Delta\Gamma = \Gamma - \Gamma_0 \propto (T - T_f)^a$]. The determination of Γ_0 and Γ_1, from the plots such as shown in Figs. 3, 4 and 6 demonstrates the universal validity of Eq. (1) with n = 1 for a large body of data, over an extended range of temperatures and frequencies, in both reentrant (Fig. 8) and spin glass (Fig. 9) alloys. In

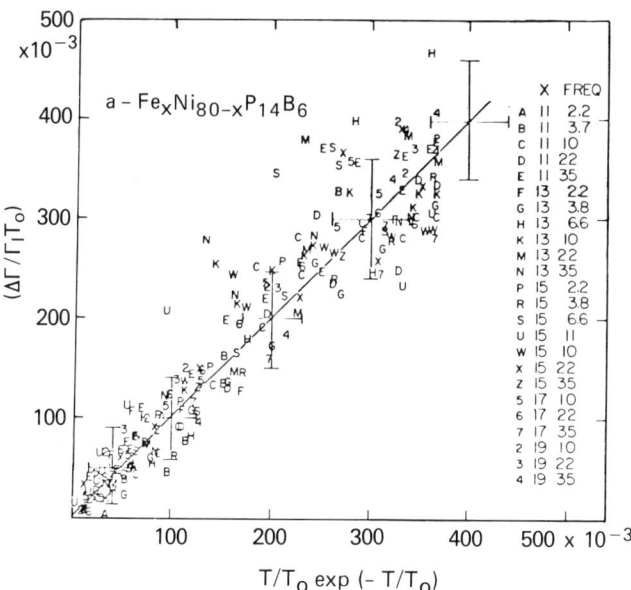

Fig. 8 Universality plot (Eq. 2 with n=1) for the reentrant alloys, a-$Fe_xNi_{80-x}P_{14}B_6$, at various frequencies.

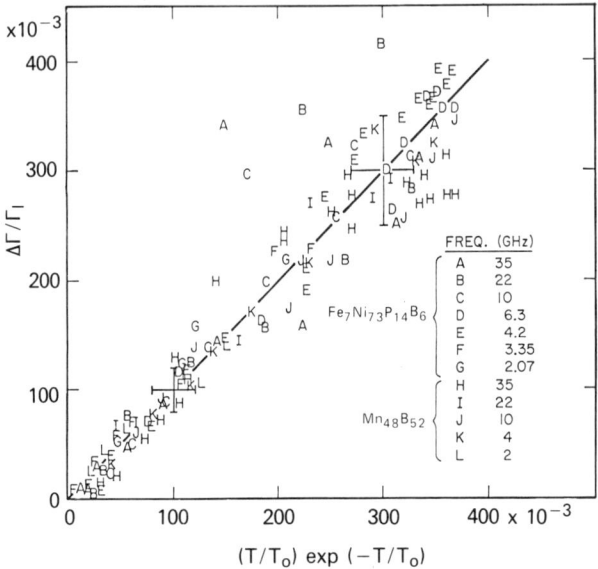

Fig. 9 Universality plot (Eq. 2 with n=1) for the spin glass alloys a-$Fe_7Ni_{73}P_{14}B_6$ and a-$Mn_{48}N_{52}$ at various frequencies.

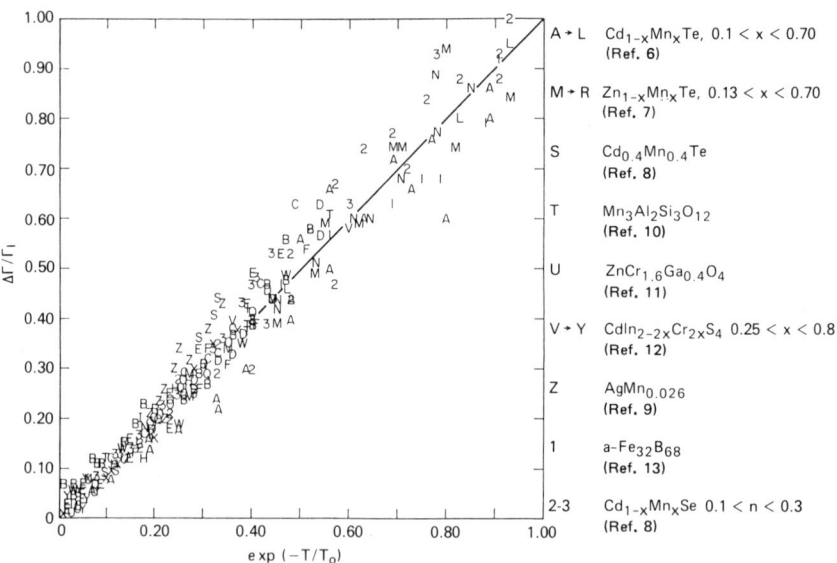

Fig. 10 Universality plot (eq. 2 with n=0) for various alloys with randomly located spins.

fact, the dynamical behavior in an enormous number of random spin systems that include, semi-magnetic semiconductors[6-8], dilute transition-metal--noble-metal alloys[9], insulating alloys[10-12], and amorphous metallic system[13], which exhibit the low-temperature rise in $\Delta\Gamma$, but not the peak at $T < T_f$, is well accounted for by Eq. (2) with $n = 0$ (Fig. 10).

A theoretical model, which may be appropriate to the universal behavior discussed above, is based on the interaction of magnetic excitations (e.g. spin waves in an infinite magnetic cluster) with the magnetic two-level systems considered to arise from finite magnetic clusters which act as relaxation channels for excitations in the infinite cluster[14,15]. The two-level systems are analogous to the tunneling centers in glasses[16] where the transitions occur, across a barrier height V, between two levels separated by an energy E. For compact clusters and near the percolateion threshold[15], the barrier height distribution follows the exponential probability function $P(V) = 1/V_0 \exp(-V/V_0)$, and leads[14] to Eq. (2) with $T_0 = V_0/\ell n (\omega\tau_0)$, in the limits $\omega\tau_0 \ll 1$ and $kT/V_0 \ll 1$, where $\tau_0 \simeq 10^{-13}$ secs is the inverse of an attempt frequency. In this model, T_0 corresponds to the maximum in the linewidth and denotes the temperature at which the finite clusters freeze. Below T_0, the linewidth decreases since the frozen clusters are no more efficient channels for excitations in the infinite cluster so that the lifetime of excitation modes increases. Though the overall observed behavior of Γ is reproduced by the model, since for $\omega/2\pi = 10$ GHz, $V_0 \simeq 5 kT_0$, the observed effects should occur at temperatures much below T_0, in contrast to the experimental data.

Next, we mention the existence of a characteristic frequency, ν_c, for the spin glass alloys at which Γ rises far more rapidly with decreasing temperature than at frequencies below or above ν_c[5]. This is shown for a-Mn$_{48}$B$_{52}$ in Fig. 11 where $\nu_c = 4$ GHz. This resonant anomaly is observed for a number of spin glass alloys (Fig. 12) and strongly points to the presence of a high "density of states" local mode in the random spin systems.

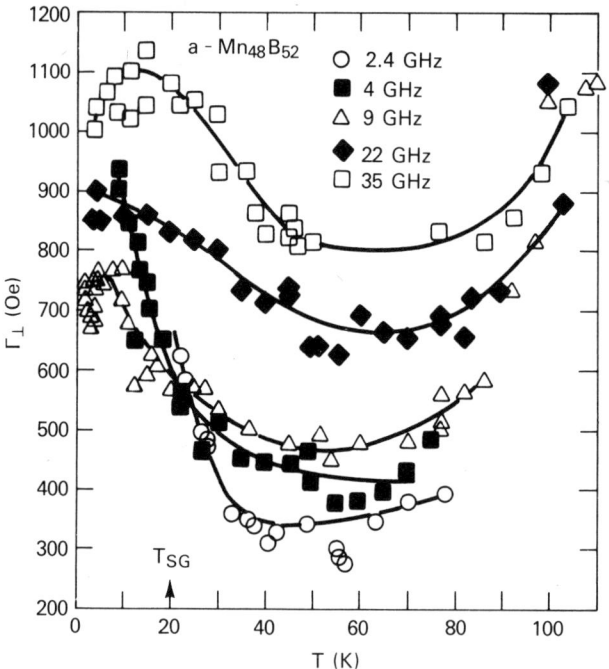

Fig. 11 Variation of linewidth with temperature and frequency in a-$Mn_{48}B_{52}$. Note the anomalous behavior at ν=4GHz (see text

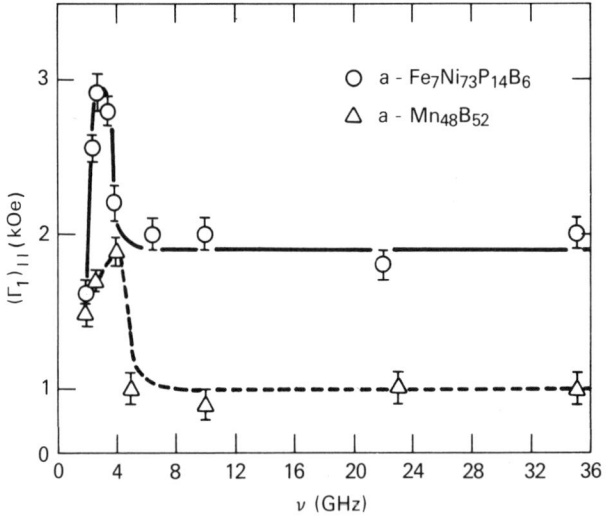

Fig. 12 Frequency dependence of linewidth in a-$Fe_7Ni_{73}P_{14}B_6$ (ν_c=3.4GHz) and a-$Mn_{48}B_{52}$ (ν_c=4GHz).

6. Non-Ergodicity

Finally, we discuss a dynamic effect that is observed in the non-ergodic behavior[2,17] seen only in a-Fe_xB_{1-x} in a narrow concentration range, $41 \leq x \leq 49$, of the reentrant alloys (Fig. 1) at all microwave frequencies. These alloys differ from others was already remarked upon at the beginning of the paper where it was noticed that the transition line between the ferromagnetic and the frozen state is nonmonotonic (Fig. 1). This is due to the small size of boron atoms which allows Fe-Fe atoms to approach each other closer than in other alloys, thus allowing a rapid change in the number of anti-ferromagnetic bonds as a function of varying concentration. Considering Ising spins on a square lattice, it is easy to demonstrate that the number of frustrated plaquettes will initially increase with the increasing number of anti-ferromagnetic bonds, but will then decrease since it takes odd number of anti-ferromagnetic bonds on an even-membered ring to lead to frustration[2,18]. This has consequences in the static (Fig. 1) as well as dynamic (Fig. 13) behavior of the alloys. The linewidth Γ as a function of concentra-

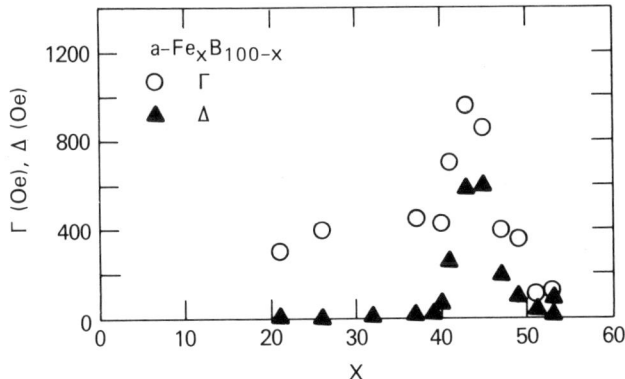

Fig. 13 Concentration dependence of linewidth and width of the hysteresis loop in a-Fe_xB_{100-x} alloys.

tion x, reflects the nonmonotomic behavior seen in Fig. 1 and furthermore follows the nonmonotonic variation of Δ, the width of the hysteresis loop[2]. More importantly, the non-ergodic behavior is seen in Fig. 14, where the resonance field for a-$Fe_{49}B_{51}$ at $\nu = 12$ GHz in the paral-

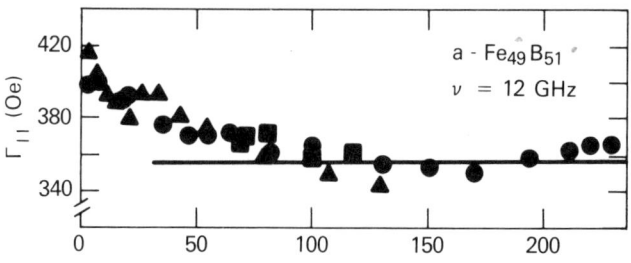

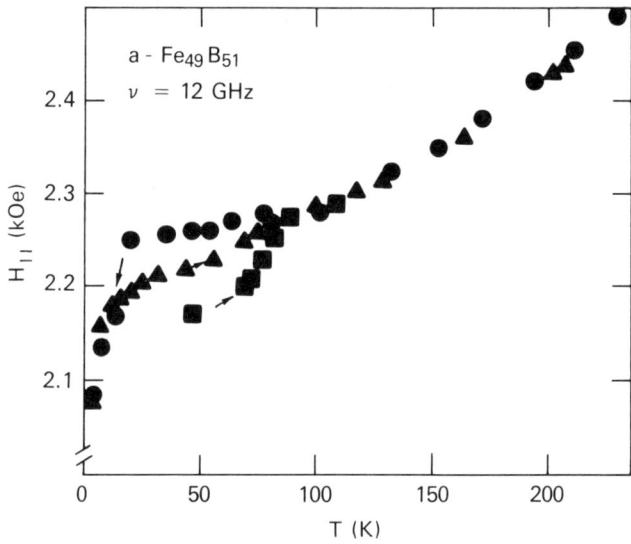

Fig. 14 Temperature dependence of linewidth (upper) and resonance field (lower) in a-$Fe_{49}B_{51}$. Non-ergodic behavior observed during thermal cycling is noted in the text.

lel geometry, has a temperature dependence that depends on the thermal history[2]. In the ferromagnetic regime (T ≳ 100K), the value of $H_{||}$ is unique. But in the reentrant regime, 4K < T < 100K, several paths result depending on the thermal cycling. The circles represent data taken while cooling from 300K; the triangles and squares denote data obtained while warming, subsequent to cooling in zero field to 4K and

48K respectively. Note that the linewidth (upper part, Fig. 14) is essentially independent of the path. Similar behavior is seen in other reentrant alloys (41 ≤ x ≤ 49) and at various frequencies. We ascribe it to non-ergodicity[2,17]. The free energy for the spin glass phase must exhibit a large number of quasi-degenerate minima separated by barriers of varying heights (a generalized version of the two-level system). It is then possible that the spin system gets 'locked' into a metastable state and is unable to 'visit' the other minima. In each of the equivalent many spin configurations, the angular dependences of the local energies are expected to be different, so that they will represent a variety of local anisotropy fields.

7. Conclusions

The spin resonance measurements have clearly demonstrated that the effects of transitions to the non-collinear magnetic phases are felt at temperatures far above the transition temperature obtained from the static measurements. The temperature and frequency dependence of the resonance linewidth and field has revealed a variety of dynamical phenomenal that include universality, the existence of characteristic frequencies and non-ergodicity. It is well established that random systems do not posses a unique length-scale[19]; this paper establishes the absence of a time-scale, for dynamics of spins, in random media.

8. References

1. K. Moorjani and J. M. D. Coey, <u>Magnetic Glasses</u>, Elsevier Science Publishers (Amsterdam), 1984.

2. D. J. Webb, S. M. Bhagat, K. Moorjani, T. O. Poehler, F. G. Satkiewicz and M. A. Manheimer, J. Magn. and Magn. Mat., 44, 158 (1984).

3. I. M. Puzei and V. C. Pokatilov, Solid State Phys (Soviet Union), 16, 671 (1974); also see S. M. Bhagat and P. Lubitz, Phys. Rev. B10, 179 (1974), and L. Kraus, Z. Frait and J. Schneider, Phys. Stat. Sol. (a) 63, 669 (1981).

4. S. M. Bhagat, D. J. Webb and M. A. Manhiemer, J. Magn. Magn. Mat., (in press).

5. M. J. Park, S. M. Bhagat, M. A. Manheimer and K. Moorjani, Phys. Rev., B (in press).

6. S. M. Bhagat and H. A. Sayad (to be published).

7. S. M. Bhagat and H. A. Sayad (to be published).

8. S. Oseroff, Phys. Rev., B25, 6584 (1982).

9. G. Mozurkewich, J. H. Elliott, M. Hardiman and R. Orbach, Phys. Rev., B29, 278 (1984).

10. J. P. Jamet, J. C. Dumais, J. Seiden, and K. Knorr, J. Magn. and Hagn. Mat., 15-18, 197 (1980).

11. D. Fiorani, S. Viticoli, J. L. Dorman, J. L. Tholence and A. P. Murani, Phys. Rev., B 30, 2776 (1984).

12. S. Viticoli, D. Fiorani, M. Nognas and J. L. Dorman, Phys. Rev., B 26, 6085 (1982).

13. K. Moorjani, T. O. Poehler, F. G. Satkiewicz, M. A. Manheimer, D. J. Webb and S. M. Bhagat, J. Appl. Phys., 57, 3444 (1985).

14. M. A. Continentino, Phys. Rev., B 27, 4351 (1983); J. Phys. C: solid St. Phys., 16, L71 (1983).

15. M. H. Cohen and M. A. Continentino, Solid State Commun., 55, 609 (1985).

16. P. W. Anderson, B. I. Halperin and C. M. Varma, Phil. Mag., 25, 1 (1972); W. A. Phillips, J. Low. Temp. Phys., 7, 351 (1972).

17. D. J. Webb, S. M. Bhagat, K. Moorjani, T. O. Poehler and F. G. Satkiewicz, Solid State Commun., 43, 239 (1982).

18. K. Moorjani, in Physics of Disordered Materials, Eds. D. Adler, H. Fritzsche and S. R. Ovshinsky, Plenum Press, New York, 1985, pp. 699.

19. For example, see various papers in the present volume.

EXCITATION AND RECOMBINATION OF
PHOTOCARRIERS IN TRANS POLYACETYLENE

Z. Vardeny and E. Ehrenfreund
Physics Department and Solid State Institute
TECHNION - Israel Institute of Technology
Haifa 32000
ISRAEL

ABSTRACT

Excitation and recombination processes of photocarriers in trans-$(CH)_x$ were measured by transient induced absorption from a fraction of a picosecond to 0.1 second. We found that intrachain excitation results in an overall neutral state with subnanosecond geminate recombination. Interchain excitation alone leads to charge separation which may result in photoconductivity. The recombination of these photocarriers is much longer involving a broad distribution of recombination times. Charge transport, however, is not dispersive and does not play a role in the recombination process.

1. Introduction

The recently developed theoretical picture of elementary excitations in trans polyacetylene $(CH)_x$[1] has created a great deal of interest both experimentally and theoretically. In this elegant picture, electronic excitations across the Peierls gap[2] are entirely different from the electron-hole pairs of conventional semiconductors. Instead, the proper description of the quasiparticles is 1D domain walls or solitons[1], which separate the degenerate ground-state structures. As a result of their translational invariance[1] solitons are thought to play the role of energy and charge carrying[3] excitations in this commensurate Peierls insulator.

To a great extent, work on photoexcited $(CH)_x$ was stimulated by the prediction of soliton photogeneration[2]. Within the framwork of the Su-Schrieffer-Heeger(SSH) model Hamiltonian[1], which contains electron phonon (e-p) interactions but does not contain electron-electron interaction, it has been shown[2] that photoexcited electron-hole (e-h) pair is unstable towards the formation of soliton-antisoliton ($s\bar{s}$) pair. Subsequently, it has been demonstrated[4] that as a consequence of the Pauli principle and charge conjugation symmetry in $(CH)_x$, the photogenerated soliton-antisoliton are oppositely charged.

The study of photoexcited $(CH)_x$ has revealed several unexpected phenomena which were not predicted by the SSH model of the soliton. Most importantly overall neutral states as well as charged excitations are observed[5-7]. This finding together with the absence (in undoped $(CH)_x$) of optical transitions at mid-gap[8], where transitions of neutral or charged solitons should have appeared according to the SSH picture[1], showed that electron correlations in $(CH)_x$ should not be ignored. Recently, more direct measurements of the electron-electron interaction such as double ENDOR[9], photomodulation of neutral soliton defects[10] and resonant Raman spectroscopy[11], have shown that the electronic gap in $(CH)_x$ is due partly to electron correlations rather than entirely due to electron-phonon interaction as in the SSH model. Under these circumstances, the nature of the photoexcitations in $(CH)_x$ may be very different from that predicted by the SSH Hamiltonian[12]. In addition, theoretical and experimental works on finite polyenes have shown conclusively[13] that the low lying excitation is a neutral state with an even symmetry rather than a singlet state with the symmetry of two bond alternation defects which is analogous to soliton pair photoproduction.

Photogeneration of charges in trans-$(CH)_x$ was demonstrated by several experiments[14-16]. Earlier results[15] on photoexcited $(CH)_x$ showed that photoconductivity exists in trans-$(CH)_x$ but not is cis-$(CH)_x$. More recently[17], it was demonstrated that a photovoltaic effect also exists in trans-$(CH)_x$. Direct evidence for charge photogeneration without external electric fields was given by the photoinduced absorption experiments[6,18].

The origin of the branching process which determines the relative photoproduction of neutral vs. charged states is not quite well understood. One possibility to explain the branching process is by the Onsager theory[19] which has successfully explained charge photoproduction in disorder materials and in molecular crystals. The difficulty with this approach is that trans-$(CH)_x$ is a quasi-1D semiconductor for which the Onsager theory based on the e-h Coulomb interaction may not be applicable. In addition, the application of this theory to 1D materials[20] results in negligible quantum-efficiency for charge photoproduction, contrary to the experiments. In order to solve this problem, it was suggested[20] that the 1D-3D interplay plays an important role in the photo-physics of conducting polymers. In the proposed model[12], <u>intrachain</u> excitation results in a neutral state while only <u>interchain</u> excitation may produce separate charges (on neighboring chains) if the interchain geminate recombination is overcome.

2. Experimental

Perhaps the best way to detect and characterize photoexcitations in $(CH)_x$ is to study their optical absorption. As a consequence of their localization they give rise to gap states in the electron and the phonon level spectrum. The scheme of our experiments is the following: We photoexcite with above-gap light, and then probe the optical absorption of the sample in a broad range from IR to visible. Essentially we obtain difference spectra, i.e. the difference in the optical absorption α of $(CH)_x$ when it contains a non-equilibrium carrier concentration and that in the equilibrium ground state. For photoexcitation we have used a variety of light sources. For steady-state measurements we have used a chopped cw (Ar^+) laser; a pulsed (10ns duration) dye laser was used to measure the carriers dynamics in the microseconds time range. For the picoseconds measurements we used the pump and probe technique with a cavity dumped passively mode-locked dye laser having 0.7 ps resolution at $\hbar\omega = 2$ eV. We measured the induced changes in α ($\Delta\alpha$) for probe beam polarization parallel to the polarization of the pump beam (↑↑) or perpendicular to it (⊥). The samples were semitransparent films (∼0.1mm thick) on KBr or sapphire substrates, isomerized to trans-$(CH)_x$ by heat treatment at 170°C.

3. Steady State Photoinduced Absorption Spectrum

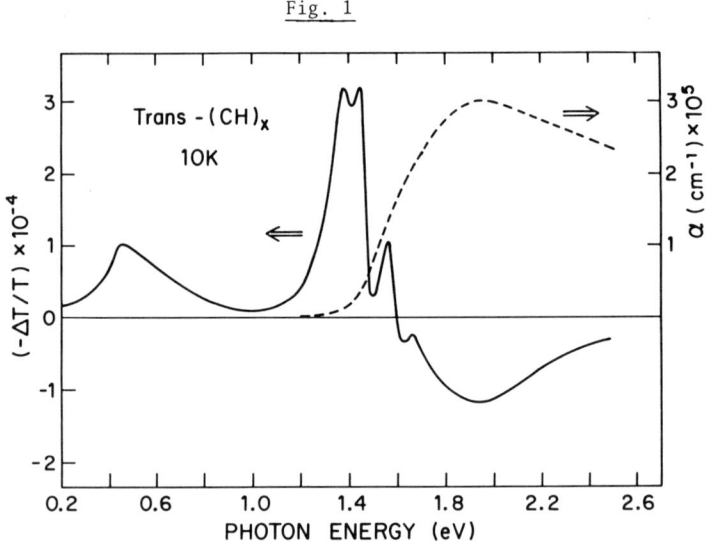

Steady state PA spectrum of trans-$(CH)_x$ at 10K. $\alpha(\omega)$ is also displayed

An overview of the steady state photoinduced absorption (PA) spectrum in trans-$(CH)_x$ at 10K, obtained with excitation of 20 mWcm^{-2} at 2.4 eV, is shown in Fig. 1; the energy dependent absorption constant $\alpha(\hbar\omega)$ is also displayed for comparison. The spectrum consists of two main PA bands with $\Delta\alpha>0$, and bleaching ($\Delta\alpha<0$) for $\hbar\omega > 1.65$ eV. The low energy (LE) PA band peaks at 0.43 eV and it originates from photoinduced charged defects. The intensity of the LE band easily saturates with the excitation power[21,22] and it is sample-dependent. Moreover it was recently demonstrated[23] in oriented $(CH)_x$ films, that the LE band is induced preferentially with light polarized perpendicular to the chains direction. Therefore, it seems unlikely that the LE band is an intrinsic excitation of the polymer chain.

The LE band is now considered[10,24] to be due to charged soliton transitions which are pushed away from mid-gap (predicted by the SSH theory) due to electron correlation.

The high energy (HE) band peaks at 1.4 eV and is due to an overall neutral state[6] which was tentatively identified as a bound ss^- pair[5-7]. This PA band does not depend on sample treatment[25] nor does it saturate at high laser illumination[26] and therefore it is intrinsic to the $(CH)_x$ chain. As seen in Fig. 1 the bleaching spectrum is in reverse to α ($\hbar\omega$) i.e. peaks at the same energy (2 eV), indicating bleaching of the interband transitions. However, the bleaching spectrum associated with the LE band peaks at about 1.45 eV[10,24] and is negligibly small at 2 eV. Therefore, the bleaching of the interband transitions is associated with the HE band alone, and it is possible to follow its dynamics simply by following the time dependence of the bleaching signal at 2 eV. This was verified by Shank et al.[26] who demonstrated that the bleaching dynamics at 2 eV and the HE band dynamics at 1.4 eV are the same. Since our picosecond measurements[27] were done at $\hbar\omega$ = 2 eV, we could follow only the HE band (i.e. the bound ss^- pair dynamics).

4. Geminate Recombination

The time dependence of the photoinduced bleaching at 2 eV for trans-$(CH)_x$ at 300 K, up to 1400 ps, is shown in Fig. 2 for both $\parallel$ and $\perp$ configurations, on logarithmical scales. $\Delta\alpha$ is induced in times less than 2×10^{-13} sec[26,27] indicating that localization occurs instantaneously. The dichroism $\Delta\alpha_\perp \neq \Delta\alpha_\parallel$ is also induced instantaneously[27] implying that the photoexcitations are preferentially polarized along the chain direction. The induced dichroism is lost ($\Delta\alpha_\perp = \Delta\alpha_\parallel$) after about 1.5 ns when the photoexcitations have diffused over sufficiently large distances that the excited $(CH)_x$ chains are random with respect to the pump polarization. Therefore the photodichroism decay was used[7,27] to evaluate the photoexcitations diffusion constant D; D was found to be[27] about $2 \times 10^{-2} cm^2 sec^{-1}$ at 300K.

Fig. 2

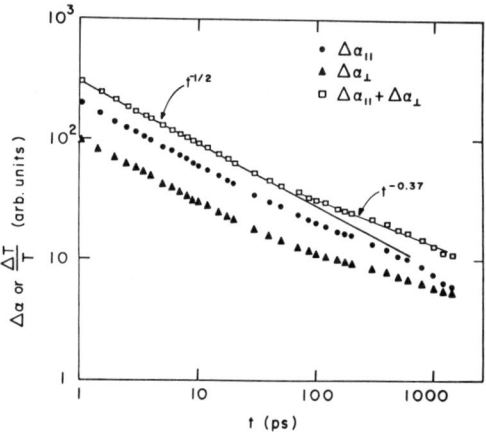

Subnanosecond decays of the polarized photo-induced bleaching at 2eV for trans-$(CH)_x$ at 300K.

Since $\Delta\alpha_\perp(t)$ gradually approaches $\Delta\alpha_\parallel(t)$ (Fig. 2), the polarization memory relaxation can be eliminated by studying the decay of $\Delta\alpha(t) \equiv \Delta\alpha_\perp(t) + \Delta\alpha_\parallel(t)$. At 300K (Fig. 2) $\Delta\alpha(t)$ decays as $t^{-\frac{1}{2}}$ up to about 50 ps and then slows to $t^{-0.37}$. At 80K, the inverse square root decay extends only to 20 ps[10]. The decay curves do not change when the excitation intensity is increased by 4 orders of magnitude[26] implying monomolecular recombination kinetics as the underlying mechanism for the decay.

The initial $t^{-\frac{1}{2}}$ decay of $\Delta\alpha$ may be due to geminate recombination of the $s\bar{s}$ pairs[10]. However, attempts to affect the decay with electric fields up to 5×10^4 V cm^{-1} were unsuccessful[27], showing that the binding of the geminate $s\bar{s}$ pair is not due to Coulomb attraction. This is consistent with the idea[12] that the HE PA band (whose decay is monitored at 2eV) is not due to a solitonic exciton i.e. a charged $s\bar{s}$ pair, but is due to an electronic configuration which might have an even symmetry as predicted for a polymer with significant electron-electron interaction[13]. Attempts were made to explain the $t^{-\frac{1}{2}}$ decay by a diffusion limited 1D recombination[7,26] where the photocarriers have to diffuse towards each other in order to recombine. However, the inverse square root behaviour is too general that one cannot specify the nature of the diffusing species or even the dimensionality of the motion. Another possibility is that the power law decay results from a broad distribution of monomolecular recombination times as described for example by the CTRW theory introduced[28] for disordered materials. Since trans-$(CH)_x$ is disordered, as was shown by the dispersion of the RRS spectrum[11] and by the distinct Urbach-tail below the interband transitions onset[8], it should not be surprizing that the disorder affects the recombination kinetics. In this case, the $t^{-\frac{1}{2}}$ decay results from a waiting time distribution $\Psi(t)$ where $\Psi(t) \sim t^{-3/2}$, a form of particular interest which was thoroughly discussed in this meeting[29].

5. Photocarriers Dispersive Recombination

Fig. 3

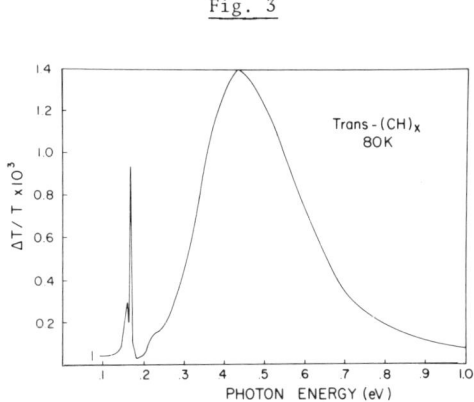

Steady state PA spectrum of trans-$(CH)_x$ at 80K from 0.1 to 1 eV.

The LE PA band is shown in detail in Fig. 3; the spectrum includes also the phonons energy range down to 750 cm^{-1}. The two relatively narrow PA bands at 1280 cm^{-1} and at 1360 cm^{-1} are due to localized vibrations which are coupled to the photoinduced charges and become ir active[18,21]. Their active vibrations and the LE band were shown previously[21] to have the same dependencies on the temperature, laser intensity and excitation profile[30]. It was concluded[21] therefore that the induced vibrations and the LE band result from the same photoinduced defects; hence the defects are charged. Later on the LE PA band was indentified[10,22,24] as due to electronic transitions from (to) the continuum bands to (from) the charged soliton level; this explains its assymmetric line-shape (Fig. 3).

The LE band is intimately related to the photoconductivity (PC) which can be generated in trans-$(CH)_x$. This was shown by measuring the PA excitation profile; it has precisely the same shape as the PC excitation profile[30]. It was concluded therefore that the photocurrent is carried by the same charges responsible for the solitonic PA band. Hence, a study of the LE PA band decay in $(CH)_x$ may provide valuable information about the photocarriers transport and recombination. This can be done by studying the time dependent PA after a pulsed laser excitation.

Fig. 4

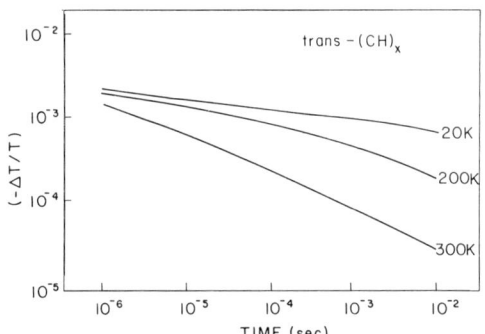

PA decays at a probe energy of 0.5 eV following 10 nonoseconds pulse excitation, for various temperatures.

In Fig. 4, we show the LE PA band decay following the pulse (∿ 10ns) excitation, for several temperatures. The measurements were done at a probe energy $\hbar\omega = 0.5$ eV and the pump pulse intensity was 40 μJ cm^{-2} at 590 nm. The curves show approximate power-law decay in time: $t^{-\beta}$, with $\beta < 1$. As is seen in Fig. 4, β increases with the temperature from $\beta \simeq 0.13$ at 20K to $\beta \simeq 0.5$ at 300K. The PA decays shown in Fig. 4 are surprisingly similar to PA decays in more conventional disorder semiconductors such as a-Si:H[31] and a-As$_2$Se$_3$[32], which were measured using similar experimental set-ups such as ours. The PA decays in these materials were attributed to diffusion limited bimolecular recombination kinetics with time dependent recombination constants. These were taken as evidence[31,32] for photocarriers dispersive transport which often occurs in disordered materials[28]. It is tempting, therefore, to explain the PA decays in (CH)$_x$ using the same ideas as in amorphous semiconductors. This was partially done earlier by Etemad et al.[15] to explain the time dependent PC which also has power law decay. In the following, we demonstrate that in spite of the similarity between the PA decays in amorphous semiconductors and in trans-(CH)$_x$, the underlying mechanism for the power law decays in (CH)$_x$ is <u>not</u> bimolecular kinetics involving dispersive transport.

For bimolecular recombination kinetics

$$\frac{dN}{dt} = -b\ N^2, \ N(t) \text{ is given by}^{31}$$

$$N(t) = N(o) / (1 + t/\tau) \qquad (1)$$

where τ is the recombination time

$$\tau = (N(o)b)^{-1} \qquad (2)$$

In Eqs. 1,2 N(o) is the initial photocarriers density and b is the bimolecular recombination constant. Eq.(2) predicts faster recombination for larger N(o). Therefore, the decays should become faster at high laser excitation intensities, as was verified for a-Si:H[33]. However, this is not the case for the PA decays in trans-(CH)$_x$.

We measured the PA strength at 0.5 eV as a function of the chopping frequency; the results are shown in Fig. 5 for two cw laser intensities which differ by a factor of 30. For the lower excitation intensity the PA signal is approximately constant up to 100 s^{-1}; this shows that the photocarriers recombination time τ is of order 10 ms for this excitation intensity. However, at the higher excitation intensity (Fig. 5), the PA strength starts decaying at about 10 s^{-1} which means τ of order 100 ms. This observation contradicts bimolecular recombination kinetics[33], since the life-time

Fig. 5

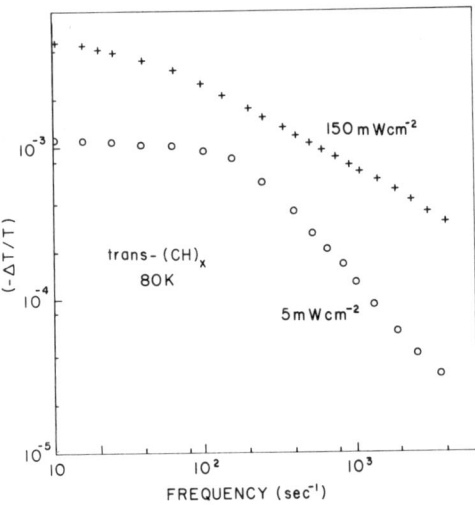

PA strength at 0.5 eV as a function of the cw laser chopping frequency, for two different excitation intensities.

increases with the excitation intensity ($\sim N(o)$) contrary to Eq. (2). We have verified that the same phenomenon occurs in the PA decays measured more directly (as in Fig. 4); the decays become slower as the pulse intensity increases. On the other hand, since the PA signal easily saturates (as is evident in Fig. 5), the linear excitation intensity dependence characteristic of monomolecular recombination kinetics could not be observed.

We show next that dispersive transport is also not involved in the recombination. The photocurrent density i(t) in a pulsed pc experiment is:

$$i(t) = N(t) \, \mu(t) \, eE \qquad (3)$$

where μ is the photocarriers mobility and E is the applied electric field.

For dispersive transport $\mu(t)$ decreases with time[28,31-33] and from Eq. 3 $i(t)$ decays considerably faster than $N(t)$. On the other hand, the PA signal $\Delta T(t)$ is:

$$\Delta T(t) = N(t)\sigma \qquad (4)$$

where σ is the optical absorption cross-section. If σ is time independent (as was verified for the LE PA band for $t > 10$ μsec) $\Delta T(t)$ essentially follows $N(t)$. For such a case by dividing $i(t)$ (Eq. 3) by $\Delta T(t)$ (Eq. 4) $\mu(t)$ can be directly extracted from the experiments. This was done for a-As_2Se_3[32] and was used as a direct proof for a time dependent mobility even when recombination sets in. On the other hand μ is time <u>independent</u> for $(CH)_x$. This was verified by J. Orenstein[34] who measured $i(t)$ and $\Delta T(t)$ on the <u>same</u> $(CH)_x$ sample. The two signals decayed together for $t > 10$ μsec thus proving that dispersive transport does not occur in $(CH)_x$.

The entire body of transient experiments can be explained, however, simply by postulating a broad distribution $G(\tau)$ of recombination times τ. This may result, for example, from a relatively narrow distribution in the photogenerated $s^+ - s^-$ distances r if τ depends exponentially on r

$$\tau = \nu^{-1} \exp(2\alpha r) \qquad (5)$$

where α^{-1} is the wave function extent and ν is a characteristic phonon frequency. A similar model was proposed by Street[35] to explain the power law decays of the geminate (e-h) photoluminescence intensity in a-Si:H. If $G(\tau)$ has a long tail as in the CTRW models: $G(\tau) \sim \tau^{-(1+\beta)}$ it is easy to show[36] that $N(t)$ decays as a power law in time $N(t) \sim t^{-\beta}$. This happens regardless of the recombination kinetics; it is true for both bimolecular or monomolecular. This explains the power law decay for $N(t)$ in $(CH)_x$ even if the recombination is not bimolecular. The longer decays observed for higher excitation intensities in trans-$(CH)_x$ (Fig. 5) may be explained in this model by postulating photogenerated s^+-s^- pairs with larger r at higher intensities so that $G(\tau)$ has longer tails (β is smaller). Since transport is not involved in the recombination process, i.e. the recombination is dispersive while μ is not, the transport mechanism is probably very different from the recombination process in $(CH)_x$. For instance, the transport may be truly 1D (e.g. by soliton intrachain diffusion[27]) while the recombination might be 3D involving interchain carriers hopping[3] from s^- to s^+. Clearly these ideas should be further tested before reaching definitive conclusions, especially more work must be done on oriented $(CH)_x$ samples[23,37].

In conclusion, we showed that the quasi 1D nature of trans-$(CH)_x$ has a very profound effect on the excitation, recombination and transport of photocarriers. Only interchain excitation result in charge photogeneration. It is therefore not surprising that charge recombination is also 3D and involves broad distribution of recombination times due to disorder. Intrachain excitations result in overall neutral states which may be described by neutral soliton antisoliton pairs which are subject to fast geminate recombination.

6. Acknowledgements

Parts of this work were done in collaboration with A.J. Heeger and J. Tauc under the Binational Science Foundation (BSF), Jerusalem, Israel. We thank J. Orenstein for many useful discussions and for sending us his preprint prior to publication. We also thank J. Klafter, A. Blumen, H. Scher and M. Silver for informative discussions during the meeting.

References

1. W.P. Su, J.R. Schrieffer and A.J. Heeger, Phys. Rev. Lett. $\underline{42}$, 1698 (1979); Phys. Rev. B$\underline{22}$, 2099 (1980); $\underline{28}$, 1138 (E) 1983.

2. W.P. Su and J.R. Schrieffer, Proc. Natl. Acad. Sci. USA $\underline{77}$, 5626 (1980).

3. S. Kivelson, Phys. Rev. Lett. $\underline{46}$, 1344 (1981).

4. R. Ball, W.P. Su and J.R. Schrieffer, J. Phys. (Paris) Colloq. $\underline{44}$, C3-429 (1983).

5. J. Orenstein and G.L. Baker, Phys. Rev. Lett. $\underline{49}$, 1043 (1982).

6. J. Orenstein, Z. Vardeny and G.L. Baker, J. Phys. (Paris) Colloq. $\underline{44}$, C3-407 (1983).

7. Z. Vardeny, Physica 127B, 338 (1984).

8. B.R. Weinberger, C.B. Roxlo, S. Etemad, G.L. Baker, and J. Orenstein, Phys. Rev. Lett. $\underline{33}$, 86 (1984).

9. H. Thomann, L.R. Dalton, M. Grabowski and T.C. Clarke, Phys. Rev. B$\underline{31}$, 3141 (1985).

10. Z. Vardeny and J. Tauc, Phys. Rev. Lett. $\underline{54}$, 1844 (1985).

11. E. Ehrenfreund, Z. Vardeny, O. Brafman and B. Horovitz, Phys. Rev. (submitted).

12. J. Orenstein, Z. Vardeny, G.L. Baker, G. Eagle and S. Etemad, Phys. Rev. B$\underline{30}$, 786 (1984).

13. B.S. Hudson, B.E. Kohler and K. Schulter, Excited States (Academic, New York, 1982) Vol. 6, p.1

14. T. Tani, P.M. Grant, W.D. Gill, G.B. Street and T.C. Clarke, Sol. State Commun. $\underline{33}$, 499 (1980).

15. S. Etemad, T. Mitani, M. Ozaki, T.C. Chung, A.J. Heeger and A.G. MacDiarmid, Sol. State Commun. $\underline{40}$, 75 (1981).

16. Y. Yacoby, S. Roth, K. Menke, F. Keilman, and J. Kuhl, Sol. State Commun. $\underline{47}$, 869 (1983).

17. B.R. Weinberger, Phys. Rev. Lett. $\underline{50}$, 1693 (1983).

18. Z. Vardeny, J. Orenstein and G.L. Baker, J. Phys. (Paris) Colloq. 44, C3-325 (1983).

19. D.M. Pai and R.C. Enck, Phys. Rev. B11, 5163 (1975).

20. A.S. Siddiqui, J. Phys. C: Solid State Phys., 17, 683 (1984).

21. Z. Vardeny, J. Orenstein and G.L. Baker, Phys. Rev. Lett. 50, 2032 (1983).

22. G.B. Blanchet, C.R. Fincher, T.C. Chung and A.J. Heeger, Phys. Rev. Lett. 50, 1938 (1983).

23. P.D. Townsend et al., Springer Series in Solid-State Sciences 63 (H. Kuzmany, M. Mehring and S. Roth editors), Springer-Verlag 1985 p.p. 50.

24. S. Etemad, G.L. Baker, C.B. Roxlo, B.R. Weinberger and J. Orenstein, Mol.Cryst. Liq. Cryst. 117, 275 (1985).

25. Z. Vardeny, E. Ehrenfreund and O. Brafman, Mol. Cryst. Liq. Cryst. 117, 245 (1985).

26. C.V. Shank, R. Yen, R.L. Fork, J. Orenstein and G.L. Baker, Phys. Rev. Lett. 49, 1660 (1982).

27. Z. Vardeny, J. Strait, D. Moses, T.C. Chung, and A.J. Heeger, Phys. Rev. Lett. 49, 1657 (1982).

28. H. Scher and E.W. Montroll, Phys. Rev. B12, 2455 (1975).

29. A. Blumen, J. Klafter, B.S. White and G. Zumofen, Phys. Rev. Lett. 53, 1301 (1984); and these proceedings.

30. G.B. Blanchet, C.R. Fincher and A.J. Heeger, Phys. Rev. Lett. 51, 2132 (1983).

31. Z. Vardeny, P. O'Connor, S. Ray and J. Tauc, Phys. Rev. Lett. 44, 1267 (1980).

32. J. Orenstein and J. Kastner, Phys. Rev. Lett. 43, 161 (1979).

33. Z. Vardeny, P. O'Connor, S. Ray and J. Tauc, Phys. Rev. Lett. 46, 1108 (1981).

34. J. Orenstein, in "Photoexcitations of Conjugated Polymers", to appear in "Conducting Polymers" ed. by T. Skotheim.

35. R.A. Street, Adv. in Physics 30, 593 (1981).

36. J. Tauc, private communication.

37. G. Leising, H. Kahlert and O. Leitner, in Ref. 23, p. 56.

Dispersive Recombination in an Epitaxial Semiconductor

M. A. Tamor

Research Staff, Ford Motor Company, Dearborn, Michigan 48121-2053

ABSTRACT

Dispersive decay of a hole photocurrent has been observed in thin films of $Pb_{1-x}Sr_xS$, an epitaxial alloy semiconductor. Electron states in $Pb_{1-x}Sr_xS$ are localized by alloy scattering, rather than topological disorder as in amorphous materials. Transient photoconductivity was measured at room temperature for $0.16 < x < 0.69$. In all samples, the photocurrent decayed as $t^{-\alpha}$ with $\alpha = 0.5 \pm 0.05$. Samples with $x = 0.16$ and 0.22 were studied as a function of temperature. At low temperature, α was linear in T. This is consistent with the multiple trapping model, in which the hole photocurrent decays as electrons detrap from exponential density of localized states. The hole current is a census of the population of trapped electrons. The temperature dependent portion of the decay persists until a bimolecular process takes over. The room temperature behavior is not understood, but may reflect the fractal nature of disordered systems.

1. Introduction

$Pb_{1-x}Sr_xS$ is a nonisoelectronic pseudobinary semiconductor alloy: Pb has 4 valence electrons ($6s^2 6p^2$) while Sr has only 2 ($5s^2$). When Sr is substituted for Pb the large difference in ionic potential moves the now empty s-states high in the conduction band.[1] In spite of the differing number of valence electrons, the material remains a semiconductor. The very large alloy scattering enormously broadens the conduction band states. For $x < 0.4$ the conduction band edge is PbS-like ($L_{2'}$ symmetry) but broadens rapidly with increasing Sr content.[2] Above $x = 0.4$ the PbS conduction band begins to "dissipate" to become discrete Pb impurity states within the SrS gap.[3] The valence band, composed mainly of S-3p states in both PbS and SrS is largely unaffected. Figure 1 shows the

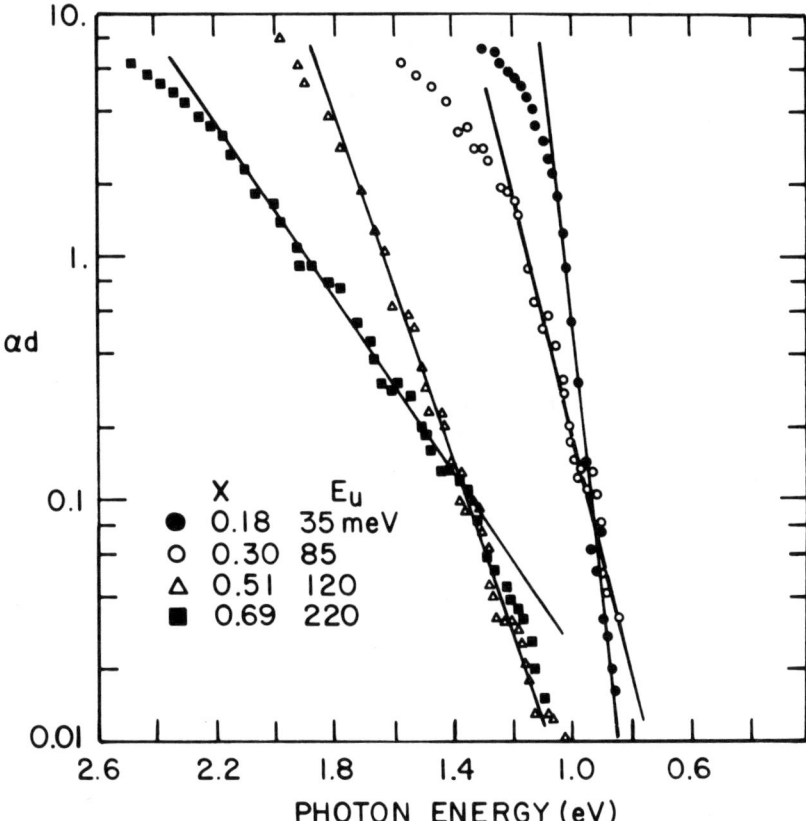

Fig. 1. Absorbtion spectra of several $Pb_{1-x}Sr_x$ films measured using photothermal deflection spectroscopy (PDS). The Urbach width varies linearly with composition.

absorption spectrum of several films measured by photothermal deflection spectroscopy. The absorption edge shows clear Urbach behavior, with the Urbach width, E_u, increasing linearly with x. It is proposed that the very strong alloy disorder in $Pb_{1-x}Sr_xS$ will result in transport and relaxation properties similar to those observed in amorphous materials. The ability to continuously vary the Urbach width over such a wide range (0 to 240 meV or more, vs. a fixed value of roughly 50 meV in a-Si and a-As_2Se_3) may be a very useful tool in the study of disordered systems.

2. Experimental Results

Films of $Pb_{1-x}Sr_x$ were produced by vacuum deposition on BaF_2 substrates. Films were 1-2 μm thick. X-ray diffraction showed the films to be essentially single crystals. A simple photo-Hall measurement, in which carrier type and density were measured with and without intense illumination, confirmed that the photocurrent was due to holes. To examine the transport properties, transient photoconductivity (TPC) was measured in the coplanar electrode geometry. Under these conditions, photoexcited carriers do not exit the sample before decaying. Thus the hole current is a census of population of trapped electrons. For amorphous materials, where the photocurrent is due to the small fraction of electrons thermally excited to conduction states, the analogous measure is the photo-induced absorbtion (PA) due to the population of trapped electrons.

TPC was measured at room temperature with 10 μs resolution for samples with x ranging from 0.16 to 0.69. In all cases, the photocurrent decayed as $t^{-\alpha}$ with $\alpha = 0.5 \pm 0.05$. Figure 2 shows decays for films with x = 0.16 and 0.69. The decay parameter, α, was independent of pulse width and intensity. The magnitude of the photocurrent did increase with excitation, ruling out a simple decay by a three particle (Auger) recombination process.

For samples with x = 0.16 and 0.22, TPC was measured with 1 μs resolution as function of temperature from 80 to 380 K. At low

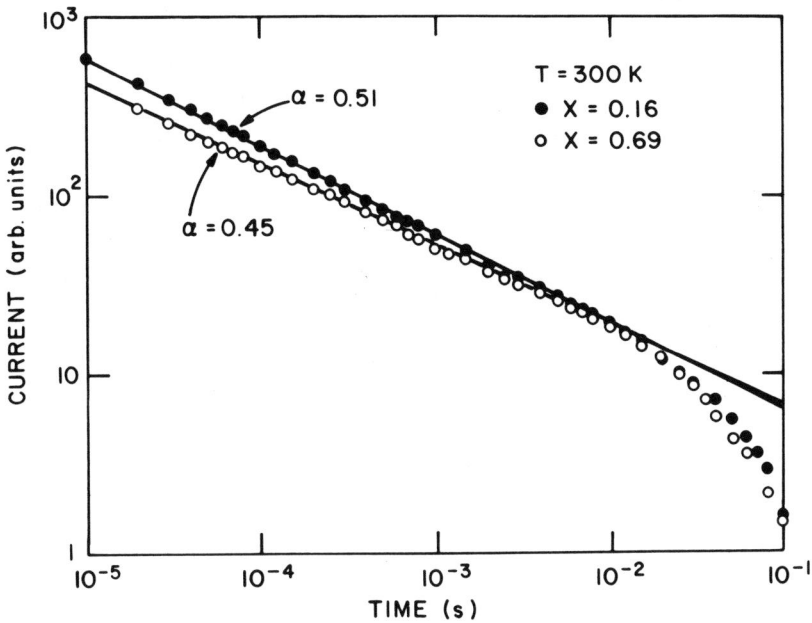

Fig. 2. Decay of the photoconductivity transient following a 10 μs Argon laser pulse. At room temperature the decay parameter, α, was 0.5 for all compositions, x. The current scales have been shifted to bring the decays into the same decade.

temperature the decay breaks into two regimes: 1) an initial decay with α_i increasing linearly with temperature and 2) a final decay with α_f independent of temperature. Figure 3 shows photocurrent decays at several temperatures in a film with x = 0.16. The temperature dependence of α_i for both compositions is shown in Figure 4. At low temperature, the slope is in excellent agreement with the relation $\alpha = kT/E_u$ predicted in the multiple trapping model.[4-6] This result is very similar to the decay of photoinduced absorbtion in a-Si and a-As$_2$Se$_3$.[6,7] The hole current decays as electrons detrap from an exponential density of localized electron states. The time and temperature dependence are consistent with the multiple trapping (MT) model originally developed for amorphous materials.[4-6] This is the first observation of a dispersive relaxation process in a crystalline semiconductor.

The mechanism of the decay at later times and higher temperature not well understood. At 80 K, the current transient eventually joined a decay curve that was independent of pulse intensity. As the pulse intensity was reduced, the current decayed with α_i until the final decay curve was reached at successively later times. This indicates that the final decay is due to a bimolecular process. The two compositions showed somewhat different behavior. For x = 0.16, the final decay showed a slope of -0.5 (Figs. 3 and 4). A t^{-1} behavior, more typical of a bimolecular decay, was only observed at very long times or very high pulse intensity. For x = 0.22 the final decay had a slope of -1 for all pulse intensities. However, the bimolecular rate coefficient, b, where $\tau_{BM} = 1/bn_o$, appeared to be several times shorter than in the x = 0.16 sample, possibly concealing the $t^{-0.5}$ decay.

3. Discussion

With the exception of the $t^{-0.5}$ behavior for x = 0.16, the decay of the TPC is entirely consistent with the multiple trapping model at early times, and a bimolecular decay at later times. Recall, however, that at room temperature the TPC decayed as $t^{-0.5}$

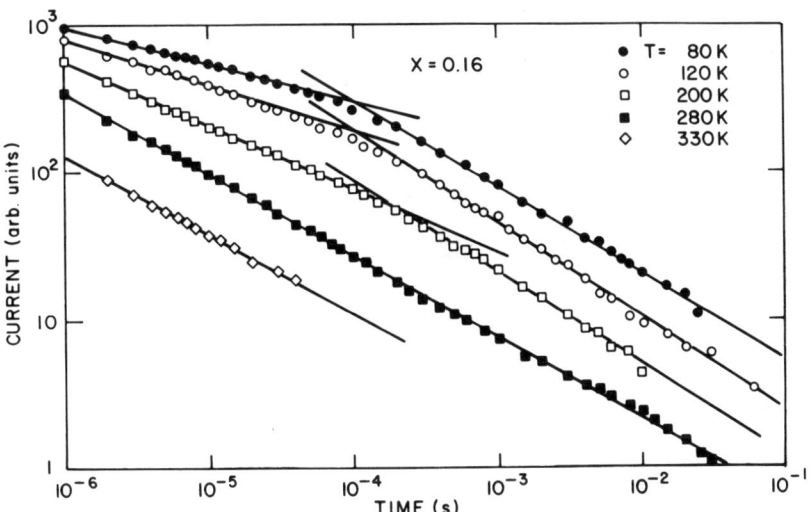

Fig. 3. Temperature dependence of the photoconductivity transient following a 1 μs pulse in a film with x = 0.16. The position of the "knee" is power dependent, indicative of bimolecular recombination at late times.

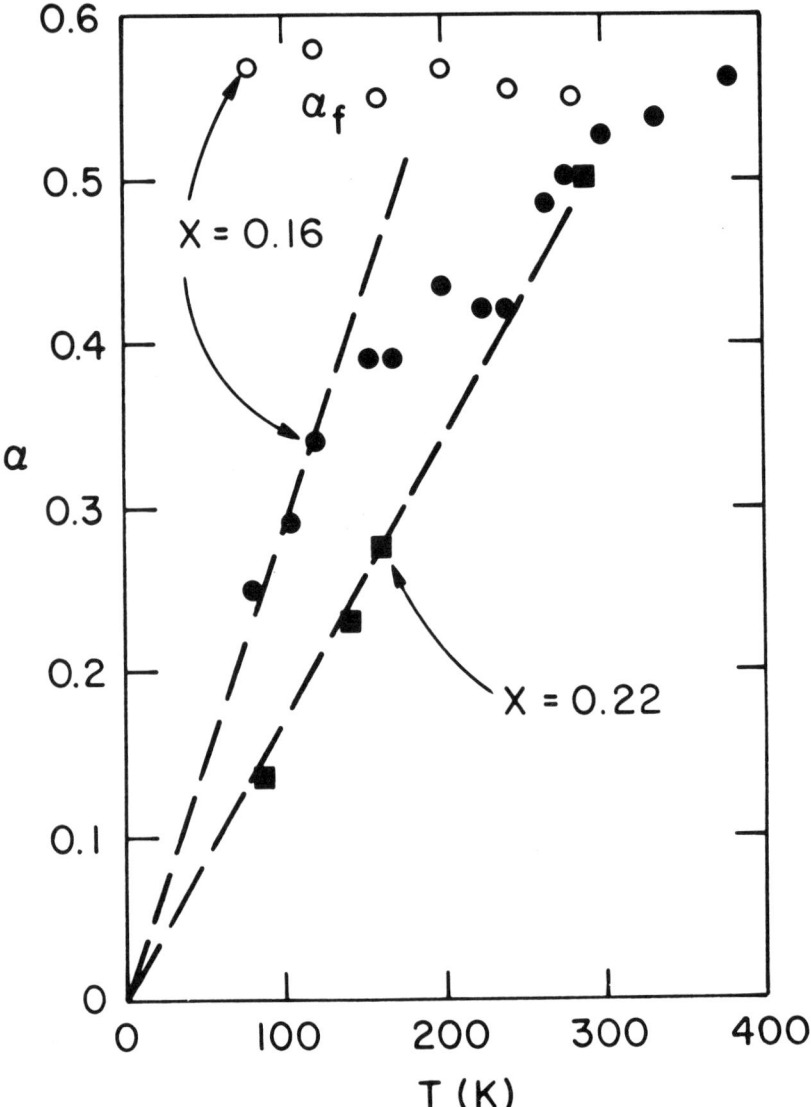

Fig. 4. Temperature dependence of the initial decay exponent, α_i. The solid squares are for a film with x = 0.22, the solid circles for x = 0.16. The open circles show α_f for x = 0.16. The dashed lines indicate the slope expected from the measured Urbach width ($\alpha = kT/E_u$).

for all compositions and was not consistent with an Auger recombination process. For x = 0.69 at 300 K, the measured Urbach width predicts α = 0.11, far smaller than the measured value. It was recently pointed out by Monroe that thermally activated hopping between localized states just below the mobility edge is indistinguishible in dynamics from multiple trapping, where carriers are excited to extended conduction states.[7] Thus a hopping model makes predictions similar to multiple trapping, and so can be ruled out. This leaves tunneling, where electrons move directly between deep lying localized states, as the mechanism by which the trapped electrons move to recombination centers. Deviations from the multiple trapping behavior in a-Si at low temperature have also been attributed to tunneling.[8]

That tunneling should appear to dominate at room temperature in these experiments may be explained simply in terms of the time scales on which the observations are made. At high temperatures detrapping from any but the deepest (and therefore sparsest) traps is very rapid. In the extreme case, the detrapping will be almost entirely finished during the light pulse. This is born out by the observation that the initial photocurrent dropped very rapidly as the temperature was raised above 250 K, even though α remained constant at 0.5 (see Fig. 4). For the low Sr content films, the experiment simply does not have the time resolution to observe the $t^{-\alpha}$ decay at high temperature. For the high Sr samples, where the Urbach width is enormous and the conduction band, if any, lies at very high energy (near the SrS band gap of 5 eV), multiple trapping will be very slow and may not enhance recombination. In this case recombination will proceed entirely by tunneling and a will not be temperature dependent.

The origin of the $t^{-0.5}$ decay of the room temperature photoconductivity is not understood. Auger (3-particle) processes are already ruled out. One intriguing possibility is the fractal nature of the eigenstates of these heavily disordered systems. Although the film is certainly a three dimensional system, the very deep trap states may have a one dimensional string-like nature (like the one dimensional rivers at the bottom of a two dimensional landscape). The recombination process in this one dimensional system will proceed with a time dependent rate constant that leads to a $t^{-0.5}$ decay. This type of behavior has already been observed in exciton recombination.[9] Preliminary theoretical results also point out the fractal nature of the eigenstates in disordered systems.[10] In summary, the initial decay of transient photocurrents in $Pb_{1-x}Sr_xS$ shows many of the features observed in amorphous materials. The decay at later times and high temperatures may indicate a new behavior also due to the disordered nature of this material.

References

1. M. A. Tamor, L. C. Davis and H. Holloway, Phys. Rev. Lett. 52, 946 (1984).

2. L. C. Davis, Phys. Rev. B 28, 6961 (1983).

3. M. A. Tamor, L. C. Davis and H. Holloway, Phys. Rev. B 28, 3320 (1983).

4. T. Tiedje and A. Rose, Solid State Commun. 37, 49 (1980).

5. J. Orenstein and M. A. Kastner, Solid State Commun. 40, 85 (1981)

6. J. Oreinstein, M. A. Kastner and V. Vaninov, Philos. B 46, 23 (1982).

7. D. Monroe, Phys. Rev. Lett. 54, 146 (1985).

8. D. Pfost and J. Tauc, Solid State Commun. 48, 195 (1983); D. Pfost, Z. Vardeny and J. Tauc, Phys. Rev. Lett. 52, 376 (1984).

9. R. Kopelman, this Proceedings.

10. C. M. Soukoulis and E. M. Economou, Phys. Rev. Lett. 52, 565 (1984); C. M. Soukoulis and E. N. Economou, Materials Science Forum, vol. 4, p.145 (Trans Tech Publications, Switzerland, 1985).

CHANGES IN ELECTRONIC TRANSPORT DURING STRUCTURAL RELAXATION OF GLASSY CHALCOGENIDES

M. A. Abkowitz
Xerox Webster Research Center
800 Phillips Rd., 0114-39D
Webster, NY 14580

ABSTRACT

Changes in electronic transport clearly associated with relaxation of the glassy structure can be identified in a wide composition range of chalcogenide glasses. Analysis of the drift mobility and deep trapping of photoinjected carriers and of the bulk thermal generation of free carriers demonstrates that in these glasses key electronic processes are controlled by specific localized electronic states in the mobility gap. Near T_g the respective populations of these states are found to vary systematically during thermostructural relaxation on an experimentally accessible time scale always tending toward a temperature dependent quasi-equilibrium value. The present observations clearly establish the thermodynamic defect origin of these key electronic gap states and simultaneously elucidate the mechanism by which relaxation impacts charge carrier transport.

1. Introduction

All good glasses exhibit physical aging at average rates which slow precipitously as temperature recedes from the glass transition region. An important consequence of thermostructural relaxation is the hysteretic behavior observed during rate heating and cooling which is represented schematically for enthalpy in figure 1. The extrapolated melt curve for enthalpy is the broken line between (3) and (1). During rate cooling the system in its super-cooled melt phase falls out of equilibrium along curve segment (3) - (4). At the freeze-in value (4) structural relaxation can no longer be observed on the time scale defined by the

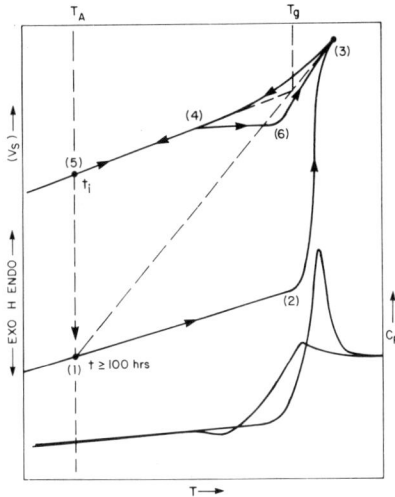

Figure 1. Enthalpy H versus temperature T during isobaric heating and cooling of what is initially a well-annealed glassy film. The large peak in heat capacity C_p corresponds to enthalpy relaxation along segment (2)-(3) and is observed only during cycle 1 heating of a well-annealed film. The small peak in C_p corresponds to the trajectory (5)-(4)-(6)-(3) described by the system in multiple cycling experiments. (See text.)

cooling rate. At lower temperature the system is frozen into the glassy state, (4) - (5), which has been determined by its prior thermal history. It is useful to define a fictive temperature T_f at the intersection of the extrapolated melt (3) - (1), and enthalpy (4) - (5), curves. T_f can be regarded as the glass transition temperature T_g associated with the glassy state defined by (4) - (5). If the glass is allowed to undergo isothermal annealing at a temperature T_A not too far below T_g then it will tend to stabilize, after evolving through a series of intermediate states, at (1) on the equilibrium line. When this annealed glass now undergoes further rate heating from (1), structural relaxation rates are far too slow to allow the system to track the melt curve so instead the system follows glass curve (1) - (2). In the glass transition region (2) - (3) relaxation rates increase precipitously and the glassy system can realize the opportunity to readjust toward the melt curve at (3). As shown in the lower part of the figure this enthalpy relaxation is typically observed as a heat capacity peak in a differential scanning calorimeter. Suppose the glass at (3) undergoes further rate cooling forcing it to track along (3) - (4) - (5) and then immediate rate heating. Under these circumstances the glass follows (5) - (4) - (6) - (3). The loop (3) - (4) - (5) - (6) - (3) is the residual hysteresis manifested at any practical scanning rate during continuous thermal ramping. Note the corresponding behavior in the smaller of the two heat capacity peaks below. The area under these heat capacity peaks is called the excess enthalpy and can be a measure of the extent of glass relaxation during isothermal annealing below T_g. The entire process depicted in figure (1) reflects a fully repeatable alteration of structure and will manifest, systematically, in many of the systems' intensive thermodynamic properties. In the following sections, it is demonstrated that in chalcogenide glasses thermostructural relaxation also has a remarkably similar impact on the transport of photoinjected carriers. Thermal cycling induces hysteresis in both drift mobility and injected carrier range (as manifested in the behavior of xerographic potentials). The latter is not, however, a universal feature of glasses which transport electronic charge. In fact in the case of chalcogenide glasses these observations bear critically on the question of the origin and nature of the electronic states which control their photoelectronic behavior.

2. Results and Discussion

(1) Drift Mobility of Injected Carriers

Drift mobility is the velocity per unit applied electric field of an excess injected charge. Drift mobilities are most commonly measured[1] by the small signal time of flight technique (TOF). In the present case 10 to 60 micron thick calcogenide films were prepared by vacuum evaporation at a rate of two microns/min. onto oxidized aluminum substrates held above the glass transition to assure vitrification. A semitransparent thin metallic film was then evaporated to complete the sandwich cell. Both substrate and top contact must block the injection of charge in the dark. Specimens were typically dark annealed at room temperature (294K) for months prior to initial measurement. To make the measurement a field is applied and shortly thereafter the specimen is illuminated with a weak pulse of strongly absorbed light. Weak photoexcitation (i.e., small signal conditions) insures that the drifting charge packet does not itself distort the field. Depending on polarity of the applied field, a hole or electron charge sheet now drifts through the specimen. The resulting current can be sensed by either an operational amplifier or a series resistor and wideband amplifier combination. This current drops to zero (at the transit time) when the transiting charge sheet reaches the collecting electrode. If the overall circuit time constant is sufficiently less than the transit time than a current mode transient, with a clearly delineated transit time is resolved providing the sheet traversal is coherent (i.e., nondispersive). Finally, a lower bound on the experimental field range is established by the requirement that the transit time be less than the bulk dielectric relaxation time. (Any excess injected carrier must transit the bulk before it is locally neutralized.) The latter is also the threshold condition for the onset of space charge limited dark injection from an ohmic contact. Thus the requirement for blocking contacts in TOF.

The hole drift mobility in a-Se illustrates several of the important characteristics generally found in a wide composition range of chalcogenides. (1) Mobilities are simply activated over a wide temperature range. (Below 273K for both electrons and holes in a-Se.)[2,3] (2) Mobilities and their activation tend to be highly reproducible in well annealed specimen films. (3) Systematic hysteresis[4] in the mobility values is induced when any specimen is thermally cycled through its glass transition region. The latter is illustrated in figure 2 for the case of photoinjected electron transport in a 3% As in Se alloy. The associated time-temperature history is as follows: the specimen which had been dark annealed for

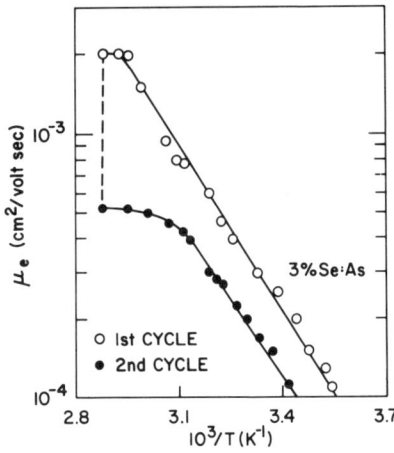

Figure 2. Log electron-drift mobility μ_e versus $10^3/T$ at 10^5 V/cm for an a-Se:As alloy film containing 3 at. % As. Open circles are measured during the thermal cycle of film previously stored at 295K for days. Solid circles are subsequent thermal scans (See text.). The dashed line depicts the drop in mobilities during cycle 1 heating.

months at room temperature was cooled to 250K then heated at 4K/min. to a temperature above its glass transition. The electron drift mobility was measured at regular intervals during this rate heating. The specimen was then immediately cooled at 10K/min. to 250K and the entire cycle repeated many times in succession. The upper data were collected during the initial (cycle 1) rate heating. As the glass transition is approached from below there is a relatively precipitous drop in mobility followed by a region in which the temperature dependence nearly vanishes. During the second and on all subsequent heating cycles the mobility retains its depressed values below the glass transition region. If however the specimen is allowed to dark anneal at 294K for hundreds of hours and is then recycled, then cycle 1 behavior is reproduced fully. The hole drift mobility behaves similarly during comparable rate heating and cooling. The behavior of the drift mobility illustrated in figure 2 is clearly aassociated with the thermostructural relaxation process depicted in figure 1. The transport behavior of the well annealed specimen and the downward mobility adjustment during the first rate heating correspond in figure 1 to trajectory (1) - (2) - (3). The behavior during subsequent thermal cycles is then associated with loop (5) - (4) - (6) - (3) - (4) - (5) and room temperature relaxation recovery with (5) - (1). Why should relaxation of the glassy structure, a process which occurs on a time scale much longer than a transit time impact transport of an excess injected carrier? The principal clue is furnished by the activated nature of the transport process which is interpreted as shallow trap controlled.[5] In the present case any excess injected carrier in its transit across the bulk undergoes multiple trapping and release. In the specimens described here the latter can be operationally distinguished from an energetically distinct manifold of deep traps. Deep trapping (as distinct from shallow trapping) causes permanent carrier loss (depletion of the carrier packet) on the time scale defined by the transit time. In its simplest form where the shallow traps form a nearly discrete manifold the shallow trap controlled drift mobility μ_d is expressed in terms of the band mobility μ_0 the number of shallow traps N_t displaced by energy E_t from the corresponding mobility edge and the effective density of states at the mobility edge N_c, as:

$$\mu_d = \mu_0 \, (N_c/N_t) \exp(-E_t/KT) \qquad (1)$$

While it is possible that the structural sensitivity could be manifest in any of all of the parameters μ_0, N_c, N_t, E_t, in (1) the absence of a resolvable change in activation and physical intuition suggests a relaxation induced variation in the population of shallow traps as the favored hypothesis. This hypothesis is in fact substantially ellaborated on and verified more directly in the measurements described in the following, which probe by the use of novel xerographic techniques the corresponding behavior of deep states (traps and emission centers) in chalcogenide glasses.

(2) Thermodynamic Defect Origin of Deep States from Dynamic Behavior of Xerographic Potentials

A relaxation induced variation in the population of traps is expected if these are thermodynamic defects. The latter means that in the equilibrium melt phase the number of such defects is determined by a defect creation energy and the temperature. Two key features are then predicted for the host glass.

(A) $N = N(T)$

In the glassy solid at a fixed (i.e. annealing) temperature the number of such defects relaxes toward the extrapolated melt value stabilizing there.

(B) Relaxation rate depends on T.

Relaxation is a complex process characterized by a distribution of relaxation times. The mean relaxation time grows precipitously longer as temperature recedes from the initial temperature range of glassy solidification.

The energy distribution of deep traps, $N(E)$, can be determined using a novel space charge (xerographic) spectroscopic technique.[6,7] Space charge buildup is induced by repeated xerographic photoinjection and subsequent bulk trapping of one sign carrier. The energy distribution of these deep traps is then determined from analysis of the isothermal trap emptying kinetics. The xerographic cycle is initiated by corona or transient contact charging to a potential V, the top surface of a film, mounted on a ground plane, which has total capacitance C. To complete the xerographic cycle the film is discharged with a flash of surface absorbed light intense enough to completely remove the surface charge and eliminate the field in the surface region. During discharge some fraction of the CV of total charge in transit becomes immobilized in deep traps. As a result the film is left with a non-zero residual potential which decays to zero on a longer time scale as thermal processes restore space charge neutrality. This decay is controlled by the underlying distribution of trap release times. It is therefore possible to calculate from the rate constants characterizing the decay of xerographic residual potential the distribution of trap energies relative to the energy of the transport state. The experiment is performed by first repeatedly cycling the film until cumulative trapping causes the residual to saturate at some value. Saturation occurs either because kinetic equilibrium is achieved between trapping and release (under fixed experimental conditions) or because all available traps have actually been filled. Temperature and cycling rate dependence can be analyzed to distinguish the former from the latter. The isothermal residual potential decay curves collected following the termination of cycling must then be deconvoluted. If the trap distribution function varies slowly with energy on a scale of (0.1-0.2)KT, then it is possible to construct the following continuum[7] approximation. After termination of cycling isothermal decay is manifest in the time dependent potential V(t) which can be represented for a specimen of thickness L and dielectric constant ε in terms of the energy distribution of occupied traps N(E,t) at time t as

$$V(t) = qL^2/2\varepsilon \int N(E,t)dE \qquad (2)$$

If $N_0(E)$ is the volume density of traps at energy E measured from the transport state then,

$$N(E,t) = N_0(E) \exp - R(E)t \qquad (3)$$

where

$$R(E) = \nu_E \exp - E/KT \qquad (4)$$

The prefactor ν_E is typically of the order of a phonon frequency. Thus

$$dV/dt = -qL^2/2\varepsilon \int N_0(E)R(E) \exp - R(E)t \, dE \qquad (5)$$

Equation (5) is simplified by recognizing that Rexp-Rt has a sharp maximum at $E = kT \ln \nu_E t$. Thus at any time t, dV/dt is dominated by the release from traps occupying a narrow energy manifold at E of width dE. If $N_0(E)$ varies slowly and if the frequency prefactor is independent of energy then a simple transformation yields

$$N_0 (E = kT\ln\nu t) = 2\varepsilon/qL^2 *t/kT*dV/dt \qquad (6)$$

where the experimental time scale has been mapped onto an energy scale.

This experimental and analytical procedure has been successfully applied to the mapping of trap distributions in a variety of glasses[6,7] in the Se:Te:As system. In the present context a key and particularly clearcut result is described in Figure (3) for hole traps in a-Se. In a-Se deep states are organized into relatively narrow and discrete submanifolds. (Progressive alloying leads to a rapid broadening of these

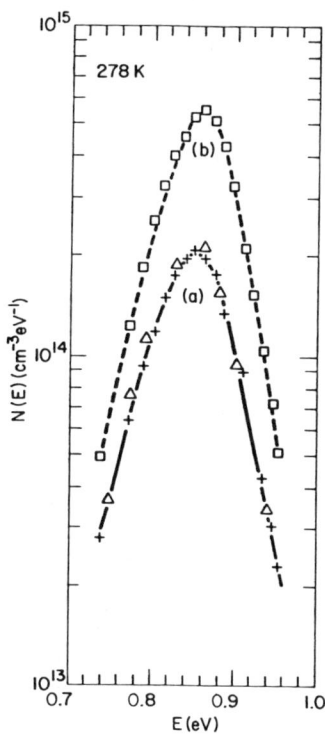

Figure 3. Hole-trap distributions computed from isothermal-residual decay curves at 278K in a 55 μm a-Se film with the following thermal history: (a) crosses, film annealed at 295K for 2 × 10³ h prior to measurement; (b) squares, film annealed at 308K for 5 h prior to measurement (a); triangles, after history (b) film is reannealed at 295K for 300 h prior to measurement. Energy is referred to valence-band-mobility edge.

states.) The spectral distribution of hole traps as a function of specific thermal history is illustrated. The crosses are for example data obtained from an a-Se film annealed for several hundred hours at 295K prior to measurement. This relatively sparse trap manifold occupies a narrow energy range centered about 0.87eV above the valence band mobility edge and is in all respects a highly reproducible feature of a-Se. The squares are obtained when the film is annealed for five hours at 308K, a temperature below T_g. The data represented by triangles was obtained after the same specimen film was reannealed at 295K for several hundred hours. In all these cases the film was quenched to 277K to arrest structural relaxation during measurement (i.e. freeze in the desired thermal history) so that both cycling and xerographic residual decay were measured at 278K. The figure demonstrates the systematic annealing behavior in the glassy state of a well delineated set of traps toward a stable temperature dependent population - precisely the expected behavior for thermodynamic defects stipulated in (A). Completely analogous behavior is observed for the deep electron traps in (ambipolar) a-Se. The above then explicitly parallels the suggested behavior of shallow traps in the process which was presumed responsible for drift mobility hysteresis.

Time resolved, strongly temperature dependent relaxation recovery behavior associated with thermostructural relaxation in the glassy state is most clearly observed in xerographic dark depletion discharge experiments.[8] The specimen, a chalcogenide film, is deposited on a conducting ground plane. The isothermal dark decay of surface voltage and its derivative is then time resolved immediately after charging. Dark decay can be driven by bulk thermal generation of carriers and by injection from the surface. The latter can be experimentally distinguished from the former. In the a-Se:Te alloys dark decay is dominated by a bulk hole depletion process. Bulk thermal generation involves the simultaneous emission of mobile holes from deep emission centers which are then immediately swept out leaving behind deeply trapped electrons. Even more specifically analysis of the resulting data[9] in a composition series of Se:Te alloys has demonstrated that as generation proceeds the bulk algebraically builds up a uniform negative space charge according to

$$\rho(t) = at^p \quad (7)$$

where p is a parameter ≤ 1 which depends on composition. Under these circumstances it can be shown that the hole depletion process in a specimen of thickness L, having capacitance C and dielectric constant ϵ, when charged to voltage V_0 is in the absence of surface loss, described by the following pair of differential equations:

$$dV/dt = -L^2/2\epsilon \; apt^{p-1} \quad \text{when } t < t_d \quad (8)$$

and

$$dV/dt = -\epsilon/2a^2 \; (V_0/L)^2 \; pt^{-p-1} \quad \text{when } t > t_d \quad (9)$$

At the critical time t_d the volume generated space charge just equals the surface charge and

$$\rho(t=t_d) = \epsilon V_0/L^2 \quad (10)$$

The depletion time t_d obeys the scaling law

$$t_d = (\epsilon V_0/aL^2)^{1/p} \quad (11)$$

Physically the depletion time is just the time required at a given temperature for the bulk to uniformly generate an experimentally predetermined amount of charge, namely that equal to the charge initially deposited on the specimen surface. It can be shown in the present case that generation is controlled by deep emission centers. Thus the depletion time will in general depend on the distribution of these states, the charging voltage, sample capacitance, physical dimensions and temperature. Figure 4 clearly demonstrates both conformity of dark decay data with the model from which equations (8) - (11) derive and the composition dependence of the parameter p. Figure 3 demonstrated directly that the number

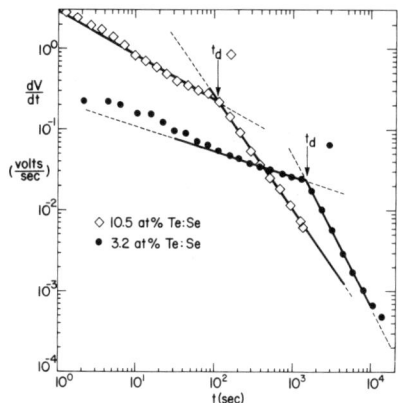

Figure 4. Rate of dark decay of surface voltage (dV/dt vs t) on a 3.2 at. % Te:Se alloy film (filled circles) on a 10.5 at. % Te:Se alloy film (diamonds). $T = 297K$, $V_0 = 100$ V, and $d = 55$ μm.

of deep traps in a-Se but not their energy distribution varied with the structural state of the glass. Similarly the corresponding invariance of drift mobility activation during thermal cycling was interpreted as a variation in the number of shallow traps but not their distribution. There is sufficient reason to also believe that while the integrated number of emission centers varies during thermostructural relaxation their spectral distribution remains fixed. Under these circumstances for any given sample charged to a fixed voltage the depletion time depends only on the integrated number of centers and temperature. With the latter in mind examine figure 5. Shown is depletion data for a 50 micron-thick, 12wt% Te:Se glassy film, charged to 100 volts. The upper part of the figure is the thermal profile imposed. The depletion time after prolonged annealing at 22°C is shown at (a). The response of the specimen to a step change in temperature from 22°C to 35°C occurs in two stages. The first stage (a) - (b) is very rapid (on the

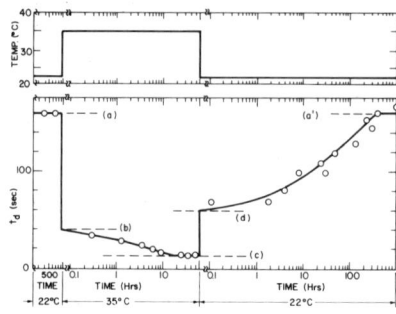

Figure 5. Time dependent variations in the depletion time t_d of a 12 wt. % Te:Se film induced by step heating and cooling. The imposed thermal history is represented schematically in the top half of the figure. (V_o = 100 V, d = 50 μm). The response is divided into three time zones. In zone 1, the specimen temperature is 22°C; in zone 2 it is stepped to 35°C. In zone 3, it is stepped back to 22°C. Shifts (a)-(b) and (c)-(d) occur under approximately isostructural conditions.

present experimental time scale) and reflects the isostructural dependence of the depletion time on temperature. The second much slower stage is the change in depletion time associated with the relaxation of glassy structure manifested through a change in emission center population. After sufficient annealing at 35°C the depletion time stabilizes at (c). The response of the film to a step decrease in temperature from 35°C to 25°C is qualitatively similar yet differs in some important details from the above. The respective asymmetry in segments (a) - (b) and (c) - (d) reflects structurally induced differences in the integrated number of emission centers which occupy a fixed distribution of finite width. Most revealing however is the significantly longer time required to stabilize the emission center population at 25°C, (d) - (a), compared to 35°C (b) - (c). The latter demonstrates precisely, the strongly temperature relaxation dependent behavior expected of thermodynamic defects in a glassy solid.

3. References

1. F.K. Dolczalck, p. 71 in "Photoconductivity and Related Phenomena", edited by J. Mort and D.M. Pai, Elsevier, New York (1976).
2. M.D. Tabak, Phys. Rev. B2 2104 (1970).
3. J.M. Marshall and A.E. Owen, Phys. Stat. Sol. A12 181 (1972).
4. M. Abkowitz and D. M. Pai, Phys. Rev. B18 1741 (1978).
5. W.E. Spear, Prof. Phys. Soc. (London) B70 669 (1957) and B76, 826 (1960).
6. M. Abkowitz and R.C. Enck, Phys. Rev. B25, 2567 (1982).
7. M. Abkowitz and J.M. Markovics, Phil. Mag. B49:L31 (1984).
8. M. Abkowitz, F. Jansen and A.R. Melnyk, Phil. Mag. B, 51, 405 (1985).
9. M. Abkowitz, G.M.T. Foley, J.M. Markovics and A.C. Palumbo, Appli. Phys. Lett. 46, 4 (1985).

MODELS FOR DYNAMICALLY CONTROLLED RELAXATION

A. Blumen[*], J. Klafter[§] and G. Zumofen[**]

[*] Max-Planck Institut für Polymerforschung, D-6500 Mainz und Lehrstuhl für Theoretische Chemie, Technische Universität, D-8046 Garching, West Germany

[§] Corporate Research Science Labs, Exxon Research and Engineering Co., Annandale, N.J. 08801, USA

[**] Laboratorium für Physikalische Chemie, ETH-Zentrum, CH-8092 Zürich, Switzerland

ABSTRACT

Relaxation in random media often follows non-exponential decay patterns. Here we present several random-walk models of parallel and sequential type which lead to non-exponential forms.

1. Introduction

Relaxation patterns in random media such as amorphous materials, mixed crystals, molecules adsorbed on surfaces or polymers, often display markedly non-exponential decay forms. Such behavior ranges from the stretched exponential expression of Kohlrausch and Williams-Watts:[1-3]

$$\Phi(t) = \exp[-(t/\tau)^\alpha] \qquad (1)$$

to algebraic long-time decays:

$$\Phi(t) \sim Ct^{-\gamma} \qquad (2)$$

Recent investigations suggested a dichotomy between parallel and sequential channels in the description of such relaxation forms.[4] Based on our previous experience on energy transport processes[5-7] we present several random-walk models which are of mixed (parallel and sequential) type and whose relaxation follows non-exponential patterns. Interestingly, models which differ in their microscopic dynamics may lead to strikingly similar decays. As an underlying theme of our study we find that such decays depend on the amount of disorder (structural, temporal and energetic). In this communication we model the structural disorder through fractals,[8] the temporal disorder through continuous-time random-walks (CTRW)[9-11] and the energetic disorder through ultrametric spaces.[12,13]

2. The Target and the Trapping Problems

An interesting dynamical problem of mixed (parallel and sequential) character obtains in reaction-diffusion situations. In systems in which the species involved interact only at short distances, for an elementary reaction to occur it is necessary that the reactants come together. Thus, the process involves a sequential aspect, since the elementary reaction step is in general preceded by the diffusive motion of the reactants. On the other hand, one monitors the development of such a reaction in a macroscopic sample: The overall relaxation is given by a parallel (ensemble) average over the microscopic events.

As a special case of reaction-diffusion problems we consider here two types of particles A and B, which perform random-walks over pre-assigned, discrete structures. In the simplest models one has one A- and several B-particles, and the A-particle is annihilated at the first encounter of a B-particle. Depending on which of the species performs the motion one distinguishes between the trapping model[6,14,15] (only the A moves), the target (scavenging) model[16-18] (only the B move) and the moving targets[19] (both species move). In these models one evaluates the survival probability of the A-particle averaged over all possible realizations of particle distributions and motions.

As recently shown by us, the target problem admits an exact solution on regular (translationally-invariant) lattices.[18] Assuming that initially the B-particles are Poisson-distributed on the lattice, and that they move independently of each other, we found the survival probability of the A-particle to be:

$$\Phi_n = \exp[-p(S_n-1)] \tag{3}$$

where p is the number density of B-particles, S_n the mean number of distinct sites visited in n-steps and Φ_n the probability that no B-particle has visited the origin in the first n steps. The quantity S_n depends on the dimension of the underlying lattice. For nearest-neighbor random walks one has for not-too-small n:

$$S_n \sim \begin{cases} n^{1/2} & d=1 \\ n/\ln(n) & d=2 \\ n & d=3 \end{cases} \tag{4}$$

and hence, at longer times and d=3 Φ_n is exponential (α being unity in Eq. (1)). On the other hand, for d=1 Φ_n shows a stretched-exponential form, with $\alpha=1/2$ in Eq. (1).

Paradoxically, the trapping problem in which only one A-particle moves is harder to handle than the target problem. In fact it turns out to be very difficult to establish analytically the relaxation $\tilde{\Phi}_n$ due to trapping. Exact expressions for trapping are known only for d=1, and one has in general to resort to simulations.[14,15] In the short and medium time domains the survival probability $\tilde{\Phi}_n$ may be expressed as a cumulant expansion. The first term in the expansion of $\tilde{\Phi}_n$ is precisely Φ_n, whereas the higher terms involve higher moments of the distribution of visited sites. Asymptotically, for very large n one finds:[20]

$$\tilde{\Phi}_n \sim \exp[-Cp^{2/(d+2)} n^{d/(d+2)}] \tag{5}$$

again a Kohlrausch-Williams-Watts form with $\alpha = d/(d+2)$. In Eq. (5) a Poisson-distribution of the B-particles is assumed: for a binominal distribution p has to be replaced by $\lambda \equiv -\ln(1-p)$.

Therefore, at long times the survival probabilities for target annihilation, Φ_n, and for trapping, $\tilde{\Phi}_n$, are <u>qualitatively</u> distinct and display different α-exponents in the stretched-exponential expression, Eq. (1). This difference unveils a fundamental aspect, namely that the course of the reaction depends on which of the species moves. This distinction is lost when reactions are treated in a pair-approximation scheme, which centers on one A- and one B-particle, and in which only the sum $D = D_A + D_B$ of the diffusion coefficients enters. In fact, in the moving target problem in which both species move, one may interpolate smoothly between the relaxation patterns $\tilde{\Phi}_n$ and Φ_n, by keeping the total D fixed and varying the relative A- and B-mobilities.[19]

Up-to-now the stochastic aspects entered only through the initial particle distributions and the random-walk realizations. In the next sections we discuss the influence of structural, temporal and energetic disorder on the relaxation forms Φ_n and $\tilde{\Phi}_n$.

3. Structural Disorder: Geometric Fractals

The basic feature of fractals is their self-similarity, i.e. the fact that they are (either individually or in their ensemble) invariant under the group of dilatation operations.[8] Many stochastically disordered systems, like linear and branched polymers, epoxy-resins, various porous materials and catalytically active surfaces[21] have been characterized as being self-similar under length-scaling. Also stochastic model structures, such as percolation clusters at criticality and aggregates constructed by diffusion-limited growth (DLA) were found to be fractal.[22] Examples for deterministically built fractals are the Sierpinski-gaskets, whose generators in d-dimensional Euclidean spaces are hypertetrahedrons consisting of d+1 hypertetrahedra of half the original sidelength.[8] One may construct many different deterministic fractals by choosing other generators: Simple changes consist in taking other scaling factors than 2 and varying the coverage of the original hypertetrahedron.[23]

Since a fractal scales with distance, one can characterize it through a parameter $\bar{d}$, the fractal dimension.[8] A by-now classical analysis by Alexander and Orbach showed that dynamical processes on fractals are not determined solely through $\bar{d}$, and that the "fracton" or spectral dimension $\tilde{d}$ is fundamental for the Laplace equation on the fractal.[24] The dichotomy between $\bar{d}$ and $\tilde{d}$ is a new aspect for fractals, an aspect not found in regular (translationally-invariant) lattices.

We are now in the position to discuss target annihilation and trapping on fractals. One may note that in the following expressions only the spectral dimension $\tilde{d}$ enters.

The analysis of the target problem on fractals parallels that on regular lattices:[18] For a <u>fixed</u> target location the relaxation is given by an expression similar to Eq. (3). Now, however, the different sites on a fractal are not identical, and an additional average over target positions is required (This average is trivial for regular lattices). Hence Eq. (3) is now only approximate:

$$\Phi_n \simeq \exp[-p(S_n-1)] \qquad (5)$$

For homogeneous fractals, such as Sierpinski-gaskets, Eq. (5) is a very good description of the true relaxation. In Eq. (5) S_n is the mean number of distinct sites visited, averaged over all starting points. One finds:[24-26]

$$S_n \sim \begin{cases} n^{\tilde{d}/2} & \text{for } \tilde{d}<2 \\ n & \text{for } \tilde{d}>2 \end{cases} \qquad (6)$$

Since $\tilde{d}$ may take any positive value, Eq. (6) interpolates smoothly between the discrete results of Eq. (4). Furthermore, Eq. (6) highlights the marginal role played by d=2. Introducing Eq. (6) in Eq. (5) leads to a stretched exponential law, with $\alpha = \tilde{d}/2$ for $\tilde{d}<2$. We note that this form allows to determine experimentally the spectral dimension of fractal objects by monitoring the relaxation due to multiple-step mechanisms in such systems.[26,27]

Turning now to the trapping problem, we again find that fractals allow to interpolate between the discrete dimensionalities of regular lattices. The short time expansion of the relaxation due to trapping has as leading term:[26]

$$\tilde{\phi}_n \sim \begin{cases} \exp(-Cpn^{\tilde{d}/2}) & \tilde{d}<2 \\ \exp(-Cpn) & \tilde{d}>2 \end{cases} \qquad (7)$$

The correction terms to Eq. (7) are, as for regular lattices, multiplicative. From our analysis of trapping on Sierpinski-gaskets, we were led to the conjecture[6,28] that the whole cumulant expansion may scale with $pn^{d/2}$ for $\tilde{d}<2$, a feature which is known to hold for $d=1$.

This scaling behavior also appears in the asymptotic (large n) form for $\tilde{\phi}_n$:[6,28]

$$\tilde{\phi}_n \sim \exp[-Cp^{2/(\tilde{d}+2)} n^{\tilde{d}/(\tilde{d}+2)}] \qquad (8)$$

In Ref. 29 we have derived Eq. (8) for $\tilde{d}<2$ by using the idea of compact exploration[30] of the available space by a random walker. The parallelity of Eqs. (5) and (8) is evident, d being here replaced by $\tilde{d}$.

4. Continuous-Time Random-Walks (CTRW)

In continuous-time one relaxes the condition that the steps of a walker may occur only at fixed time intervals.[9-11] One introduces thus a waiting-time distribution $\psi(t)$, which gives the probability density that the time between steps equals t. The simplest situation obtains for a memoryless-process, in which the probability of remaining at a given site during the time interval t is exponential:

$$\psi^P(t) = b\exp(-bt) \qquad (9)$$

Eq. (9) leads to a Poisson-process, hence the superscript P. Other waiting-time forms $\psi(t)$ display long-time tails:[9-11]

$$\psi(t) \sim 1/t^{1+\gamma} \qquad \gamma>0 \qquad (10)$$

Such expressions may be envisaged to arise from a superposition of Poisson-type laws:[11]

$$\psi(t) = \frac{1-q}{q} \sum_{j=1}^{\infty} q^j b^j \exp(-b^j t) \qquad (q<1,\ b<1) \qquad (11)$$

Evidently, Eqs. (10) and (11) scale with time, the scaling parameter being γ and $\ln q/\ln b$, respectively.

An interesting situation obtains for $\gamma<1$ (or, equivalently, for $b<q$), for which the first moment of the waiting-time distribution

$$\tau_1 = \int_0^{\infty} t\psi(t)\,dt \qquad (12)$$

diverges. In this case the properties of the CTRW differ qualitatively from the corresponding simple random-walk behavior.

Thus, the mean number $S(t)$ visited in time t depends on whether τ_1 is finite or not.[7,9-11] For $\tau_1<\infty$

$$S(t) \sim \begin{cases} t^{\tilde{d}/2} & \text{for } \tilde{d}<2 \\ t & \text{for } \tilde{d}>2 \end{cases} \qquad (13)$$

i.e. Eq. (13) reproduces the findings of Eq. (6), with n being replaced by t/τ_1. For an infinite first moment, $\gamma<1$ in Eq. (10) one has:[7]

$$S(t) \sim \begin{cases} t^{\gamma \tilde{d}/2} & \text{for } \tilde{d}<2 \\ t^{\gamma} & \text{for } \tilde{d}>2 \end{cases} \qquad (14)$$

We expressed Eqs. (13) and (14) in terms of the spectral dimension $\tilde{d}$, since the fractal approach is general, and includes the regular lattices as special cases ($d=\tilde{d}$). Interestingly, the geometrical and temporal scaling aspects, represented by $\tilde{d}$ and by γ, combine multiplicatively in the exponent of $S(t)$, Eq. (14). Hence the two processes subordinate.[6-8]

Turning now to the target annihilation through walkers whose steps are taken in continuous-time, one may show that the relaxation is given through:[31]

$$\Phi(t) = \exp[-pS(t)] \qquad (15)$$

where $S(t)$ is given through Eqs. (13) or (14). Here again we note the appearance of a Williams-Watts stretched exponential form. Now the parameter α in Eq. (1) may take any value between 0 and 1, if one chooses γ in Eq. (14) accordingly. This is true even for regular lattices in arbitrary dimensions, so that the CTRW provides an adequate extension for the Glarum-model of spin relaxation in glasses,[32] as pointed out by Shlesinger and Montroll.[11,16]

Due to the multiplicative connection of γ and $\tilde{d}/2$ in Eq. (14), which stems from subordination, a measured stretched-exponential parameter α may be modelled by an infinity of (γ, $\tilde{d}/2$) pairs. Monitoring only the target relaxation, one is not in the position of differentiating between the temporal disorder, exemplified by γ, and the spatial disorder, given by $\tilde{d}$. Such a distinction may be drawn in the trapping problem, since here the CTRW forms show different relaxation patterns in the short- and long-time regimes.[7]

The short-time behavior of the decay law due to trapping, $\tilde{\Phi}(t)$, may again be expressed in the form of a cumulant expansion, by viewing the different realizations of the waiting-times between the individual steps as an additional temporal average. The leading term in the cumulant expansion is[7]

$$\tilde{\Phi}(t) \simeq \exp[-pS(t)] \qquad (16)$$

The approximate relation, Eq. (16), holds for small trap concentrations, $p \ll 1$ and for short times. Remarkably, the long-time behavior of $\tilde{\Phi}(t)$ is determined mainly by $\psi(t)$.[7] The relaxation follows the behavior of:

$$\Psi(t) \equiv 1 - \int_0^t \psi(t')dt' \qquad (17)$$

which gives the probability that no step has occured until t. Thus, for an algebraic decay, as in Eq. (10), one has $\psi(t) \sim t^{-\gamma}$, and a detailed analysis shows that:[31]

$$\tilde{\phi}(t) \sim \begin{cases} t^{-\gamma}/(p^{2/\tilde{d}}) & \text{for } \tilde{d}<2 \\ t^{-\gamma}/p & \text{for } \tilde{d}>2 \end{cases} \quad (18)$$

Therefore at long times the relaxation pattern for trapping under algebraic waiting-time forms is algebraic, Eq. (2), and does not follow a stretched-exponential expression. Similar behaviors as displayed by Eq. (18) occur for all waiting-time distributions with long-time tails, and are due to avoided crossings of the decay patterns. For such distributions the asymptotic form (8) does not show up. The trapping problem allows to differentiate between the temporal and the spatial scaling aspects: One can determine γ from the time dependence of Eq. (18) and $\tilde{d}$ from its concentration dependence.

5. Energetic Disorder: Ultrametric Spaces (UMS)

Up-to-now we considered all steps to neighboring sites to be equivalent. In fact, in realistic situations neighboring sites may be separated by energy barriers, whose height may be random. Then a given activation energy allows a walker to visit only a subset (cluster) of sites in the neighborhood of the starting point, the cluster being separated from the other sites by barriers higher than the prescribed activation energy. One may then classify the sites through the energy required to reach them:[12] To such a classification corresponds a ultrametric space (UMS).[12,13]

An example for a UMS is the set of <u>tips</u> of a finite Bethe-lattice. A distance between two sites may be defined as being the minimal number of branches one has to walk on the tree in order to get from one tip to the other. Such a distance has a physical meaning as being proportional to the energy required to get from one site to the other.

It is straightforward to verify that the so-defined distance $d(x,y)$ satisfies the strong triangle inequality:[13)]

$$d(x,y) \leq \max(d(x,z), d(y,z)) \qquad (19)$$

for all sites x,y,z of the UMS. Furthermore, specifying a value $\tilde{E}$ for the activation energy leads to the partition of the UMS into a set of disjoint clusters, where any two points of the same cluster are separated by barriers of energy lower than $\tilde{E}$, and any two points belonging to different clusters by barriers higher than $\tilde{E}$. Here we take for simplicity the barrier heights to be hierarchically distributed, so that all consecutive energy levels differ by Δ, and assume that the branching ratio z is constant over the whole Bethe lattice.

A qualitative argument allows now to determine $S(t)$, the mean number of distinct sites visited during t. For thermally activated jumps, the intersite transition rates are given by $w \exp(-j\Delta/kT)$, where $j\Delta$ is the barrier height to be surmounted, T the temperature, and w a unit rate. During the time interval

$$e^{j\Delta/kT} \leq wt_j < e^{(j+1)\Delta/kT} \qquad (20)$$

z^j points of the UMS are accessible to the walker. Hence

$$z^j \sim z^{(kT/\Delta)\ln(wt_j)} = (wt_j)^\gamma \qquad (21)$$

where we set

$$\gamma \equiv (kT/\Delta)\ln z \qquad (22)$$

For $\gamma<1$, z^j increases more slowly than t_j and the walker explores practically all accessible points. Therefore, using this idea of compact exploration:

$$S(t) \sim (wt)^\gamma \sim t^\gamma \qquad (\gamma<1) \qquad (23)$$

On the other hand, for $\gamma>1$, z^j increases more rapidly than t_j, and the mean number of distinct sites visited stays proportional to t_j. Hence

$$S(t) \sim t \qquad (\gamma>1) \qquad (24)$$

One may now compare Eqs. (23) and (24) with the similar expressions for CTRW, Eqs. (13) and (14), for $\tilde{d}>2$. In fact it is this parallelity which motivated our use of the letter "γ" in the definition (22).

From numerical simulations on several UMS we have established that Eqs. (23) and (24) are indeed well obeyed.[33] Furthermore, numerical results allowed us to determine the relaxation behavior in the trapping and target annihilation problems on UMS.

The target problem shows again as exact decay law[31,33]

$$\Phi_n = \exp[-p(S_n-1)] \qquad (25)$$

where n is the number of steps. This is, with Eqs. (23) and (24), a stretched-exponential decay, in which the parameter α of Eq. (1) is a function of T. Depending on temperature one has a crossover from exponential decays, $\alpha=1$, for large values of T ($kT \ln z > \Delta$) to stretched-exponential forms for small T ($kT \ln z < \Delta$). This is an extremely interesting finding for experimental purposes: In the case of a hierarchical energy disorder one may switch through the marginal behavior at $\gamma=1$ ($kT \ln z = \Delta$) through a simple change of temperature. Evidently, such a qualitative dependence of the decay pattern on T may be used as a diagnostic for energy randomness.

For the trapping problem one may express the short-time relaxation in terms of a cumulant expansion,[31,33] whose first term is given by the right-hand side of Eq. (25). For $\gamma \lesssim 1$, say $\gamma \simeq 0.5$ one finds that the higher terms in the expansion lead to a relaxation form, whose exponent is a power-series in pn^γ. A comparison to fractals, Eq. (7), shows another possible interpretation for γ on UMS, namely of being up to a factor of 2 an "effective" spectral dimension[34] $\gamma = \tilde{d}/2$. A word of caution: This spectral dimension is via Eq. (22) temperature-dependent. But with this <u>proviso</u> the analogy may be pushed quite far. For a fixed temperature the relaxation laws on UMS parallel closely the findings for fractals,[31,33-36] both in the short-time regime (as discussed) and also at long times. On UMS one has as long-time relaxation due to trapping:

$$\tilde{\Phi}(t) \sim \exp[-C p^{\gamma/(\gamma+1)} n^{1/(\gamma+1)}_j] \tag{26}$$

which is identical to Eq. (8) for $\gamma = \tilde{d}/2$.

6. Summary

In this communication we have shown that non-exponential decay patterns may arise from several, microscopically distinct, processes. Thus, stretched exponential relaxation forms show up in trapping and in target annihilation, when the underlying systems are regular lattices, fractals or ultrametric spaces and where the individual steps may occur either at fixed time intervals or in continuous time (CTRW). Thus, in order to be able to specify the underlying mechanisms to an experimental finding, one needs additional information to the decay pattern. Such information may be, for instance, the microscopic structure of the material under investigation or the temperature dependence of the relaxation.

Acknowledgments

We are thankful to Prof. K. Dressler for continuous encouragement and help. The support of the Deutsche Forschungsgemeinschaft and of the Fonds der Chemischen Industrie is gratefully acknowledged by AB.

References

1) R. Kohlrausch, Ann. Phys. (Leipzig) 12, 393 (1847)
2) G. Williams and D.C. Watts, Trans. Faraday Soc. 66, 80 (1970)
3) G. Williams, Adv. Polym. Sci. 33, 59 (1979)
4) R.G. Palmer, D.L. Stein, E. Abrahams and P.W. Anderson, Phys. Rev. Lett. 53, 958 (1984)
5) A. Blumen and G. Zumofen, J. Chem. Phys. 77, 5127 (1982)
6) J. Klafter, A. Blumen and G. Zumofen, J. Stat. Phys. 36, 561 (1984)
7) A. Blumen, J. Klafter, B. White and G. Zumofen, Phys. Rev. Lett. 53, 1301 (1984)
8) B.B. Mandelbrot, "The Fractal Geometry of Nature" (W.H. Freeman and Co., San Francisco, 1982)
9) H. Scher and M. Lax, Phys. Rev. B7, 4491; 4502 (1973)
10) H. Scher and E.W. Montroll, Phys. Rev. B12, 2455 (1975)
11) E.W. Montroll and M.F. Shlesinger, in "Nonequilibrium Phenomena II: From Stochastics to Hydrodynamics", J.L. Lebowitz and E.W. Montroll eds. (North Holland, Amsterdam, 1984)
12) A.D. Gordon, "Classification" (Chapman and Hall, London, 1981)
13) W.H. Schikhof, "Ultrametric Calculus" (Cambridge Univ. Press, 1984)
14) G.H. Weiss and R.J. Rubin, Adv. Chem. Phys. 52, 363 (1983)
15) G. Zumofen and A. Blumen, Chem. Phys. Lett 88, 63 (1982)
16) M.F. Shlesinger and E.W. Montroll, Proc. Natl. Acad. Sci. USA 81, 1280 (1984)
17) S. Redner and K. Kang, J. Phys. A17, L451 (1984)
18) A. Blumen, G. Zumofen and J. Klafter, Phys. Rev. B30, 5379 (1984)
19) A. Blumen, G. Zumofen and J. Klafter, J. Physique (1985), in press
20) M.D. Donsker and S.R.S. Varadhan, Comm. Pure Appl. Math. 28, 525 (1975) and 32, 721 (1979)
21) J.M. Drake and J. Klafter, J. Lumin. 31-32, 642 (1984)
22) F. Family and D.P. Landau eds. "Kinetics of Aggregation and Gelation" (North Holland, Amsterdam, 1984)
23) R. Hilfer and A. Blumen, J. Phys. A17, L537; L783 (1984)
24) S. Alexander and R. Orbach, J. Phys. Lett. 43, L625 (1982)
25) R. Rammal and G. Toulouse, J. Phys. Lett. 44, L13 (1983)
26) A. Blumen, J. Klafter and G. Zumofen, Phys. Rev. B28, 6112 (1983)
27) J. Klafter and A. Blumen, J. Chem. Phys. 80, 875 (1984)
28) G. Zumofen, A. Blumen and J. Klafter, J. Phys. A17, L479 (1984)
29) J. Klafter, G. Zumofen and A. Blumen, J. Phys. Lett. 45, L49 (1984)
30) P.G. de Gennes, C.R. Acad. (Paris) Ser. II, 296, 881 (1983)
31) A. Blumen, J. Klafter and G. Zumofen, "Models for Reaction Dynamics in Glasses", in "Optical Spectroscopy of Glasses", I. Zschokke-Gränacher Ed., (Reidel, Dordrecht, 1986)
32) S.H. Glarum, J. Chem. Phys. 33, 639 (1960)
33) A. Blumen, J. Klafter and G. Zumofen, J. Phys. A, (1986), in press
34) S. Grossmann, F. Wegner and K.H. Hoffmann, J. Phys. Lett. 46, L575 (1985)

35) A.T. Ogielski and D.L. Stein, Phys. Rev. Lett $\underline{55}$, 1634 (1985)
36) B.A. Huberman and M. Kerszberg, J. Phys. $\underline{A18}$, L331 (1985)

Fractal Nature of Geometric and Energetic Disorder: Exciton Kinetics in Crystalline Films, Glasses and Membranes

Raoul Kopelman
Department of Chemistry
The University of Michigan
Ann Arbor, Michigan 48109

Heterogeneous kinetics are shown to differ drastically from homogeneous kinetics for both steady state and transient (relaxation) conditions. For the elementary reaction A + A → Products the transport-limited heterogeneous reaction rate is proportional to $[A]^X$ or $t^{-h}[A]^2$, where $(X - 1)(1 - h) = 1$ and $0 < h < 1$ (compared to simply $[A]^2$ for the homogeneous reaction). The random walk exponent $f = 1 - h = (X-1)^{-1}$ is given by $\beta d_s/2$ where d_s is the spectral dimension of the real fractal space and 2β that of the effective fractal potential energy space. These relations are borne out by supercomputer simulations as well as by experiments. Experimentally, $\beta=f(T)$ and approaches 1 when $kT > W$, where W is the inhomogeneous linewidth in the case of triplet exciton fusion. This anomalous, transport limited, exciton kinetics is studied in naphthalene crystals and films and in naphthalene embedded polymeric glasses, polymeric membranes, filters and other porous media.

1. **INTRODUCTION**

Transport and reactions in fractal-like media have been of much recent interest [1-13]. Exciton transport on <u>percolation clusters</u> was studied experimentally a decade ago[14,15] and has been analyzed and simulated in terms of random walks on percolation clusters [16-18] even before the popular de-Gennes ant first started its drunken walk in a labyrinth [19], and before fractals were introduced into percolation [20]. Our first method[16] of studying mobile exciton transport was based on exciton - trap "reactions".

The rate coefficient k(t) of this reaction was related to the random walk efficiency: k(t) ∝ dS/dt where S is the range of the random walker (mean number of distinct sites visited)[18,21]. Evesque and Duran[2] used the same approach for the reaction of mobile excitons with trapped excitons (heterofusion) while Klymko and Kopelman[1,22,23] removed the traps altogether and monitored the annihilation (homofusion) reaction between mobile excitons. The steady-state and relaxation measurements gave anomalous results[23] (e.g. reaction orders of 20 rather than 2). The classical ("euclidean") picture had to be abandoned in favor of a fractal model. This model of geometric disorder appears to have general applicability to disordered media. It is not only complementary to the temporal-disorder (or energetic disorder) model of Scher, Lax, Montroll and Shlesinger[24-28], but the two models can be combined as suggested theoretically by Klafter et al.[8] and via simulations by Anacker et al.[29-31].

We have studied triplet exciton annihilation on naphthalene aggregates which are embedded inside pores[32] (of various membranes or vycor glass), cavities[33] (in polymeric glasses), grain boundaries[34] and on random clusters in random alloys[1,12,35]. The common denominator of all these systems is the non-classical or fractal-like annihilation kinetics. The annilhilation rate coefficient k is not a constant but rather described by the time-dependent form: $k \propto t^{-h}$, where h is a heterogeneity exponent (0 < h < 1) that vanishes (h = 0) only for homogeneous samples, giving back the classical form: k = const. We also measure the dependence of h on temperature. The effective spectral dimension of the fractal-like medium is given by: $d_s = 2(1-h)$.

2. FRACTAL-LIKE EXCITON KINETICS IN POROUS GLASSES AND MEMBRANES (with S. Parus and J. Prasad)

The sample preparation of the porous polymeric membranes and filter papers involved soaking in solutions of naphthalene (zone refined, in spectroscopic grade hexane). The porous vycor (7930 Corning glass) samples were prepared by naphthalene sublimation in vacuum. The original vycor glass was treated with acids, water and alcohol. The samples were excited at low temperatures by a 1600 W xenon arc lamp through a monochromator set at 310 nm. Spectra were collected via a 1 meter JY double monochromator (with EMI 9816 QB PMT and PAR 1109 photon counter). The delayed fluorescence decays were obtained by shuttering the excitation beam and using various filters with an EMI 9781R phototube and a PAR 4202 signal averager.

The principle behind the approach taken here is the relation of the instantaneous exciton annihilation probability to the site visitation efficiency. The instantaneous exciton annihilation is $d\rho/dt$ where $\rho(t)$ is the exciton density. The fraction of exciton population annihilated (in time dt) is $\rho^{-1} d\rho/dt$. This fraction is linear in the density ρ (binary collision theory argument). Hence the normalized instantaneous exciton annihilation probability k is given by:

$$k = \rho^{-2} d\rho/dt \quad . \tag{1}$$

Classically k is time independent ("rate constant" in chemical language). It has been shown that k is proportional to the visitation efficiency ϵ of a random walker[6,18,21] (the time derivative of the mean number of <u>distinct</u> sites visited):

$$k \sim \epsilon \sim t^{-h} \qquad 0 \leq h \leq 1 \qquad (2)$$

where $h = 0$ for 3-dimensional Euclidean spaces and lattices (classical result) but for $d_s < 2$, e.g. fractal domains,[1,2]

$$h = 1 - d_s/2 \qquad h > 0 \quad . \qquad (3)$$

The delayed fluorescence (F) feeds directly on the triplet exciton fusion: $F(t) \sim d\rho/dt$. The phosphorescence (P) is due to the natural decay of the triplet exciton population: $P(t) \; \rho(t)$. From Eqs. 1, 2 we get:

$$F/P^2 \sim t^{-h} \quad . \qquad 0 \leq h \leq 1 \qquad (4)$$

We note that under special circumstances, where one of the two fusing excitons is trapped ("heterofusion"), and is present in large abundance (relative to the freely moving excitons), one gets,[1,2] instead of the binary reaction (1) ("homofusion"), a pseudo-unary reaction:

$$k = \rho^{-1} d\rho/dt \qquad (5)$$

and thus

$$F/P \sim t^{-h} \qquad (6)$$

We note that for both (4) and (6) one has at early times ($P \sim \rho \sim$ const.):

$$f \sim t^{-h} \quad . \qquad t \to 0 \qquad (7)$$

The fluorescence and phosphorescence spectra of

naphthalene in these environments show broadened bands, physical trap emissions, excimer emission and radical (1-hydronaphthyl) emission. All these are characteristic features of disordered aromatic solids. Spectra of naphthalene in vycor depend on temperature. In the membrane and filter paper samples the temperature does not affect the emission spectra. A weak phosphorescence (relative to fluorescence) intensity is common to all samples. This may be, in part, because of the triplet-triplet annihilation (fusion). See Figs. 1,2.

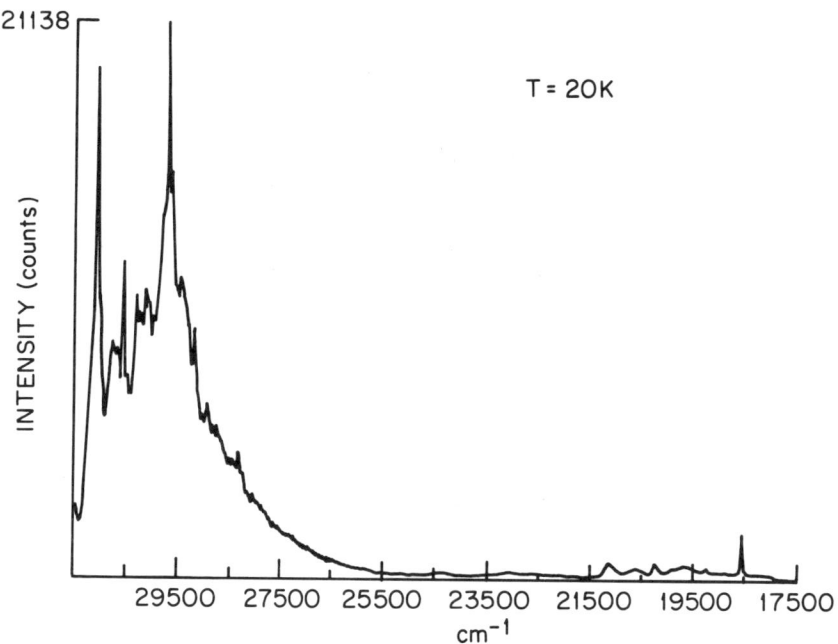

Fig. 1. Emission spectra of naphthalene embedded in nylon membrane.

Our experimental results are shown in Figs. 3 and 4. The heterogeneity exponent h is extracted from the slopes of ln k vs ln t. The values so obtained for h are listed in Table 1, along with values for the effective spectral dimension d_s of the fractal-like medium calculated from the relation $d_s = 2(1-h)$. We note that for all samples at low temperatures $h > 0$. This implies that all samples exhibit a fractal-like behavior at these low temperatures. We also note that for the vycor sample $d_s = 1.1$. This value does not change much with increasing temperature. On the other hand, we observe (Table 1) some drastic temperature effects on the membrane and filter paper values; notably the cellulose filter paper approaches a classical behavior at T = 80K, i.e., $h = 0$.

We show below, using vapor-deposited naphthalene, that random energetic disorder can also result in fractal-like kinetics (and thus in an effective h and d_s). Moreover, geometric and energetic effects do superimpose[34] in a way that is analogous to the "subordination principle" of Klafter, Blumen and Zumofen.[8] In the latter, the spectral dimension d_s is effectively reduced by a constant factor accounting for a "continuous time random walk" (CTRW) with an anomalous hopping time distribution (infinite second moment). We note that this CTRW model has been shown to be equivalent to models of energetic disorder.[8,30,31] We can thus separate the roles of geometric and energetic disorder via temperature studies. In the limit where the thermal energy is large compared to the inhomogeneous line-broadening (which we observe spectroscopically), the effects of the energetic disorder are minor and the geometric effect dominates. Table 1 shows that both in glass (synthetic) and cellulose (natural) <u>filter papers</u> the triplet excitons of the naphthalene embedded in the pores move anomalously (at low temperatures) due to energetic disorder. However, it appears that while the glass filter paper has a pore network with a real fractal-like structure (geometric spectral dimension of 1.6) the natural filter paper has $h = 0$, which is consistent with $d_s \geq 2$, i.e. a non-fractal (Euclidean) pore network (which would be more efficient for biological

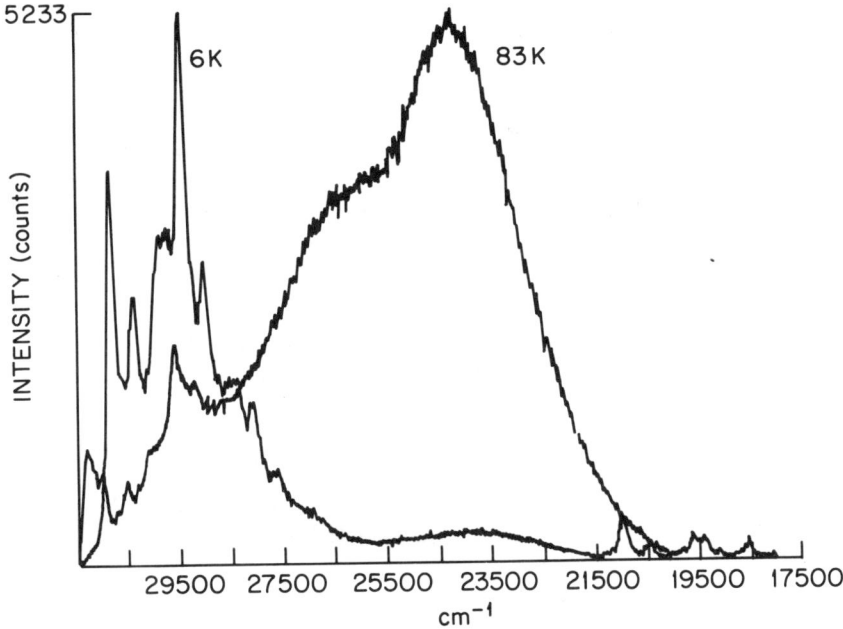

Fig. 2. Emission spectra of naphthalene embedded in vycor at 6K and 83K.

Table 1

Sample	Pore Size(μ)	T (K)	Exponent(h) Homofusion	Spectral dim.(d_s')
Acetate (GA8) Membrane†	0.2	4	0.16	1.7
		80	0	3
Acetate (GA1) Membrane†	0.5	4	0.47	1.1
Acetate (GA3) Membrane†	1.2	4	0.44	1.1
Nylon (B214) Membrane†	0.2	4	0.21	1.6
Glass Filter Paper	0.6	6	0.42	1.2
		80	0.20	1.6
Cellulose Filter Paper	0.6	6	0.33	1.3
		80	0.06*	1.9 - 3
Vycor Porous Glass	0.003	6	0.44	1.1
		58	0.5	1

* Heterofusion (a homofusion model gives a negative value)
† Gelman Sciences (courtesy of A. Korin and M. Kraus)

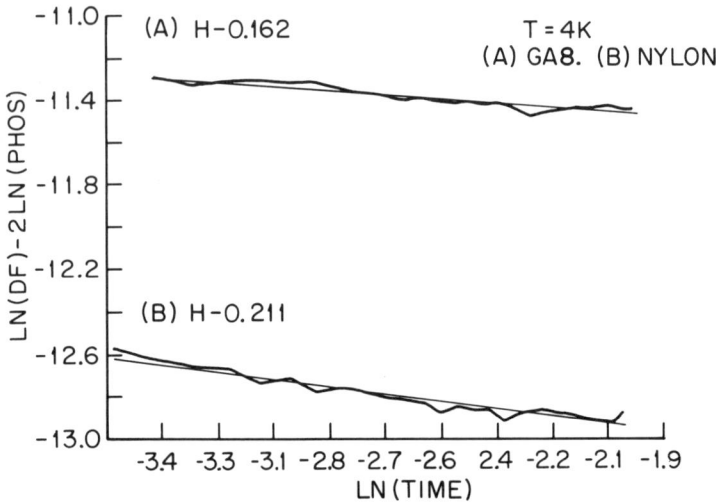

Fig.3. Annihilation rate coeff., k = (DF)/(Phos) vs. ℓn t on a ℓn-ℓn plot for polymeric membranes. H is the heterogeneity exponent h.

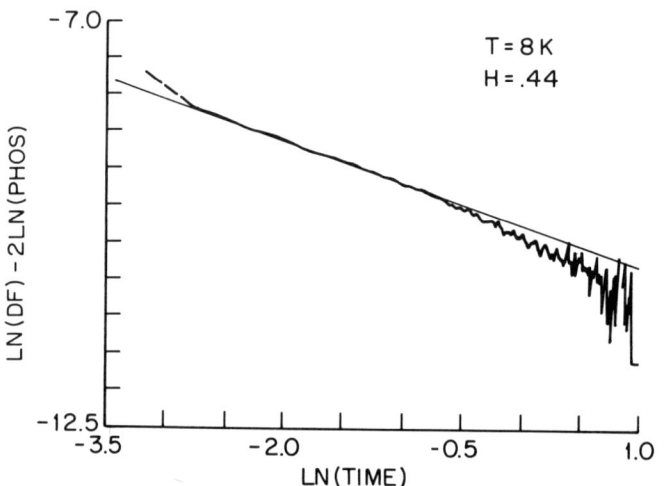

Fig. 4. Same type of plot as in Fig. 3 for vycor porous glass.

diffusion processes).

The low spectral dimension value (d_s = 1.1 ± 0.2) for our sample of porous vycor (which supposedly is the same as that used by Drake et al.[36] is consistent with a number of literature claims. Even et al.[37] derived a value d_f = 1.7 for the fractal dimension of their porous vycor sample (not necessarily the same as ours) and interpreted it in terms of a percolation backbone. This would result in $1 < d_s \leq 4/3$. Drake et al. [36] have chosen to model their porous glass sample with a 3-dimensional percolation cluster (d_f = 2.5). This gives a d_s value of about 1.3.

We estimate that the corresponding length-scale for exciton migration is on the order of 1000A (0.1μ). We note that the nominal pore sizes (Table 1) give upper limits, rather than mean values, for the membranes and filter papers. Due to the much faster exciton migration in Euclidean (ordered) spaces (the bulk migration is at least two orders of magnitude more efficient than migration in fractal-like domains) we believe that our data on these samples characterize the more numerous but narrower microporous networks, rather than the nominal pore sizes. The only exception is the porous vycor glass sample whose pore size is a mean value based on molecular adsorption measurements.[36]

In conclusion, the studies of naphthalene embedded in polymeric membranes, filter papers and vycor glass samples have demonstrated a fractal-like exciton annihilation kinetics. Temperature studies establish both the subordination of geometric and energetic disorder and a technique for assessing the relative roles of geometric and energetic constrains. The network spectral dimension is usually in the range $1 < d_s < 2$, but some samples are consistent with d_s = 1 and others with d_s = 3. Pore size studies of polymeric membranes might characterize the crystalline domain sizes in deposited films.

3. EXCITON TRANSPORT IN NAPHTHALENE-DOPED POLY(METHYLMETHACRYLATE): SPATIAL AND ENERGETIC DISORDER (with E. I. Newhouse)

Polymer backbones have been assigned fractal properties.[20] One might expect the spaces between them ("cavities", "pores") to also be fractal-like, and thus dopants in a polymer may be forced into a fractal environment. We have chosen naphthalene: poly(methylmethacrylate) (PMMA) ("plexiglass") for such experiments, as the spectroscopic properties of naphthalene in other phases are very well known, and thus meaningful comparisons can be made between the new data to be presented here and those previously reported for other systems of naphthalene.

Zone-refined and potassium-fused naphthalene (Aldrich) was used. PMMA (Polysciences, "medium molecular weight") was reprecipitated from dichloromethane (Baker, spectrograde) by hexane (Baker, hplc grade), and vacuum dried. Samples prepared by solvent casting from this polymer behaved the same as samples made from unreprecipitated polymer. Other samples were made by polymerizing freshly distilled methylmethacrylate (Scientific Polymer Products) with an appropriate amount of naphthalene dissolved in it. 2,2-azodi-isobutyronitrile (AIBN) was used as initiator and the evacuated sealed tube heated to 60C for 24-36 hours. Spectra of these samples were indistinguishable from those made by solvent casting. PMMA is a brittle polymer. These samples, however, are not brittle, due to the plasticizing effect of the dopant.

All spectroscopic measurements were made in an immersion cryostat at 77K unless otherwise noted, because the samples turned brown after about 1/2 hour ultraviolet illumination at room temperature. Excitation was with a

1600W xenon arc lamp, appropriately filtered. A custom-built (Karl Lambrecht Corp.) relay lens was used to match the f number of the 1 m Jarrell-Ash monochromator. A cooled ITT F-4013 phototube and PAR 1120 photon counter were used to detect the emission at right angle to the excitation. Spectra were recorded on an LSI 11/03 microcomputer. Phosphorescence and delayed fluorescence decays were recorded with the same optical train using out-of phase shutters in the excitation beam and in front of the monochromator. Only the first half of the monochromator was used, as high rejection of scattered light was not necessary. Detection was with an EMI 9781R phototube and PAR 4202 signal averager, whose output was stored on the microcomputer. Fluorescence decays were excited with the frequency doubled output of a Rhodamine 6G solution pumped with a Molectron UV1000 nitrogen laser. Decays were detected with the EMI phototube and a PAR gated integrator (model 164) and boxcar averager (model 162). Due to electrical noise from the laser, the microcomputer could not be used. Instead, decays were recorded with a chart recorder and subsequently digitized.

Table II: Naphthalene triplet lifetimes (5000A)

Concentration (%)	τ (4K) (sec)	τ (77K) (sec)
20	1.9	0.8
10	2.2	1.0
5	2.1	1.7
1	2.3	2.0

Table III: Triplet lifetime as a function of wavelength (77K)

Wavelength (A)	τ (sec)
4800	0.8
4750	0.9
4700	1.6

The phosphorescence lifetime in samples doped only with

naphthalene is concentration dependent (see the next two figures and Table II). If one selects the detection wavelength of the phosphorescence in the 20% sample, the observed lifetime <u>increases</u> the more toward the blue one observes, as evident in Table III. Thus, the emission spectrum represents a distribution of sites with varying intermolecular interactions or a distribution of domain sizes. Some sites, with limited excitation exchange interaction with other sites, have an emission spectrum characteristic of dilute naphthalene systems, and a long emission lifetime. In other regions of the sample there is aggregation of varying degrees, resulting in a red-shifted emission spectrum, and a shorter lifetime, because of triplet-triplet annihilation and/or trapping. The phosphorescence lifetime of all these samples lengthens at 4K (see Table II), implying that triplet energy transport is a thermally activated process that has ceased at 4K. The phosphorescence decay is "bi-exponential", (Figure 5) with the lifetimes of the two "components" indicated on the Figure. The deviation from a single exponential decreases with concentration, and the phosphorescence decay from a 5% sample can be fit with a single exponential.

While the delayed fluorescence can also be fit to two exponentials, one usually expects a linear $I^{-1/2}$ vs. time behavior at short times and an exponential decay at long times with a decay time 1/2 that of the phosphorescence decay time. Note that the slow component of the delayed fluorescence decays much more rapidly than at twice the slow phosphorescence decay rate. This can be explained as being due to the presence of domains with varying amounts of naphthalene-naphthalene interactions, with triplet-triplet annihilation occurring only on those domains with the strongest interactions (shortest lifetimes). (So far, we are describing the system as an ensemble of different regions, but all with Euclidean topology).

One expects annihilation on a fractal to have a lifetime behavior reflecting the spectral dimension, d_s:

$$I_{df} \propto \rho^X,$$

where I_{df} is the delayed fluorescence intensity, ρ is the exciton density, and $X = 1+(2/d_s)$. Specifically the ratios observed in these samples correspond to $d_s = 1.4$ (18% naphthalene) and $d_s = 1.2$ (8.4% naphthalene).

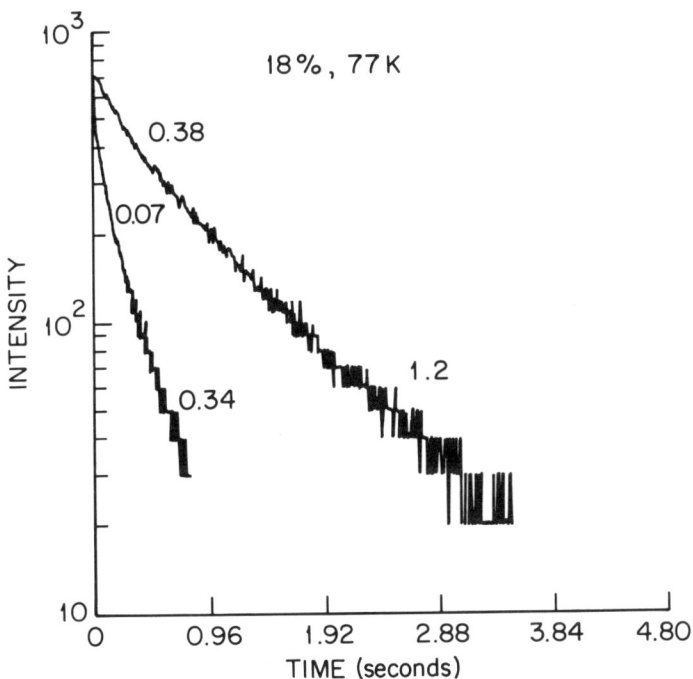

Fig. 5. Phosphorescence and delayed fluorescence decays for 18% naphthalene samples at 77K. The upper curve is phosphorescence, the lower fluorecence. The numbers are the decay time values obtained from a least squares fit to a bi-exponential function. The "steps" reflect the voltage resolution of the A/D converter used to record the data.

We were interested to check for further evidence of

sample heterogeneity. A plot of $\log(I_{df}/I_{ph}^2)$ vs. $\log(\text{time})$ (Fig. 6) shows that these samples are heterogeneous, as expected from the phosphorescence decays. (For homogeneous samples these plots are horizontal lines). The "slopes" are 0.2-0.3, corresponding to d_s of 1.6-1.4, for all the data shown in this Figure. A plot of the type shown in Fig. 6 will be linear only as long as the excitons can meet and annihilate. Thus, for finite clusters, one expects to see deviations from linearity as in Fig. 6, when the number of clusters with more than one exciton on them begins to decline significantly.

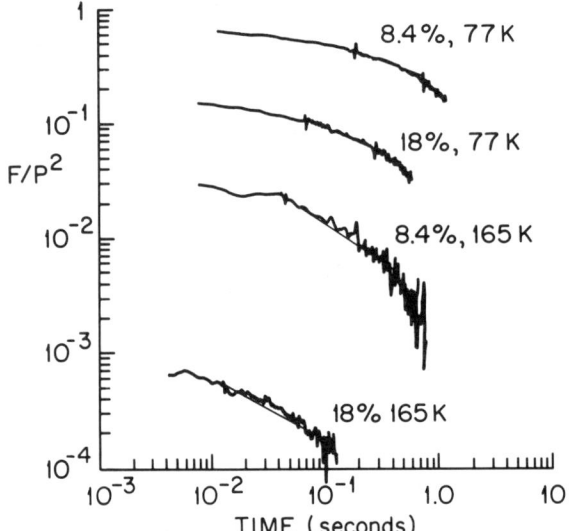

Fig. 6 Test for sample heterogeneity effects on triplet transport at two temperatures and two concentrations. The concentrations and temperatures are given on the Figure. The superimposed lines indicate the slopes of the fit and are plotted only over the region used for the fit. The values range from 0.2 to 0.3. The kinetics are clearly nonclassical, as a horizontal line is expected for classical kinetics, and is clearly not observed here. The values of the ordinate are arbitrary, but the time in seconds can be read directly from the abscissa.

In summary, the fluorescence emission from these samples is broad, indicating a wide distribution in site energies in this polymeric medium. The lifetime is the same as for naphthalene crystals. The phosphorescence spectra are naphthalene concentration dependent and are extremely broad. This is a dynamic process which disappears at 4K. The phosphorescence spectrum of each concentrated sample represents a superposition of spectra from sites with different environments, or with different domain sizes. In the phosphorescence of concentrated samples, "spectral diffusion" of excitons to sites with low energy environments is seen (triplet exciton energy relaxation). The fluorescence and phosphorescence spectra both reflect the energetic disorder in the naphthalene. The phosphorescence lifetime is also concentration dependent, unlike the singlet lifetime. Triplet transport is a thermally activated process which ceases at 4K. The non-exponential nature of the phosphorescence decay may indicate that there is a distribution of sites. That is further evident from the abnormally fast decay of delayed fluorescence, which implies that only sites with shorter phosphorescence lifetimes allow triplet-triplet annihilation. these observations may also be explained by describing the system as fractal-like with a spectral dimension of about 1.5. It has been noted that the behavior of samples consisting of regions with different Euclidean geometry may be fractal-like (provided that they contain one-dimensional bottlenecks). The delayed fluorescence data can be used to extract an annihilation rate constant which is at least 100 times smaller than in naphthalene crystals, indicating that the regions supporting triplet migration are not crystalline, in spite of the fact that they seem to have a well-aggregated structure in the two most concentrated samples studied. $\mathrm{Log}(I_{df}/I_{ph}^2)$ vs. log(time) plots demonstrate the heterogeneity by deviating drastically from the classical picture. Forcing a fit with a fractal model yields slopes of 0.2-0.3 and d_s of about

1.3-1.6. The kinetics are thus seen to reflect the geometric disorder, yielding information complementary to the emission spectra, which reflect primarily the energetic disorder. For triplet energy transport the medium is effectively heterogeneous, with some parts apparently not transporting energy at all. Those regions which do transport do so much less readily than pure crystals. The transport topology is fractal-like, but only as a first approximation.

4. EXCITATION TRANSPORT KINETICS IN VAPOR-DEPOSITED NAPHTHALENE (with L.A. Harmon)

Naphthalene of extremely high chemical purity was necessary in this work to isolate the effects of deliberately introduced structural and energetic disorder. Commercially obtained naphthalene was zone-refined (100 passes), fused with potassium metal, and subsequently zone-refined (500 passes). Only the center third of material thus prepared was used for experiments. Disordered naphthalene samples were prepared by evaporation onto quartz substrates from a miniature oven in the sample chamber of a Janis Research Co. liquid helium dewar. The oven was mounted in a teflon cage and the substrate assembly attached to a sample rod. Before deposition, the cryostat, oven and substrate were cooled to 20-30K, following which the sample chamber was pumped out to 74 cm vacuum (measured with a mechanical gauge). While the substrate was held in the optical path 4 inches below the oven, the oven was heated to about 35 C. The substrate was raised into position directly opposite the oven mouth and held there typically for two minutes. The range of oven temperatures used was 20-50 C. During the deposition process, the substrate temperature increased by 20-30 degrees because there is no heat sink in the evacuated chamber. Within the range of oven and substrate temperatures used, no effects were observed on the spectra of the resulting samples. After deposition was complete, the quartz substrate with its vapor-deposited naphthalene was lowered into the optical path, vacuum in the sample chamber was broken with helium gas and the system recooled to liquid helium temperatures.

Samples were annealed by warming the sample chamber to 150K over a 30 minute interval and maintaining that temperature for 5-6 hours. Naphthalene is volatile and sublimes even at cryogenic temperatures. Consequently the thermal energy requisite for annealing also drives the competing sublimation process. Visual inspection of samples in situ sometimes revealed nonuniformity in sample thickness after annealing. Spectra were unchanged by cycling between 100 and 5K. Samples held at 120K did not appear to anneal significantly. Excitation was by 310 nm radiation from a 1600 W Xenon arc lamp passed thorough a .25m Jobin-Yvon monochromator with 2mm slits. Emission spectra were collected through a Jobin-Yvon 1m double monochromator with an EMI 9816QB photomultiplier and digitized with a PAR model 1109 photon counter. To obtain delayed fluorescence decays, the excitation beam was shuttered. The delayed emission was passed through two Corning 7-54 glass filters which transmit the UV and eliminate phosphorescence. Time-resolved intensities were monitored by an EMI 9781R photomultiplier whose output was averaged with a PAR model 4202 Signal Averager. The output photomultiplier was protected from lamp emission and prompt fluorescence by a shutter in the emission beam triggered to open 3 msec after the excitation shutter closed. Each decay consists of 1024 points at 1 msec intervals, typically averaged over 100 scans. Scans were repeated at intervals of 10 or 20 seconds to allow the triplet excitations to equilibrate during and after illumination. All experimental data were recorded on an LSI-11/03 laboratory computer; fitting of experimental delayed fluorescence decays was performed on an Ahmdal 470 at the University of Michigan.

A tentative description of the physical structure of these samples, consistent with the spectroscopic results,[34] is sketched in Figure 7. Reasonably ordered regions exist in which fluctuations of molecular coordinates from single crystal values are relatively small and random; these regions give rise[34] to spectrum "I". Forming the boundaries of such regions are areas of much greater structural disorder, acting as lower-dimensional, energetically-disordered systems, which are characterized[34] by an effective bandwidth of about 100 cm^{-1}, observed in Spectrum II. We observe that the excimer emission does not rise dramatically until thermalization allows transport within regions II and therefore place the majority of excimer-forming sites in type II regions. For similar reasons, the 1-hydronaphthyl(1-HNR) radical is also assumed to be produced mainly in type II regions.

Figure 7. Sketch of proposed domain structure of vapor-deposited napthalene. Unshaded areas represent nearly-crystalline, Type I regions. Shaded areas represent highly disordered, Type II regions. The relative widths of the type II regions have been exaggerated. (R) 1-HNR; (E) Excimer.

We note that there is no information regarding the concentration of sites which are capable of excimer or radical formation.

Table IV. Summary of expressions for $I_{df}(t)$. γ is the incomplete gamma function.

$h = 0$	
$m=1$	$k_{ann} \cdot \rho_0 \exp\{-(k_{ann}+k_1)t\}$
$m=2$	$(\frac{k_1^2}{k_{ann}})\{(1 + \frac{k_1}{k_{ann}\rho_0})e^{k_1 t} - 1\}^{-2}$
$h \neq 0$	
$m=1$	$k_{ann}\rho_0 t^{-h}\exp\{-\frac{k_{ann}t^{1-h}}{1-h} - k_1 t\}$
$m=2$	$k_{ann}^2 \rho_0^2 t^{-h} e^{-2k_1 t}\{\frac{k_{ann}}{k_1^f}\rho_0 \gamma(f,k_1 t) + 1\}^{-2}$

We have selectively examined the kinetics of triplet excitation transport in type II regions of our vapor-deposited naphthalene by monitoring triplet-triplet exciton annihilation in the millisecond time domain. In crystalline naphthalene, this process occurs on a microsecond or shorter time scale: annihilation observed at longer times is due to transport which has been slowed by the presence of disorder.

The kinetics of triplet-triplet annihilation in structurally disordered naphthalene is examined through analysis of the time-dependence of the resulting delayed fluorescence. We start from a rate equation for the triplet population density, $\rho(t)$, incorporating both annihilation and monomolecular decay:

$$-\frac{d\rho}{dt} = k_{ann} t^{-h} \rho^m + k_1 \rho \quad . \tag{8}$$

In this expression, the molecularity, m, of the annihilation reaction has been left unspecified to include both binary (m = 2 and unary (m = 1) processes. Although annihilation is a strictly binary reaction, pseudo first-order kinetics may be observed if one excitation belongs to a population whose density is time-independent, as for example a constant population of long-lived excited traps. The form of k(t) from equation (2) has been retained; k_1, the monomolecular decay rate, is equal to the inverse natural lifetime of the excitation in the medium. We note that eqn. (8) contains as a special case an annihilation rate <u>constant</u> if h = 0. Equation (8) has been solved[34] for the four possible cases: m = 1 or 2, h = 0 or h > 0.

The delayed fluorescence intensity, $I_{df}(t)$, is proportional to the first term in (8):

$$I_{df}(t) \propto k_{ann} t^{-h} \rho^m(t) \quad . \tag{9}$$

Explicit expressions for $I_{df}(t)$ are found by substitution of $\rho(t)$ obtained from solutions of (8). The resulting expressions for $I_{df}(t)$ are summarized in Table IV. At early times, these expressions reduce to t^{-h} (for h > 0); at long times we find simple exponential behavior with rate constants equal to $k_{ann} + k_1$ (m = 1, h = 0) or $k_1 m$ (m = 1 or 2, h > 0).

A simplex procedure[34] has been used to fit experimental delayed fluorescence decays by finding the set of parameters [$(k_{ann} \rho_0)$, h, k_1] in each expression of Table

V which best describes the experimental decay. The product $(\rho_0 k_{ann})$ was treated as a single parameter. The fitting procedure was found to be insensitive to bias by initial guesses of the parameter values. The exponent, h, proved to be the primary factor in determining the quality of the fit. By fixing k_1 and fitting only $(\rho_0 k_{ann})$ and h, we found that an increase in k_1 from 0.4 to 0.6 sec^{-1} resulted in (at most) a 15% increase in the calculated value of h. This is reasonable in light of the fact that the decays are fit over only about 0.5τ.

Table V. Comparison of different models in fitting delayed fluorescence.

	k_{ann}	$k_1(\tau)$	h	x^2
h = 0				
m = 1	6.56	2.38 (0.42)	0	4.82
m = 2	7.61	0.00054 (1851)	0	1.09
h ≠ 0				
m = 1	0.29	1.13 (0.88)	0.49	0.0051
m = 2	0.13	0.65 (1.5)	0.49	0.0051

The results of fitting a single decay are summarized in Table V and shown graphically in Figure 8. Fits with a time-independent rate coefficient, i.e. h = 0, are compared to those in which h is allowed to vary. The magnitudes of the corresponding x^2 values for fits with h = 0 suggest that a time-dependent rate coefficient is necessary to describe the experimental decay. This is substantiated by comparison

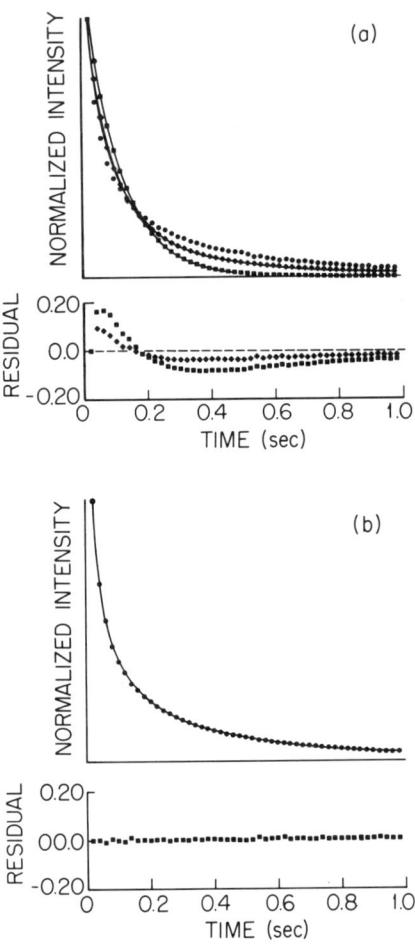

Fig. 8. Delayed fluorescence decay of vapor-deposited naphthalene with fits.
(a) <u>Time independent rate coefficient</u>. Circles: experimental data. Solid line with squares: best fit, m = 1. Solid line with diamonds: best fit, m = 2.
(b) <u>Time-dependent rate coefficient</u>. Circles: experimental data; solid line: best fit, m = 2. Residuals are plotted.

of Figure 8a with Figure 8b. The best fits obtained with $h = 0$, i.e. $k(t)$ = constant, are shown in Figure 8a, with the residuals plotted below. We see here that a time-independent rate coefficient cannot describe the experimental decay. It is clear that a time-dependent rate coefficient ($h = 0$) is necessary to reproduce the observed time dependence of the delayed fluorescence. Fits obtained with $m = 1$ and $m = 2$ are indistinguishable in Figure 8b. However, unrealistic values of the natural lifetime are necessary to fit with a unary model ($m = 1$). We therefore conclude that the annihilation process in these regions is primarily occurring between excitations from a single population. We note that the value of h obtained is independent of m, indicating that $k(t)$ is the dominant factor in the decay over this time regime.

These results were verified by repeated experiments with other vapor-deposited naphthalene samples. All of the qualitative spectral and kinetic features proved to be reproducible, and h was found to vary between 0.45 to 0.6. These values of h are comparable to those obtained from low concentration isotopic mixed naphthalene crystals[14] and are indicative of a highly ramified medium. The values of h were found to be unchanged after annealing; this confirms that the transport observed on this time scale occurs in type II regions since they are affected little by annealing.

In conclusion, our work reported here has demonstrated that disordered naphthalene samples prepared by vapor-deposition at low temperatures are heterogeneous and contain both near-crystalline and highly disordered domains. This method has been shown to provide reproducible model systems for the study of the effects of deliberately introduced disorder on excitation transport. We have been able to completely describe the delayed fluorescence, arising from triplet-triplet annihilation in the disordered regions of our vapor-deposited naphthalene samples, by rate expressions incorporating both annihilation and natural decay. The annihilation process in these samples is well-described by a time-dependent rate coefficient, $k(t)$ t^{-h}, with h near 0.5.

5. ISOTOPIC ALLOY CRYSTALS (with P. W. Klymko and L. W. Anacker)

A recent analysis[38] of the annihilation kinetics of triplet excitons in percolating clusters of naphthalene-h_8 in naphthalene-d_8 single crystals gives a refined value of h = 0.36 ± 0.03 and d_s = 1.28 ± 0.06. An analysis of the older exciton trapping data of Ahlgren[10] provided[39] critical percolation exponents (β = 0.13 ± 0.05 and γ = 2.1 ± 0.2) from which the fractal dimension d_f can be derived: $d_s = d(\gamma + \beta)/(\gamma + 2\beta)$. We note that the triplet exciton transport in naphthalene crystals is confined to the ab plane[39], giving a euclidean dimension d = 2. A comparison of theory, simulation and experiment is given in Table VI. Note that no free parameters were used to obtain d_f and d_s in Table VI. The striking agreement of both the

Table VI: Experimental vs. Theoretical Dimensions

d=2	Percolation Theory	Percolation Simulation	Experiment	Classical Theory
d_f	1.89583...*	1.895±0.002+	1.89±0.03	2
d_s'	1.26±0.02**	1.25±0.02++	1.28±0.06	2

*ref.40. +ref.41. **refs.3,4,9. ++ ref.9

fractal and the spectral experimental dimensions with theory and/or simulation (Table VI) is strong support for the fractal-like kinetics as opposed to classical kinetics. The naphthalene crystal has been called "the H atom of molecular crystals". It appears that the isotopic mixed naphthalene crystal (at the critical concentration) is "the H atom of fractal materials". It is the only system that is fractal from atomic lengths (10^{-10} to finger-size length scales (10^{-2}m).

6. REACTION KINETICS ON FRACTAL AND EUCLIDEAN STRUCTURES WITH ENERGETIC DISORDER: RANDOM WALK SIMULATIONS (with L. W. Anacker)

Systems with spatial and energetic disorder are related to those with temporal disorder [8,28,42]: "All the randomness of space is incorporated into choosing the appropriate waiting-time distribution [28]," $\psi(t)$. This type of temporal disorder is often referred to as continuous time random walk [8,25,26,44]. A fractal distribution of event times [28] is generated when the mean time between events is infinite while the median time is finite; this occurs [8,28,42] for any distribution $\psi(t)$ with a long-time tail of the form $t^{-1-\lambda}$ where $0<\lambda<1$. The superposition of temporal disorder on a system with spatial disorder or on fractal structures [8] requires d, d_f, d_s, and λ to characterize the dynamics. Energetic disorder, as considered here, introduces a temperature-dependent fractal-like distribution of waiting times with a Boltzmann-weighted probability for moves to sites of higher energy [31,45,46]. A walker moves from site i to one of its z nearest-neighbor sites with probability

$$P_{ij} = \begin{cases} z^{-1}\exp(-(\epsilon_j - \epsilon_i)/kT) & \epsilon_j > \epsilon_i \\ z^{-1} & \epsilon_j \leq \epsilon_i \end{cases} \quad (10a)$$

or remains on site i with probability

$$P_{ii} = 1 - \sum_{j=1}^{z} P_{ij}. \tag{10b}$$

Note the asymmetries: $P_{ij} \neq P_{ji}$ and therefore, unlike the continuous time random walk [26], $\langle t_{ij} \rangle \neq \langle t_{ji} \rangle$.

Transport is measured by the mean number of distinct sites visited, $\langle S(t) \rangle$, at time t. For several types of disorder $\langle S(t) \rangle$ is surprisingly well described by a simple power law [5,46]

$$\langle S(t) \rangle \propto t^f \qquad t \to \infty \tag{11}$$

in which:

$$f = \begin{cases} 1 & d_s > 2 & \text{(spatial)} \\ d_s/2 & d_s < 2 & \text{(spatial)} \\ \lambda & 0 < \lambda < 1 & \text{(temporal)} \\ \lambda d_s/2 & d_s < 2 & \text{(spatial + temporal)} \\ f'(T) & & \text{(energetic)} \end{cases} \tag{12}$$

Higher order terms exist, but for the most part do not contribute significantly.

Consider now diffusion-limited chemical reactions with a bimolecular decay mechanism for the A + A reaction. When two particles meet they react instantaneously, and the rate of reaction is governed strictly by the rate of transport in the system considered (see eq. 1):

$$-\frac{d\rho}{dt} = k(t)\rho^2. \tag{13}$$

This corresponds to the Smoluchowski-type equation:

$$-\frac{d\rho}{dt} = k_0 \left(\frac{dS}{dt}\right) \rho^2 \tag{14}$$

which can be written as

$$\frac{d\rho^{-1}}{dt} = -\rho^{-2}\frac{d\rho}{dt} = k_0 \left(\frac{dS}{dt}\right) \tag{15}$$

or integrated to obtain

$$\rho^{-1} - \rho_0^{-1} = k_0(S(t) - S_0) \qquad (16)$$

where $\rho_0 \equiv \rho(t=0)$ and $S_0 \equiv S(t=0)$.

Random walk simulations were used to study transport and diffusion-limited chemical reactions on 8th order planar Sierpinski gaskets (9843 sites), and on square (512^2 sites) and simple cubic (64^3 sites) lattices. All simulations on Euclidean lattices imposed periodic boundary conditions while the three vertices of the largest triangle on the Sierpinski gasket formed reflective boundaries. Lattice site energies for each run were assigned when a walker tried to visit a site with a previously unassigned site energy; once a site energy was assigned it retained that fixed value. The direction of motion for every walker at each time increment was decided by a uniform pseudo-random number generator; time was incremented after all walkers attempted to move. A new lattice was generated for each run.

Initial conditions for single walker simulations set $S_0 \equiv S(N=0) = 1$. Single walker simulations on the Sierpinski gasket started each random walker at a random initial site; on Euclidean lattices, the walker started at the center of the lattice. Reacting random walker simulations set $\rho_0 \equiv \rho(t=0) = 0.1$ with an initial spatial distribution of walkers assigned using a uniform pseudo-random number generator.

Unitless temperature parameters were used for each energy distribution: $T" = kT/W$ with a maximum allowed energy difference of W. Exponential site energies were assigned from a normalized distribution with mean 1; discrete energies were used in such a way that energies below 0.001 were not distinguished from energies of 0.001 and energies above 10 were assigned an energy of 10, i.e. W = 10. A normalized Gaussian distribution was used with mean 0, standard deviation $\sigma = 1$, and an energy cut-off of 3σ, i.e. $W = 6\sigma$; energies outside the range $(-3\sigma, 3\sigma)$ were included but they were not distinguished from energies of $\pm 3\sigma$. The uniform energetic distribution, $T" = kT/W$, referred to here corresponds to the random site energies being uniformly distributed from 0 to 1: $W=1$. Pseudo-random number generators used were: RANF on the CYBER 205 at Colorado State University for uniform distributions; and FUNIF, FNORM, and FEXP on the Amdahl 5860 at The University of Michigan for the uniform, Gaussian and exponential distributions, respectively.

Results reported in this section strictly concern single walker transport on the planar Sierpinski gasket. Three types of pseudo-random energetic distributions were considered: exponential, Gaussian, and uniform. These simulations consisted of 1,000 realizations for each reduced temperature shown in Table VII.

TABLE VII

Unitless Temperatures: $T"$
Single Walker on a Planar Sierpinski Gasket

Uniform: $T" = .10, .15, .25, .50, 1.$
Gaussian: $T" = .03, .05, .06, .07, .08, .09, .1, .15, 1.$
Exponential: $T" = .01, .03, .05, .10, .20, 1.$

1000 realizations at each $T"$; 4000 time increments.

Log-log plots of $<S(N)>$ versus N for exponential, Gaussian and uniform distributions of energetic disorder are shown in Fig. (9). In each case, the curves become fairly linear for large N in good agreement with Eq. (11). At high temperatures, $f(T") = d_s/2$ as expected. Lower temperatures illustrate effects of energetic disorder on a fractal; this is comparable to the superposition of temporal disorder on a fractal ("subordination") [8,28]. The effect of subordination is most clearly seen in Fig. (9a), the exponentially distributed site energies, where each curve becomes linear in the limit of large N. Similar behavior is seen qualitatively for the Gaussian and uniform distributions, Figs. (9b) and (9c), respectively; however, these distributions may be more sensitive to the asymmetric hopping probabilities.

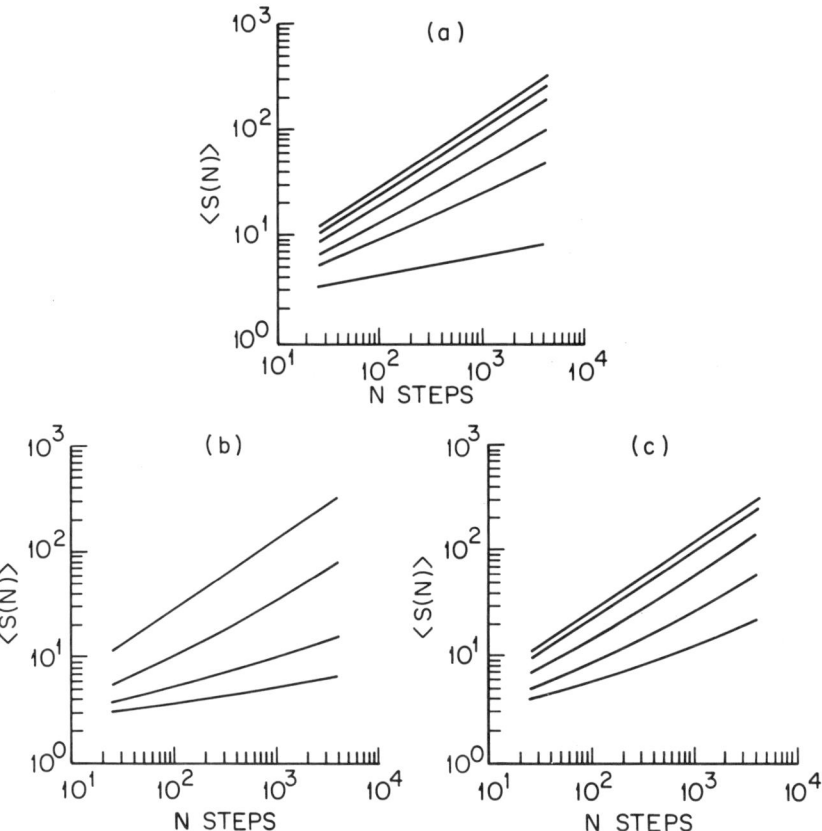

Fig.9 Mean number of distinct sites visited, <S(N)>, by a single random walker on an 8th order planar Sierpinski gasket with a) exponential [T" = .01, .03, .05, .1, .2, 1.0], b) Gaussian [T" = .03, .05, .1, 1.0] and c) uniform [T" = .1, .15, .25, .5, 1.0] distributions of random site energies. Bottom to top curves correspond to increasing T".

We now combine results from single random walker simulations and reacting walker simulations. Reacting walker simulations allowed only one walker to occupy each site at any given instant. Walkers were placed on the lattice such that each walker had an equal probability of landing on any unoccupied site. When a walker tried to move to a site already occupied, it was removed without further perturbing the system, i.e. $A + A \rightarrow A$. For the isoenergetic perfect ($P_{ij} = P_{ii}$) Sierpinski Gasket and for each T" shown in Table VIII, simulations consisted of 100 to 1000 realizations for single random walkers and 10 to 20 runs for reacting random walkers.

TABLE VIII

Unitless Temperatures: T"
Single Walker and Reacting Walker Simulations

Square Lattice*
  Uniform: T" = 0.05, 0.06, 0.07, 0.08, 0.09, 0.1

Planar Sierpinski Gasket**
  Uniform: T" = 0.10, 0.15, 0.20

Simple Cubic Lattice*
  Uniform: T" = 0.05, 0.055, 0.06, 0.065, 0.07, 0.1
  Gaussian: T" = 0.05, 0.066, 0.083, 0.166

---

* 10,000 time steps
** 4,000 time steps

For the simple cubic lattice, reacting walker simulations on the perfect lattice included 120 realizations. Reacting walker simulations with a Gaussian distribution of site energies were performed on a smaller scale; 10 runs were performed for each T", and these simulations initially placed 1563 walkers on a 25^3 site lattice. In Fig. (10), $<\rho^{-1} - \rho_0^{-1}>$ is plotted versus $<S - S_0>$ for the simulations on the simple cubic lattice with uniform site energies. The curves are found to be roughly linear as expected from Eq. (16).

As shown by the Sierpinski gasket simulations of single walker transport, the basic power law relation for the number of distinct sites visited by a random walker, Eq. (11) is a good first-order approximation even for

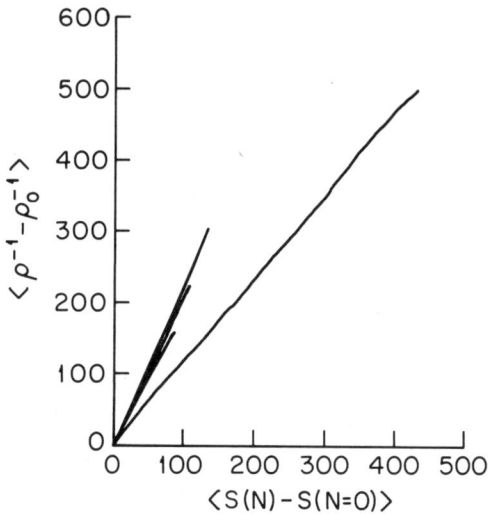

Fig. 10
$\langle \rho^{-1} - \rho_0^{-1} \rangle$ versus $\langle S(N) - S(N=0) \rangle$ for the simple cubic lattice with a uniform distribution of random site energies. Shortest to longest curves correspond to increasing temperatures: $T'' =$ 0.06, 0.065, 0.07, and 0.1.

energetically disordered systems with temperature-dependent Boltzmann weighting factors. The major difference between the exponential, Gaussian, and uniform distributions of site energies is seen in the crossover behavior: it is most apparent for the uniform distribution and to a lesser degree for the Gaussian distribution. Simulations of the exponential distribution do not show a clear crossover, but it cannot be ruled out on the basis of these simulation results.

The stochastic approach (Smoluchowski-type) adopted here relates the macroscopic time-dependent rate coefficient to the microscopic transport of individual walkers, Eq. (16). It is a fairly good approximation even for energetically disordered systems. Eq. (16) provides a scaling approach for moving from single walker simulations to reacting walker simulations. This approach works well over a fairly broad time interval and is relatively insensitive to the detailed nature of the system considered. Furthermore, it emphasizes that the rate-determining factor in these diffusion-limited reactions is indeed due to transport constraints imposed by the media on each individual walker.

7. SUMMARY

We have established a correlation between fractal-like exciton annihilation kinetics and the geometrical constrains and/or energetic disorder of a number of samples containing pure naphthalene. We have shown a behavior analogous to that of isotopic mixed naphthalene crystals (percolation clusters), in contrast to the behavior of perfect, pure naphthalene crystals. Furthermore, for isotopic alloys at the critical percolation concentration, we have demonstrated a genuine fractal behavior, with excellent values for the fractal and the spectral dimensions.

8. ACKNOWLEDGMENT

Supported by NSF Grant DMR 8303919 and NIH Grant 2R01 NS80116-16.

9. REFERENCES

1. P. W. Klymko and R. Kopelman, J. Phys. Chem. 87 4565(1983).

2. P. Evesque and J. Duran, J. Chem. Phys. 80 3016(1984).

3. A. Aharony and D. Stauffer, Phys. Rev. Lett. 52 2368(1984).

4. S. Alexander and R. Orbach, J. Phys. (Paris) Lett. 43 L625(1982).

5. R. Rammal and G. Toulouse, J. Phys. (Paris) Lett. 44 L13(1983).

6. P. G. de Gennes, C. R. Acad. Sci. Ser. A296 881(1983).

7. D. Ben-Avraham and S. Havlin, J. Phys. A15 L691(1982).

8. J. Klafter, A. Blumen and G. Zumofen, J. Stat. Phys. 36 561(1984).

9. P. Argyrakis and R. Kopelman, Phys. Rev. B29 511(1984).

10. D. C. Ahlgren, Ph.D. Thesis, The University of Michigan (1979).

11. J. S. Newhouse, Ph.D. Thesis, The University of Michigan (1985).

12. R. Kopelman, P. W. Klymko, J. S. Newhouse and L. W. Anacker, Phys. Rev. B29 2164(1984).

13. L. W. Anacker and R. Kopelman, J. Chem. Phys. 81 6402(1984).

14. R. Kopelman, E. M. Monberg, F. W. Ochs and P. N. Prasad, J. Chem. Phys. 62 292(1975).

15. R. Kopelman, E. M. Monberg, F. W. Ochs and P. N. Prasad, Phys. Rev. Lett 34 1506(1975).

16. R. Kopelman, Topics in Applied Physics 15, 247(1976).

17. J. Hoshen and R. Kopelman, J. Chem. Phys. 65, 2817(1976).

18. P. Argyrakis, Ph.D. Thesis, The University of Michigan (1978).

19. P. C. de Gennes, La Recherche, 7 919(1976).

20. R. B. Pike and H. E. Stanley, J. Phys. A14, L169(1981).

21. R. Kopelman and P. Argyrakis, J. Chem. Phys. 72, 3053(1980).

22. P. W. Klymko and R. Kopelman, J. Lumin. 24/25, 457(1981).

23. P. W. Klymko and R. Kopelman, J. Phys. Chem. 86,

3686(1982).

24. E. W. Montroll and H. Scher, J. Stat. Phys. 9, 101(1973).

25. H. Scher and M. Lax, Phys. Rev. B7, 449(1973).

26. H. Scher and E. W. Montroll, Phys. Rev. B12, 2455(1975).

27. M. F. Shlesinger, J. Stat. Phys. 10, 421(1974).

28. M. F. Shlesinger, J. Stat. Phys. 36, 639(1984).

29. L. W. Anacker, R. Kopelman and J. S. Newhouse, J. Stat. Phys. 36 591(1984).

30. L. W. Anacker and R. Kopelman, Phys. Rev. B (in press).

31. L. W. Anacker and R. Kopelman, 1985 Proceedings of the 2nd International Conference on Unconventional Photoactive Solids, Ed. H. Scher. Plenum Press (1986).

32. J. Prasad, S. J. Parus and R. Kopelman, 1985 Proceedings of the 2nd International Conference on Unconventional Photoactive Solids, Ed. H. Scher. Plenum Press (1986).

33. E. I. Newhouse and R. Kopelman, 1985 Proceedings of the 2nd International Conference on Unconventional Photoactive Solids, Ed. H. Scher. Plenum Press (1986).

34. L. A. Harmon and R. Kopelman, 1985 Proceedings of the 2nd International Conference on Unconventional Photoactive Solids, Ed. H. Scher. Plenum Press (1986).

35. L. W. Anacker, P. W. Klymko and R. Kopelman, J. Lumin. 31/32, 648(1984).

36. J. M. Drake et al., this volume.

37. U. Even, K. Rademann, J. Jortner, N. Manor and R. Reisfeld, Phys. Rev. Lett. 42 2164(1984).

38. P. W. Klymko, Ph.D. Thesis, The University of Michigan (1984).

39. A. H. Francis and R. Kopelman, Topics in Applied Physics 49, 241(1981).

40. D. Stauffer, Introduction to Percolation Theory, Taylor and Francis, London (1985).

41. J. S. Newhouse, J. Hoshen and R. Kopelman (unpublished).

42. J. T. Bendler and M. Shlesinger, Macromolecules, 18 591(1985)

43. E. W. Montroll and G. H. Weiss, J. Math. Phys. 6 167 (1965)

44. J. Klafter and R. Silbey, Phys. Rev. Lett. 44 55 (1980)

45. L. W. Anacker, R. Kopelman and J. S. Newhouse, J. Stat. Phys. 36 591 (1984).

46. P. Argyrakis, L. W. Anacker and R. Kopelman, J. Stat. Phys. 36 561 (1984).

MONTE-CARLO SIMULATIONS -- A MEANS TO GAIN
INSIGHT INTO DIFFICULT TRANSPORT PROBLEMS IN RANDOM MEDIA

M. Silver

Department of Physics and Astronomy
University of North Carolina
Chapel Hill, NC 27514

and

B. Ries and H. Bassler
Fachbereich Physikalische Chemie
Philips Universitat
Marburg, D-3550 F.R. Germany

ABSTRACT

Theories of transport and energy relaxation in random media are difficult to document experimentally because the systems are generally not well characterized. To ameliorate this problem, we have performed a series of monte-carlo simulations on hopping through various energy distributions of hopping sites which can give insight into the physical processes characterized by the random media. The result of a number of these simulations are presented.

1. Introduction

One of the reasons for this conference is that transport in disordered media is very complicated because the experimental systems are not well characterized. As a consequence, theoretical models are not easily documented. To ameliorate this problem we have performed a series of monte-carlo simulations on a variety of transport problems. Generally we have studied the hopping transport through various energy distributions of hopping sites. The advantage of this approach is that while it does not give analytic or general solutions, it does yield information on well characterized systems.

In this paper we will briefly review the monte-carlo method specifically as it applies to electronic and excitonic hopping transport. (The multiple trapping problem is considerably easier because only energy disorder is involved; spatial separation of sites is not important and therefore this method will not be outlined.) We focus on a cubic lattice in which the energy of each site is a random function of some probability distribution, usually gaussian. Spatial randomness is approximated since we in principle allow for jumps to all sites.

In this short paper we cannot discuss all the problems that have been explored. Instead, we will concentrate on a few interesting problems which are:

(1) equivalence between the multiple trapping and hopping model

(2) E-hopping in a gaussian distribution of hopping sites with electric field, temperature and width of the distribution as parameters.

(3) hopping in a dilute system for comparison with a percolation (cluster) approach and

(4) geminate recombination

Included will be a short discussion of the interpretation of some related problems including a possible explanation of the Meyer-Neldel rule. (The Meyer-Neldel rule relates to the general experimental observation that the pre-factor in an activated process is a function of the activation energy.) The monte-carlo method applied to transport problems is in fact very simple. One, first maps a probability function of some variable v onto uniform random number space. (In practice for computational efficiency we often use non uniform random number space such as exponential or gaussian depending upon the problem being considered.) Schematically then we have

(1) $$P(v)\, dv = P(R)\, dR.$$

where $P(v)$ is the probability per unit variable v and $P(R)$ is the probability per unit random number.

For illustrative purposes taking $P(R) = 1$ we have

(2) $$\int_0^{v_e} P(v)\, dv = R_e$$

Consequently, R_e specifies a specific event v_e. Equation (2) in general can easily be inverted and the specific event v_e is related to R_e through:

(3) $$v_e = F(R_e)$$

where $F(R_e)$ is obtained from the inversion of equation (2). Consequently, if one chooses a random number R_e then this designates the event v_e. One can see this more graphically by showing curves for specific distributions: a) fig. (1a) $P(v) = \frac{1}{\tau} e^{-t/\tau}$ and where the variable v is time, t, and in fig. (1b) where $P(v) = $ const.

The first curve (1a) is used to calculate a sequence of jumps, for example, $P(t) = \nu_0 \exp[-\nu_0 t]$ or by inverting as in equation (3) $t_e = -\frac{1}{\nu_0} \ln(1 - R_e)$. The sequence of jumps time $\sum t_{ej} = \sum -\frac{1}{\nu_0} \ln(1-R_{ej})$. Where the R_{ej} are a sequence of selected random numbers.

The locations of a point on a lattice is chosen for example with all R's between 0 and $\frac{a}{L}$ specifies sites between 0 and $\frac{a}{L}$. Random numbers between $\frac{a}{L}$ and $\frac{2a}{L}$ specify sites between $\frac{a}{L}$ and $\frac{2a}{L}$ etc. Once the site is chosen, its internal properties such as ε_i are determined from another random number and the probability function $P(\varepsilon_j)$ which may be gaussian or exponential etc.

By a sequence of randomly selected numbers we specify the motion of a particle under our pre-selected conditions in a given lattice which is

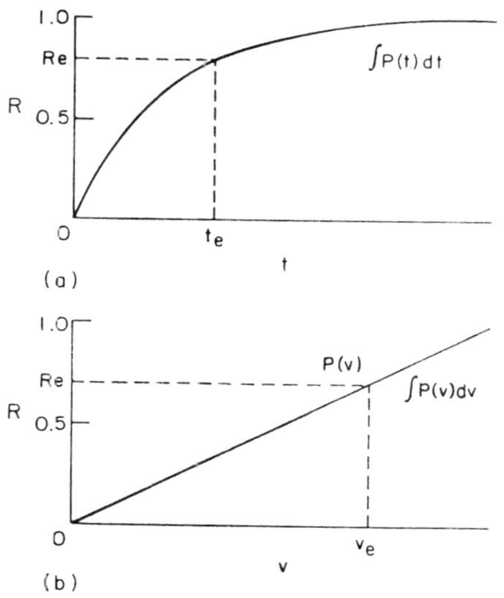

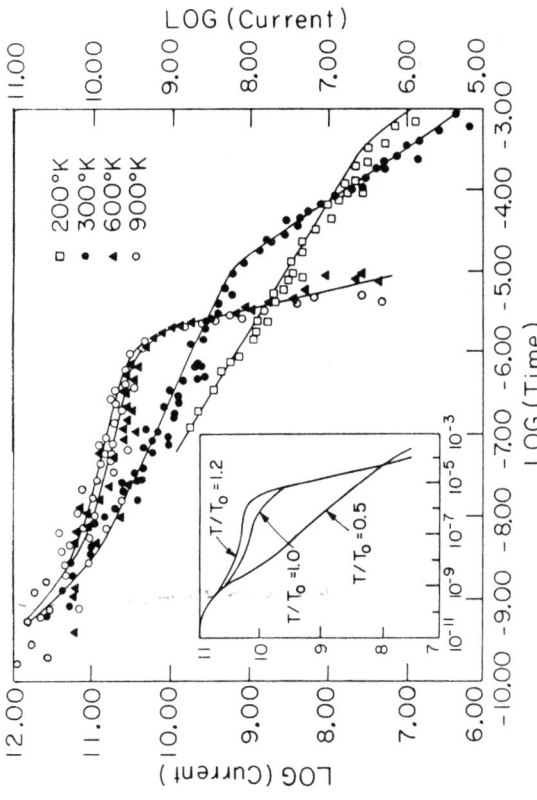

Figure 1. Schematic diagram showing relationship between the random numbers R and the probability variable. (a) shows the case where $P(t) = \nu e^{-\nu t}$. The selection of the random number R_e specifies the time of the event t_e. (b) shows the case $P(v) = $ const.

Figure 2. Dispersive transport due to hopping in an exponential distribution band tail states with $T_0 = 504°K$. The insert shows multiple dispersive transport.

well characterized. We start a particle either at a random site or at a pre-selected one. The probability that it will jump from this site i to another site j is

$$(4) \qquad P_{ij} = \frac{\nu_{ij}}{\sum_{l} \nu_{il}}$$

Where ν_{ii} is set equal to zero when the lifetime is infinite.

Obviously $\sum P_{il} = 1$ and therefore from curve (1b) one can assign a range of random number R_{ij} which when chosen specify the site to which the particle jumps. The time for the jump is obtained from figure (1a) which means:

$$(5) \qquad t_{ij} = (-1/\sum \nu_{il}) \ln(1 - R_{ij})$$

In order to calculate t_{ij} we must specify ν_{ij}. We choose

$$(6) \qquad \nu_{ij} = \nu_o \, e^{-2\gamma|r_j - r_i|} \, e^{-(\varepsilon_j - \varepsilon_i)/kT}$$

for $\varepsilon_j > \varepsilon_i$ and

$$(7) \qquad \nu_{ij} = \nu_o \, e^{-2\gamma|r_j - r_i|}$$

for $\varepsilon_j < \varepsilon_i$ and consequently $P_{ij} \neq P_{ji}$. It should also be noted that equation (7) implies that phonon emission required if $\varepsilon_j \ll \varepsilon_i$ does not limit the jump rate.

The computation proceeds by recording the position of each particle as a function of time until the particle dies (according to another probability function) or by reaching a pre-specified killer site. For each computer experiment we have generally used 50 lattices each containing more than 69,000 sites. We follow the motion of at least 20 moving particles in each lattice and only then do we average to obtain $\vec{r}(t)$, $\langle r^2(t) \rangle$, $\langle |r(t)| \rangle$, $\langle r^2(t) \rangle^{1/2}$. $N(\varepsilon,t)$, $\langle \varepsilon(t) \rangle$ and therefore $D(t)$ and $\mu(t)$ from $D(t) = \langle r^2(t) \rangle/t$ and $\mu(t) = \langle |\dot{r}(t)| \rangle/E$

2. **Results**

A) comparison between multiple trapping and hopping in an exponential energy distribution of sites.

Dispersive transport, motion in which the spreading of a pulse of carriers is non-gaussian, was first shown to result from a broad distribution of event times by Scher and Montroll[2] using Continuous Time Random Walk (C.T.R.W.). Soon after dispersion was demonstrated using monte-carlo (a technique first used in transport by M. Silver, K. S. Dy and J. L. Huang[3] for multiple trapping).[4] Later Tiedje and Rose[5] and Orenstein and Kastner[6] gave a simple physical picture of the dispersive nature of multiple trapping for a broad energy distribution of traps based on a demarcation energy approach E_d, $[(\varepsilon_c - \varepsilon_d) = kT \ln (\nu_0 t)]$, above which carriers are in quasi equilibrium and below which they have an occupation governed by their trapping constant. Because a quasi equilibrium is envisioned, Silver Schonherr and Bassler[7] pointed out that the process should not depend upon the transport mechanism and therefore, hopping through an exponential distribution of sites should be as dispersive as multiple trapping through the same distribution. This was demonstrated by Monte-Carlo and the results are shown in figure 2. The insert shows the multiple trapping results while the main figure gives results for hopping. These monte-carlo results clearly show that hopping through an energy distribution of sites does give rise to dispersion, contrary to the conclusion of Marshall and Sharp[8] (In a time-of-flight experiment Adler and Silver[9] showed by monte-carlo that dispersion can be observed without a broad energy distribution of hopping sites but with a spatially random but equal energy distribution). More recently Monroe[10] applied the simplified quasi equilibrium approach to hopping through an exponential energy distribution of hopping sites. Monroe was able to distinguish two regimes: (1) an early time solution where dispersion results from jumps only to lower energy sites and (2) a later time regime where hopping up to higher energy states as well as jumps to lower energy sites are involved. Regime 1 gives a current which decays faster than in regime (2) where the current decays as $t^{\alpha-1}$ similar to that obtained by Silver, Schonherr and Bassler[7]. Monroes predicted that the time to reach the regime (2) is given by:

(8) $$t_s = \frac{1}{\nu_0} \exp[3T_0/T]$$

where t_s is the separation time, ν_0 is the characteristic pre-factor in the jump frequency and T_0 characterizes the width of the energy distribution of hopping sites. However, according to the simulations performed by Silver[7] et al. this should occur at approximately 2×10^{-10} s for $T = 200°$ $\nu_0 = 10^{13}$ s^{-1} and $T_0 = 504°$K. However, no such time separating the two regimes was found in the monte-carlo simulations. The reason for this was that for the parameters used in the simulations the calculated t_s is less than the shortest possible hop time which was 2×10^{-9}s for the parameters used by Silver, Schonherr and Bassler.[7] The expression for the shortest hop time is $t_{sh} = \frac{1}{\nu_0} \exp[2 \gamma N_t^{-1/3}]$ where $2\gamma (N_t)^{-1/3}$ was chosen to be 10 and N_t was the density of hopping sites. One can generalize this result because t_s must always be longer than the shortest hop time. Consequently, the separation time will only be observed if $t_s > t_{sh}$ or

(9a) $$3T_0/T > 2\gamma/N_t^{1/3}$$
or
(9b) $$T < 3T_0 N_t^{1/3}/2\gamma$$

In our simulation from Equation (9b) $T < 150°$. The lowest temperature we studied was $200°$K. Consequently we did not observe the short time hopping down regime. For a-Si:H where a typical set of values might be $N_t = 10^{19}$cm^{-3}, and $2\gamma = 1/5$Å suggests that the temperature $T < 90°$. In other words unless the temperature was significantly less than $90°$K normal dispersive transport would be expected. Of course if N_t was larger than 10^{19} and/or $1/\gamma$ was larger than 10Å, then for temperatures somewhat larger than $90°$ the short time hopping down the regime followed by a longer time dispersive regime would be observed.

The similarities between hopping and multiple trapping is clear; except at short times, the time dependence of the current do not distinguish the mechanism of transport.

3. Hopping in a Gaussian Band Tail

Several interesting observations have been found for the case of hopping in a gaussian energy distribution of states. Perhaps the most interesting result is related to the dispersive behavior as a function of applied electric field. For times shorter than the first transit time, the current decays approximately as $t^{-1+\alpha}$ where the dispersive parameter α depends upon the energy width of the hopping sites. (recall that for a pure exponential $\alpha = T/T_o$.)[4] We explored the electric field dependence of α because it would be expected that the effective width of the hopping sites would decrease with applied field. The higher states would be lowered in energy in the direction of the field increasing the jump rate. The hopping rate to the lower energy states would not be affectd because hopping down in our model is not energy dependent. As a consequence, the effective α would increase and dispersion would tend to decrease with increased field. This expectation was found and the results are shown in figure 3. These monte-carlo results show an apparent increase in α (decrease in 1-α) with an increase in electric field. The change in α with electric field has another effect: The transit time t_d can no longer be represented by a power law of the form $t_d \propto (1/E)^{1/\alpha}$ as one expects[5] for a constant α. This result is shown in figure 4 where the log t_d vs log E. deviates from a straight line at high fields (just where α changes with E, see figure 3) Such a change in the form of t_d vs E at high field was found to resemble the Poole-Frenkel Behavior, $\ln t_d \propto (1/E)^{1/2}$. These results are shown in the insert in figure 4. The Poole-Frenkel effect is generally found in systems where the localized sites are charged when empty. In our simulations no charged centers were included. Thus, the similarity in behavior leads to difficulty in interpreting results and caution should be exercised in interpreting experimental results with regard to the dominant transport mechanism.

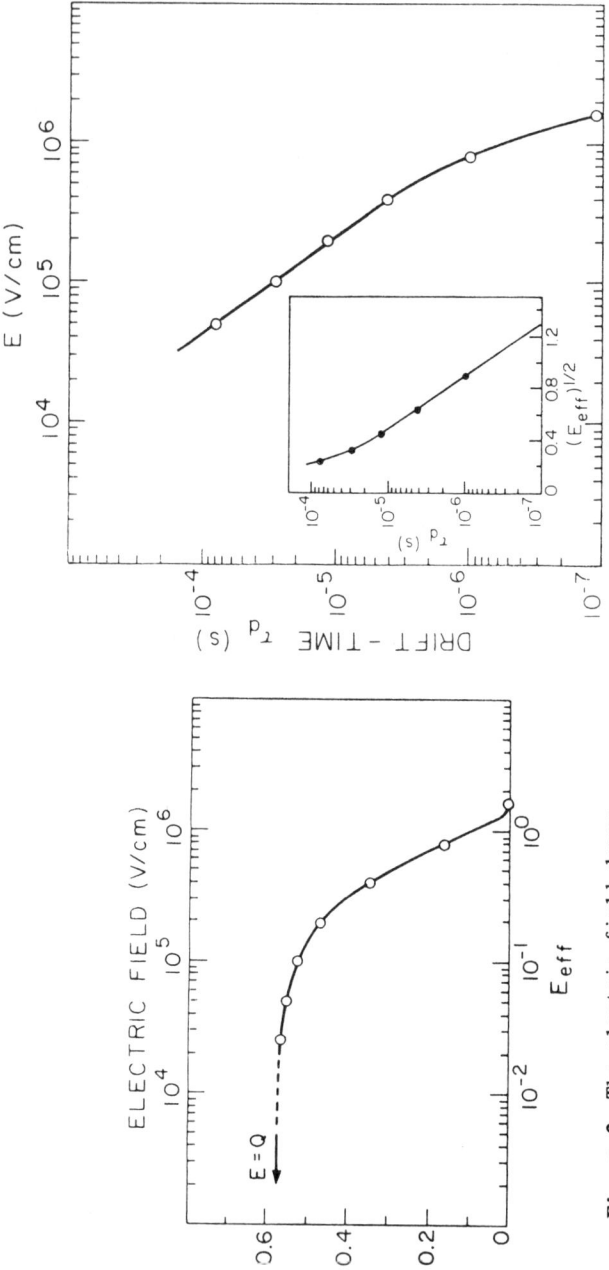

Figure 3. The electric field dependence of α where $E_{eff} = eEa/\sigma$, a is the hopping distance and σ is the energy width of the distribution of hopping sites.

Figure 4. Log of drift time vs log E (or $E_{eff} = eEa/\sigma$). Notice that at high E, there is a deviation from straight line behavior implying a change in α with field. The insert shows the same data plotted as log t_d vs $(E_{eff})^{1/2}$. There is a considerable range over which the slope is constant similar to the Poole-Frenkel effect.

The gaussian distribution shows another interesting feature which is that the apparent mobility (transit time, t_d or long time current i_∞) does not follow the normal Arrhenius plot ($\ln t_d \propto 1/T$) but in fact our montecarlo results show that a much better fit to the data is $\ln t_d \propto (1/T)^2$. We show these results in figure 5. The best fit is seen by the straight line of $\log i_\infty$ vs $1/T^2$ whereas the graph of $\log i_\infty$ vs $1/T$ is curved.

There are two features of these results that are worth noting:
(1) When the log of the intercept at $T = \infty$ is plotted as a function of $\log(C-C_o)$ one obtains a straight line implying a percolation controlled transport (e.g. $\ln \mu_o$ vs $\ln(C-C_o)$ gives $\mu_o \propto (C-C_o)^{1.8}$ where C_o is a critical concentration $\approx .15$) and C is the concentration. On the other hand, the same data when plotted $\ln \mu_o$ vs $1/C^{1/3}$ also results in a straight line implying that transport is controlled by the average hopping distance. These results are shown in figure 6. This remains a dilemma because it is important to experimentally distinguish these two different transport mechanisms. Lemus and Hirsch[12] have attempted to clarify this issue experimentally. However, to date the question is still unresolved and further work is needed.

The other feature of interest (2) is that a gaussian distribution of hopping sites can result in an apparent Meyer-Neldel rule; i.e. $\mu_o \propto E_a$ where E_a is the activation energy when the log μ is plotted vs. $1/T$. This is seen in figure 7 where some data on mobility in TNF:PVK is plotted[10]. (The activation energy was varied by the applied field.) Notice that the graphs when plotted vs $1/T$ intersect at a finite T. Therefore, the intercepts at $T = \infty$ will vary with the slope of the curves giving the Meyer-Neldel rule, $\mu_o \propto E_a$. However, even though the plots of $\ln \mu$ vs $1/T$ are reasonably straight, when the same curves are plotted vs $1/T^2$, the intercept occurs at $T = \infty$. As a consequence we suggest that the Meyer-Neldel rule can result from transport through a gaussian distribution of hopping sites. The Meyer-Neldel rule has been very confusing and perhaps this suggestion can be documented and help clear up this issue. It's importance cannot be over-stated because so much fundamental information is generally derived from the $T = \infty$ intercept.

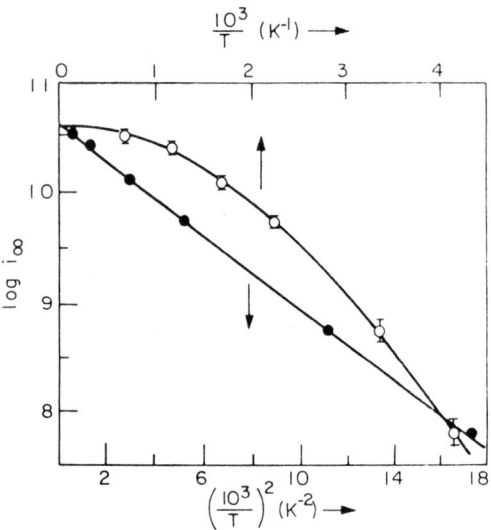

Figure 5. Log i_∞ vs $1/T$ and $1/T^2$. Notice that straight line behavior is obtained vs $1/T^2$.

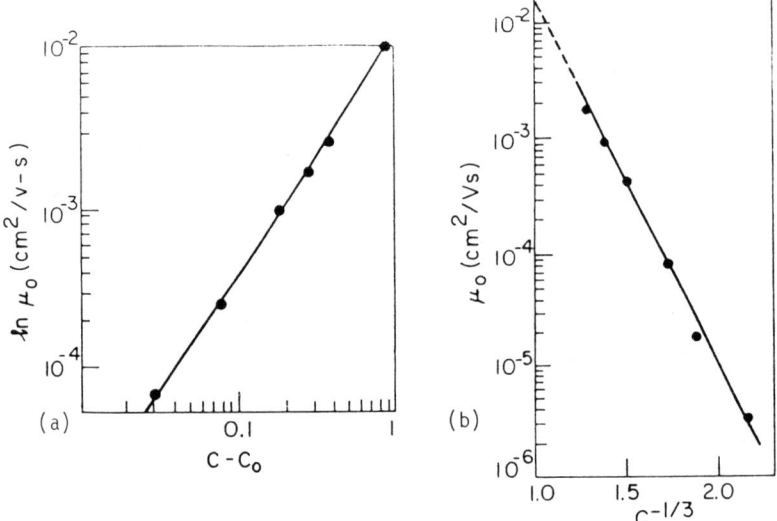

Figure 6. (a) Log μ_0 vs loc $(C-C_0)$ and (b) log μ_0 vs $C^{-1/3}$. Notice both curves give straight lines.

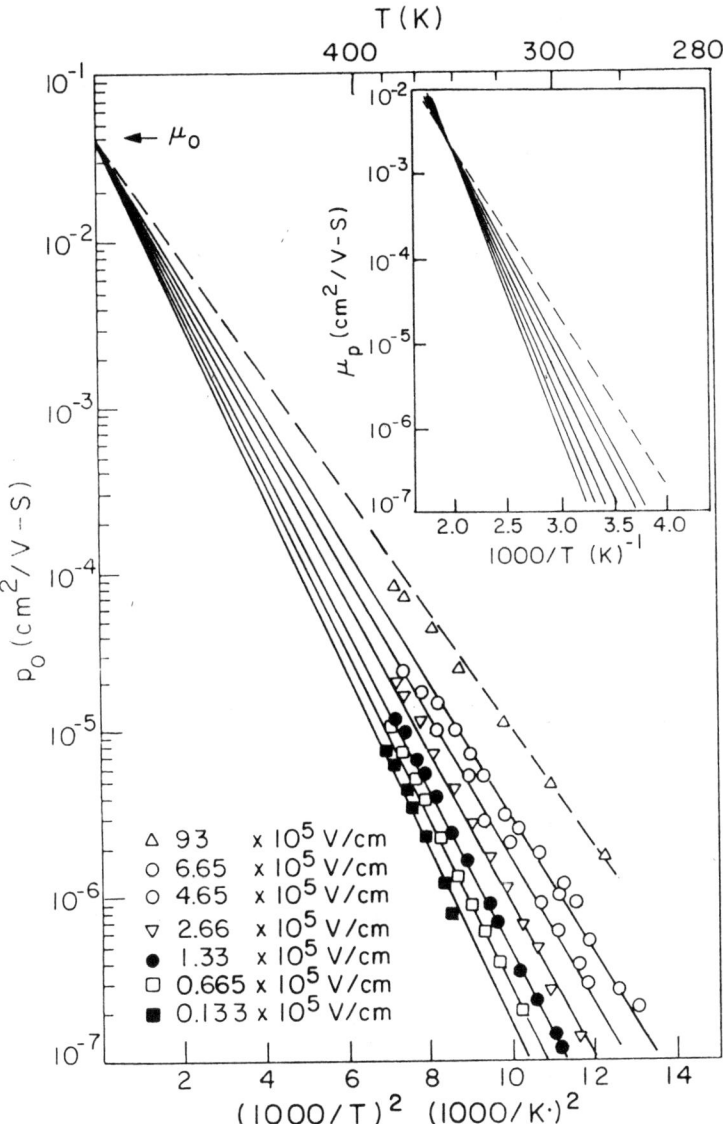

Figure 7. Log μ_o vs $1/T^2$ for various electric fields. (The electric field changes the average hopping energy thereby giving different activation energies.) Notice that all the curves intersect at $T = \infty$. The insert shows that same data plotted vs $1/T$. The intersection is at a finite temperature.

4. Geminate Recombination

The general ideas related to geminate recombination were first formulated by Onsager[12] who basically showed that the probability for two charged carriers to escape from each other depended upon the coulomb barrier that they experienced. Consequently, if two charge carriers are generated a distance r_o apart and have little kinetic energy then their escape probability P_{esc} is given by:

$$(10) \qquad P_{esc} \propto \exp[+ eV(r_o)/kT] = \exp[- r_c/r_o]$$

where $r_c = e^2/4\pi\varepsilon\varepsilon_o kT$. [An externally applied electric field will modify this simple result by lowering the effective barrier thereby giving a larger P_{esc} than indicated by equation (10)]. All of the ramifications of the steady state results cannot be discussed here. Instead we focus on some early time dependence. The time dependence of P_{esc} is best represented by the recombination rate. What is interesting is that the recombination rate does not rise instantaneously and only then decays as a power law. This result is shown in figure 8. This result is expected since the pairs are produced r_o apart and they cannot recombine until they diffuse together. (The longer time power law decay also results from the diffusive motion.)

In these simulations all downward jumps are allowed and the life time of the pair at $r = 0$ is zero. On the other hand, Orlowski and Scher[14] measured and calculated recombination in a-Si:H. Their results are shown in figure 9. They show an almost instantaneous decay of the recombination which they can fit to a model. In their model they limit upward jumps to kT in energy and downward jumps to $\hbar\omega_p$ where ω_p is an optical phonon frequency and they do not include any Coulomb interaction.[13] The difference between the results of the two models is clear and should be distinguishable experimentally. Since the difference between the two models is related to the Coulomb interaction and energy loss restrictions, the question arises under what circumstances do each of the two models apply? This is an important question which also requires further study.

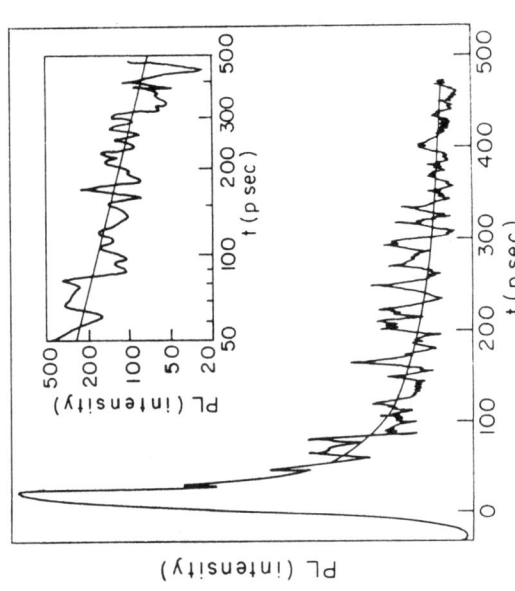

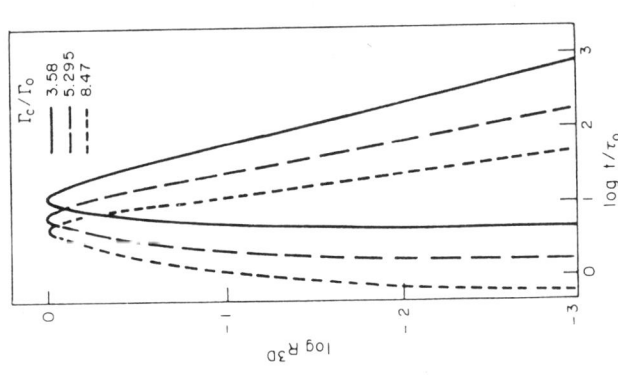

Figure 8. Log R vs log t/τ_o for various values of r_c/r_o. Where R is the recombination rate and T_o is a characteristic time. Typically it takes ~ 10^2 hops for R to reach its maximum.

Figure 9. Luminesence vs time. The figure shows an almost instantaneous rise of the luminescence (recombination rate), of order or a few ps and then a power law decay. The insert shows the same data plotted as log PL vs log t.

5. Summary

In this paper we presented the method and some results of monte-carlo simulations on transport properties of random media. The equivalence between hopping and multiple trapping was shown not only for long times but for all times provided that $3T_o/T > 2\gamma(N_T)^{-1/3}$. The field dependence of α, the dispersion parameter, is demonstrated and the results make the I-V characteristics resemble a Poole-Frenkel effect. Finally, several remaining problems were illustrated: (1) it is difficult to distinguish between percolation effects and average hopping behavior; better data is needed so that the applicability of the two regimes can be delineated better and (2) the time dependence in recombination processes are predicted to be different depending upon whether one includes Coulomb interactions and energy matching conditions. Experimental results in organic materials are consistent with simulation results where the Coulomb terms are included but downward in energy jumps are not inhibited. On the other hand, results in the pico-second regime in a-Si:H are best described in the absence of a Coulomb term and downward jumps are limited to energy differences no more than $\hbar\omega_p$.

There is no doubt that these simulations procedures are available for checking theory and gaining insight into the processes involved in experiments in random media.

References

1. H. Bassler, Phys. Stat. Sol. (b)<u>107</u>, 9 (1981).
2. H. Scher and E. W. Montroll, Phys. Rev. B<u>12</u>, 2455 (1975).
3. M. Silver, K. S. Dy and J. L. Huang, Phys. Rev. Lett. <u>37</u>, 1360 (1971).
4. M. Silver and L. Cohen, Phys. Rev. B<u>15</u>, 3276 (1977).
5. T. Tiedje and A. Rose, Sol. Stat. Comm. <u>37</u>, 49 (1981).
6. Orenstein and M. Kastner, Phys. Rev. Lett. <u>46</u>, 1421 (1981).
7. M. Silver, G. Schonherr and H. Bassler, Phys. Rev. Lett. <u>48</u>, 352 (1982).
8. J. Marshall and A. C. Sharp, J. Non-Cryst. Sol. <u>35-36</u>, 99 (1980).
9. J. Adler and M. Silver, Phil. Mag. <u>45</u>, 307 (1982).
10. D. Monroe, Phys. Rev. Lett. <u>54</u> 146 (1985).
11. W. D. Gill, J. Appl. Phys. <u>43</u>, 5033 (1972).

12. S. J. Santos Lemus and J. Hirsch, to be published.
13. L. Onsager, Phys. Rev. 54, 554 (1938).
14. T. E. Orlowski and H. Scher, Phys. Rev. Lett. 54, 220 (1985).

MOLECULAR MOTION IN GLASSY POLYCARBONATE

Alan Anthony Jones, Paul T. Inglefield, J.F. O'Gara and A.K. Roy
Jeppson Laboratory, Department of Chemistry
Clark University, Worcester, MA 10610

Motion in polymeric glasses is greatly reduced both in amplitude and rate relative to either the rubbery or dissolved state.[1] In rubbers and solutions of random coil macromolecules, the backbone bonds sample nearly all directions in space over relatively short times, certainly less than microseconds. In the glass or near the glass transition, backbone bonds do not reorient isotropically even on the time scale of seconds but the presence of considerable local motions is still evident in mechanical and dielectric relaxation studies.[1]

The polycarbonate of bisphenol A (BPA-PC) pictured in Figure 1 displays especially large dielectric and mechanical loss peaks well below the glass transition. Mobility in this glass exceeds the norm for polymeric materials and this fact has attracted many investigators. One would like to characterize the time scale, energetics and geometry of the motion. Dielectric and dynamic mechanical relaxation measurements yield broad loss peaks with an apparent activation energy of about 50 kJ/mole.[2-3] Considerable speculation concerning the geometry of the motions responsible for the loss peaks accompanied the dielectric and mechanical studies.[2] However these techniques are not capable of determining this aspect of motion relative to the structure of the repeat unit of the polymer.

Geometry of Motion from Solid State NMR

In the past five years, solid state NMR has been able to determine the geometry of motion relative to the chemical structure of the macromolecule.[4-10] Large magnetic interactions such as dipolar, shielding and quadrupolar lead to distinctive line shapes when the motion is restricted or anisotropic. Since these magnetic interactions can be precisely related to the local structure, the presence of motion can be be set into the context of the repeat unit geometry.

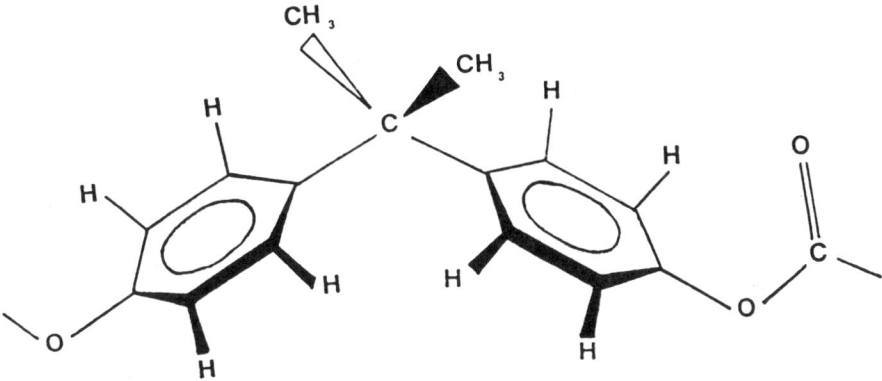

Figure 1. The repeat unit of the polycarbonate of bisphenol A (BPA-PC).

This advance has provided detailed information on local motion in glasses but in the end raises new questions as it answers the old.

The first indication of the geometry of local motion in BPA-PC came from a proton line shape study of a closely related structural analogue.[4] Here the proton-proton dipolar interaction between the neighboring protons on the phenylene groups is monitored by the presence of a Pake doublet. This distinctive line shape feature would collapse if a vector drawn between the two neighboring protons is reoriented by the onset of local motion. The Pake doublet persists in an undiminished form until the glass transition indicating no reorientation of the proton-proton internuclear direction. This direction is parallel to the effective backbone bond consisting of the phenylene group indicating that this virtual bond is not reoriented.

Proton spin-lattice relaxation in the same polymer shows the presence of considerable rapid, large amplitude motion.[11] If the phenylene protons are moving yet the corresponding backbone bond is not reorienting, the motion must be a rotation about that backbone bond. In chemists' parlance, the rotation would be about the C_1C_4 axis where C_1 is the phenylene carbon attached to the oxygen and C_4 is the phenylene carbon furthest from C_1.

This defines one aspect of the motion but the rotation needs further specification. Is it free rotation about the C_1C_4 axis or rotation by jumps between minima or a restricted rotation-libration? The first answer came from deuterium quadrupole line shape NMR[5] which indicated it was jumps between two minima separated by 180° plus restricted rotation in the bottom of each of the two minima. This result was substantiated by carbon-13 chemical shift anisotropy line shape results which can definitely distinguish between the three possibilities mentioned.[6] Carbon-13 proton dipolar line shapes are also consistent with the 180° jump or π flips plus restricted rotation.[9] These NMR experiments provide a description of the geometry of phenylene group motion not available from any other technique.

Both deuterium and carbon-13 proton dipolar[9] line shape studies on the methyl group of BPA show this entity to be rotating about the three fold symmetry axis. The motion is independent of phenylene group flips. Also the methyl group line shapes show the presence of at best only a little wiggling besides the rotation about the symmetry axis. Thus the bisphenol-A unit including the two phenylene groups, the two methyl carbons and the bridge carbon is not moving as a whole. The phenylene groups move in a two fold potential about their symmetry axis and the methyl groups move in a three fold potential about their symmetry axis.

The motion of the carbonate unit, $-O-\overset{\overset{O}{\|}}{C}-O-$ remains to be defined. Here an apparent inconsistency between NMR line shape data and dielectric data arises. Clearly the carbonate unit must be reorienting to account for the presence of the appreciable dielectric loss. However carbon-13 chemical shift anisotropy line shape data[12] on the carbonate carbon shows the presence of very little motion. This point must be addressed later in an overall motional model.[13]

Time Scale of Local Motion

Motions in polymers are known to have complex character in time which is quite apparent in the broad, skewed loss peaks usually observed in macromolecular glasses.[1] Such broad loss peaks could arise in at least two different manners.[14] First the motion of a given polar group at a certain spatial position in the glass could be highly non-exponential. The motion of other polar groups of the same chemical type but at different spatial locations could be moving with a comparable non-exponential correlation function. This situation is referred to a homogeneous by NMR spectroscopists.[14]

An alternative source of broad loss peaks or relaxation minima can be imagined. Suppose an individual polar group at a particular location in a glass reorients in a Debye like fashion characterized by a simple exponential correlation function. However other polar groups at other locations also relay with simple exponential time constants but the time constants are different.[14] The various time constants at various locations in the glass arise from the inhomogeneous nature of the glass at a microscopic level. In particular different polar groups at different locations experience different packings or intermolecular environments leading to different time scales for reorientation. This situation is referred to as inhomogeneous and leads to broad loss peaks just as the earlier homogeneous description.

Mechanical loss, dielectric loss and proton spin-lattice relaxation cannot distinguish between the origin of the distribution. However carbon-13 spin-lattice relaxation and the solid state line shape experiments can distinguish between the two possibilities[7,14,15] and now definitively point to a predominately inhomogeneous distribution in polycarbonate.

To quantitatively summarize the inhomogeneous distribution of relaxation times in polycarbonates, two forms of correlation functions have been used. Spiess[16] has employed a log normal distribution which extends over about three decades in time to account for a variety of deuterium line shape experiments. An apparent activation energy of about 40 kJ describes the temperature dependence and some change in breadth accompanies change in temperature.

Jones[11,15] et al. has employed the fractional exponential correlation function to summarize proton spin-lattice relaxation, carbon-13 spin-lattice relaxation and carbon-13 chemical shift anisotropy line shape data. A fractional exponent (β) of 0.15 is required to fit the proton data[15,17] and is consistent with the carbon data. The fractional exponent is determined from the temperature dependence of spin-lattice relaxation data since insufficient data was available to fix the exponent at different temperatures. This fixed β was capable of matching the published deuterium data[7,15] though more detailed results from this technique are forthcoming.

The Jones group has also presented a quantitative analysis of the temperature dependence of the amplitude and time scale of the restricted rotation based on a correlation function developed by Gronski.[15,18] The amplitude of the restricted rotation is found to increase with the square root of temperature which is plausible. However the amplitude extrapolated to absolute zero is large which is difficult to interpret. The time scale of the restricted rotation is contained in a rotational diffusion constant which varies linearly with temperature and falls in the range of 10^{-8}s.[17] This time scale is characteristic of phenylene group libration and is much longer than infrared vibrational oscillation.

A More Complete Motional Model

An NMR spectroscopist might be tempted to stop with the motional interpretation developed thus far. This would include methyl group rotation, phenylene group π flips plus libration and little motion of the carbonate or the bisphenol unit as a whole. While this view satisfies the NMR data it is not easily reconciled with the mechanical and dielectric data. A relaxation map[6,11,15] constructed from spin-lattice relaxation minima, line shape coalescence, dielectric and mechanical maxima shows these phenomena to be linked phenomenologically in time. Such a map is shown in Figure 2 but the motions identified by NMR will not lead to large mechanical or dielectric loss.

The discrepancy between an apparently static carbonate carbon based on line shape data and the virtual necessity of carbonate reorientation for dielectric loss has been mentioned. Similarly, the only large amplitude motions noted are anisotropic rotations of symmetric groups, methyl and phenylene; and such motions do not lead to significant shear loss as is observed.

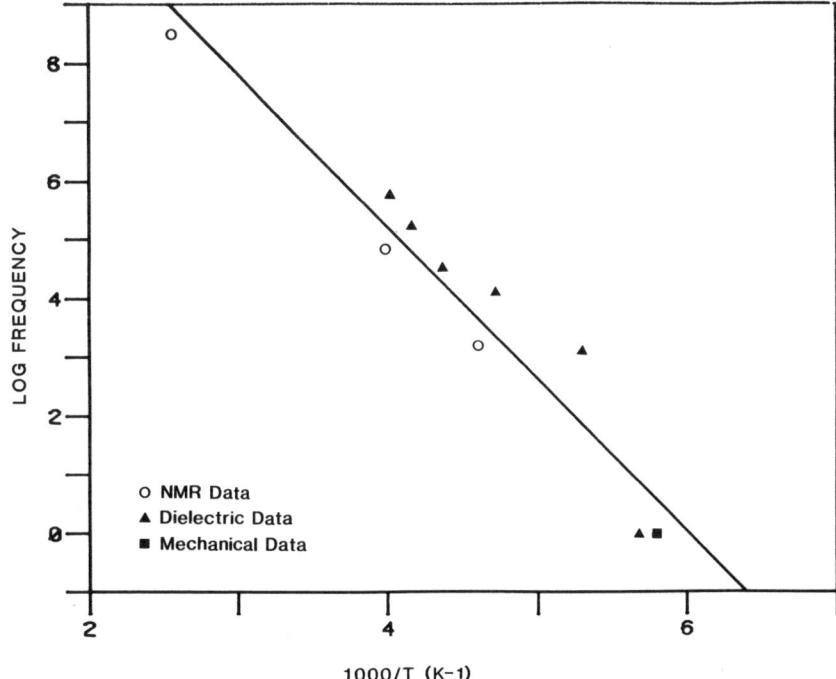

Figure 2. Log frequency versus inverse temperature or relaxation map. The highest frequency NMR point is the 90 MHz proton T_1 minimum, the next highest, 43 KHz $T_{1\rho}$ minimum and the lowest, the average position of CSA line shape coalescence. The open circles are maxima of dielectric loss curves taken at different frequencies and the square of the dynamic mechanical loss peak. The positions of all points have an associated uncertainty of the order of 10 degrees because of the broadness of loss peaks and relaxation minima.

Given the phenomenological linkage of the various relaxation experiments, the geometry of motion given by NMR, and the need to account for dielectric and mechanical loss, a motional model is proposed.[13] The model is constructed to be consistent with observations but unfortunately the data in hand do not "prove" the validity of the model.

The proposed local motion focuses on the conformation of the carbonate group. This unit is believed to reside predominantly in the trans-trans conformation.[19] As part of the motional model, the existence of a few defect conformations of either cis-trans or trans-cis forms are proposed. A cis-trans or trans-cis conformation with respect to the carbonate unit is found to be only somewhat higher in energy from quantum mechanical calculations on diphenyl carbonate.[20] Also the barrier for conformational change from cis-trans or trans-cis is found to be the lowest rotational backbone barrier in polycarbonate from the same calculations.

The motion proposed is a conformational exchange between a cis-trans or trans-cis and a neighboring trans-trans unit. This process is displayed in Figure 3. the pathway of the conformational exchange is a sequential set of rotation of the cooperative type suggested by Helfand.[21-24] The carbonate bonds which rotate are indicated by asterisks in Figure 3 and the backbone phenylene groups undergoing rotations are labelled with numbers.

The phenylene groups effectively execute π flips which is consistent with NMR line shape data. However in this motional picture, the π flips are part of a more complex limited segmental motion. The carbonate group changes shape in changing from trans-trans to cis-trans or trans-cis and the dipole moment should change in magnitude as well as orientation. These characteristics should lead to mechanical and dielectric loss linked in time to the π flips.

The carbonate carbon chemical shift anisotropy line shape does not collapse in the presence of this motion because a large population of trans-trans conformations is interchanged with a small population of cis-trans or trans-cis conformations. Even at high temperature, the average line shape in the rapid exchange limit will be close to the trans-trans line shape thereby displaying little indication of the motion. However since defect diffusion is involved, many carbonates can be reoriented by one defect leading to significant dielectric relaxation.

The BPA unit as a whole is not appreciably reoriented during the conformational interchange save for π flips which is again consistent with NMR line shape data. The BPA unit is translated which could lead

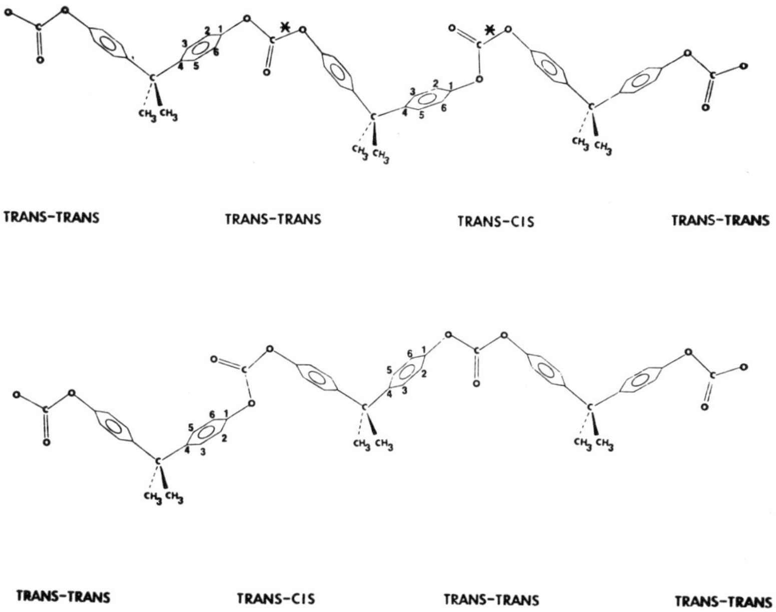

Figure 3. BPA polycarbonate chains. The top chain fragment is the initial state and the lower chain fragment is the final state. The carbonate CO bonds with asterisks indicate points of bond rotation. The phenylene rings undergoing flips in association with the CO bond rotations are numbered. The correlated conformational change from the top chain to the lower chain involves two neighboring carbonate groups and is produced by the CO rotations which interchange the trans-trans and trans-cis conformations. Note that the choice of a trans-cis unit in the figure is arbitrary. If a cis-trans unit were used, the other phenylene rings would be flipped so over a period of time all rings could be flipped as the cis-trans and trans-cis conformations diffuse along the chain.

to a bulk mechanical loss which has been observed in BPA-PC.

The interchange of a few cis-trans or trans-cis conformations with neighboring trans-trans conformation diffuses the cis-trans or trans-cis units along the predominantly trans-trans backbone. This defect diffusion character is related to several experimental observations.

First at high temperatures, all phenylene groups undergo π flips. If defect diffusion is sufficiently rapid on the line shape experiment time scale, all phenylene groups will appear to flip by either a cis-trans to trans-trans interchange or a trans-cis to trans-trans interchange. However as temperature is lowered in the glass, some carbonate units will no longer be able to participate in the interchange process because some local environments present too high a barrier. The phenylene groups near these units will appear rigid while other phenylene groups near carbonates with lower intermolecular barrier to conformational interchange will still appear as mobile. This would lead to the observed inhomogeneous character of the line shape collapse at intermediate temperatures. At quite low temperatures, little conformational interchange at any carbonate unit would result in the rigid line shape limit observed below $-100°C$.

Intermolecular interactions must contribute significantly to barrier heights since the observed activation energy in the glass is 50 kJ/mole. In solutions involving low viscosity solvents, the barriers to rotation or segmental motion is in the range of 10 to 15 kJ/mole in agreement with isolated chain calculations.[25-26] Thus the 35 to 40 kJ increase in apparent activation energy must be ascribed to intermolecular interactions allowing for the rationalization of the inhomogeneous character of line shape collapse just presented.

The defect diffusion character of the conformational interchange process involving a distribution of barrier heights can be identified with one of the derivations of fractional exponential correlation functions.[27-29] If the proposed motional model is correct, then the use of the fractional exponential correlation function is more than a convenient mathematical form but is also a physically sensible form.

In this motional model, phenylene group libration and methyl group rotation while present are not key aspects of the dynamics. Similarly, the presence of some low amplitude chain oscillation or wiggling is also a secondary aspect.

Remaining Questions

The defect conformations have not been directly observed so they remain as only postulated entities. The defect diffusion character of the motion may not be the source of the inhomogeneous distribution of relaxation times. Rather, the distribution of relaxation times may result from packing differences at individual motional sites. In this picture there is no diffusion from site to site but each site reorients with its own time scale. If this were the case, the fractional exponential function is just a convenient mathematical form and not connected with a physical derivation.

The defect diffusion conformational interchange process is shown in Figure 3 for an extended chain conformation. This extended chain conformation is not likely with other rotational angles placing one phenylene group out of the plane of the paper with respect to the other. These rotations will lead to the random coil character appropriate for polycarbonate. However whatever the relative disposition of the phenylene rings in one BPA unit, the interchange of carbonate conformations can still take place leaving the BPA unit motionless. Thus the particular conformational sequence shown in Figure 3 is convenient to draw but not a requirement of the motional model. A detailed analysis of the action of conformational interchange for other conformations associated with relative rotations of the phenylene groups in a BPA unit will be pursued with computer graphics.

The analysis of phenylene group libration in the data treatment and its role in the motional model is simplistic.[18] All phenylene groups are treated as thought they undergo the same librational amplitude. However the librational amplitude is governed by intermolecular interactions since rotation is found to be four fold or higher from calculations[10,26] and dilute solution spin-lattice relaxation.[25] If intermolecular effects dominate, a glassy matrix should lend to a distribution of libration amplitudes as well as a distribution of barrier heights. The failure of the line shape simulation to produce a reasonable intercept for the amplitude of libration at absolute zero may reflect the single amplitude approximation.

Further experimental work and analysis of the consequences of the motional model are required. At this stage it can be regarded as an interesting proposal or conjecture which may prompt particular lines of inquiry.

Acknowledgement

This research was carried out with the financial support of the National Science Foundation Grant DMR-790677, of National Science Foundation equipment Grant No. CHE 77-09059, of National Science Foundation Grant No. DMR-8108679, and of U.S. Army Research Office Grant DAAG 29-82-G-0001.

References

1. N.G. McCrum, B.E. Read and G. Williams, Anelastic and Dielectric Effects in Polymeric Solids, Wiley, New York (1967).

2. A.F. Yee and S.A. Smith, Macromolecules (1981) 14, 54.

3. S. Matsuoka and Y. Ishida, J. Polym. Sci., Part C (1966) 14, 247.

4. P.T. Inglefield, A.A. Jones, R.P. Lubianez and J.F. O'Gara, Macromolecules (1981) 14, 288.

5. H.W. Spiess, Colloid. Polym. Sci. (1983) 261, 193.

6. P.T. Inglefield, R.M. Amici, J.F. O'Gara, C.-C. Hung and A.A. Jones, Macromolecules (1983) 16, 1552.

7. H.W. Spiess, in Advances in Polymer Science, Vol. 66, H.H. Kausch and H.G. Zachmann, eds., Springer Verlag, Berlin (1985).

8. J.F. O'Gara, A.A. Jones, C.-C. Hung and P.T. Inglefield, Macromolecules (1985) 18, 1117.

9. J. Schaefer, E.O. Stejskal, R.A. McKay, W.T. Dixon, Macromolecules (1984) 17, 1479.

10. J. Schaefer, E.O. Stajskal, D. Perchak, J. Skolnik and R. Yaris, Macromolecules (1985) 18, 368.

11. A.A. Jones, J.F. O'Gara, P.T. Inglefield, J.T. Bendler, A.F. Yee and K.L. Ngai, Macromolecules (1983) 16, 658.

12. P.M. Kenricks, M. Linder, J.M. Hewitt, D. Massa and H.V. Isaacson, Macromolecules (1984) 17, 2412.

13. A.A. Jones, Macromolecules (1985) 18, 902.

14. J.I. Kaplan and A.N. Garroway, J. Mag. Res. (1982) 49, 464.

15. A.K. Roy, A.A. Jones and P.T. Inglefield, submitted to Macromolecules.

16. E.W. Fischer, L.P. Hellman, H.W. Spiess, F.J. Horth, A. Ecarius and M. Wherle, Mackro. Mol. Chem. Suppl. (1985) 12, 189.

17. J.J. Connolly, A.A. Jones and P.T. Inglefield, to be submitted to Macromolecules.

18. A.K. Roy, A.A. Jones, P.T. Inglefield, J. Mag. Res. (1985) 64, 441.

19. D.D. Williams and P.J. Flory, J. Polym. Sci. A-2 (1968); 6 (1945).

20. J.T. Bendler, Ann. N.Y. Acad. Sci. (1981) 371, 229.

21. E. Helfand, J. Chem. Phys. (1971) 54, 4651.

22. E. Helfand, Z.R. Wasserman and T.A. Weber, Macromolecules (1980) 13, 526.

23. C.K. Hall and E. Helfand, J. Chem. Phys. (1982) 77, 3275.

24. T.A. Weber and E. Helfand, J. Phys. chem. (1983) 87, 2881.

25. J.F. O'Gara, S.G. Desjardins and A.A. Jones, Macromolecules (1981) 14, 64.

26. A.E. Tonelli, Macromolecules (1972) 5, 558.

27. M.F. Shlesinger, E.W. Montroll, Proc. Natl. Acad. Sci. (USA) 1984, 81.

28. M.F. Shlesinger, J. Stat. Phys. (1984) 36, 639.

29. A. Blumen, G. Zumufen, J. Klafter, Phys. Rev. B (1984) 30, 5379.

RELAXATION AND RECOVERY

OF GLASSY POLYCARBONATE

John T. Bendler, D.G. LeGrand and W.V. Olszewski

Polymer Physics and Engineering Branch
General Electric Corporate Research and Development
Schenectady, New York 12301

ABSTRACT

Anelastic response in carefully prepared samples of isotropic, stress-free glassy polycarbonate is studied by stress-relaxation followed by simultaneous birefringence and strain-recovery measurements. The kinetics of relaxation and recovery are well-described by the Kohlrausch-Williams/Watts (KWW) decay law. The kinetic parameters are independent of thermal history and strain, while the unrelaxed and relaxed moduli are strong functions of both. The results are interpreted in terms of small regions of high-energy cis-trans carbonate conformational defects trapped in the glass during the temperature drop from above Tg. The kinetics of relaxation (KWW) is controlled by the distribution of free-energy barriers seen by the diffusing defects as they move toward frozen chain segments in their vicinity, while the amount of relaxation is determined by the total number of defects trapped-in from the melt or generated by the applied strain.

(I.) INTRODUCTION

Polycarbonate is an example of an engineering thermoplastic, a class of amorphous materials much in favor in automotive, appliance and structural applications requiring light-weight combined with high-modulus, toughness and heat-resistance. A consequence of the increased production and use of amorphous polymers in load-bearing applications is renewed attention to their mechanical properties, both at low-strain where creep, relaxation and fatigue are important, as well as in the high-strain failure range. Though considerable practical knowledge of the polymer solid-state exists, theoretical interest and understanding has developed very slowly until recently. A surge of activity directed towards such diverse disordered materials as spin glasses, metallic glasses, icosahedral crystals, amorphous Si, ceramics, dielectrics and polymers has helped to identify common features and research issues;

a. Broad distributions of relaxation times producing memory effects and hysteresis.

b. Non-equilibrium nature of the glassy state.

c. Non-linearity of stress/strain and other thermodynamic relationships.

d. Scarcity of suitable (reproducible) experimental data or data of sufficient accuracy and range to analyze and isolate [a-c].

e. Mathematical, analytical and computational complexity associated with theoretical models and data processing.

More serious, perhaps, than any of the above difficulties was the apparent lack of general theoretical methods to describe complex, highly-random systems possessing both structural and energetic disorder. One purpose of this conference is to describe theoretical progress achieved in several areas of condensed matter physics as a consequence of systematic study of "models of disorder", such as percolation, scaling and fractals. Certainly the random-walk model of polymer chain configurations in solution and bulk is one of the best examples of a successful theory of spatial molecular disorder. The time-dependent generalization, reptation, has in turn gone a long way towards rationalizing large-scale chain motions in solution and melts. Nevertheless, few attempts to extend reptation to the glass transition or solid-state have been made since the basic microscopic motions in the glass are almost certainly local rotatory-oscillatory events with only occasional segment or conformer diffusive motion. The absence of an approximate model for the solid-state spectrum has been a severe obstacle to progress since all glass-state properties are time- or frequency-dependent.

Fortunately, this situation has improved, due largely to the discovery that the fractional exponential relaxation function

$$\phi_\alpha(t) = e^{-(\frac{t}{\tau})^\alpha} \qquad 0 < \alpha < 1 \qquad (1)$$

provides a useful empirical description of many relaxation phenomena in the glass state of polymeric and non-polymeric materials. The function $\phi_\alpha(t)$ was first introduced by R.Kohlrausch [2] and in recent times by Williams and Watts [3] in connection with dielectric relaxations. Moynihan [4], Struick [5], Patterson [6], Ngai [7] and others have shown its application to thermal, optical, magnetic, electrical and mechanical phenomena. Ngai [7], Shlesinger and Montroll [8], Bendler and Shlesinger [9], and Palmer et. al. [10], have discussed theoretical models. All data does not fit the Kohlrausch-Williams/Watts function (KWW) of eq 1 very well, and Jonscher, Dissado, Hill and co-workers [11] and Macdonald [12] discuss alternate functional forms. Klafter and Shlesinger [13] have recently shown the mathematical relationship between three quite different physical models which lead to KWW. In particular, it seems that a common ingredient is a scale-invariant set of relaxation times in the problem. Bendler and Shlesinger [9,14] discuss how defect hopping over a distribution of free energy barriers can generate such a relaxation-time distribution in a disordered glass.

The present article reports stress relaxation and strain/birefringence recovery data obtained on amorphous polycarbonate at room temperature nearly nearly 130 °C below the glass transition, T_g = 149 °C. The work was initiated in an effort to learn whether birefringence recovery kinetics was related to mechanical recovery in polycarbonate, and if either could be modeled using the KWW decay law. In addition to providing positive answers in both cases, the results show a surprising lack of dependence of glass-state kinetics on thermal history or strain, and a strong dependence of relaxation strength on both. This leads us to propose that relaxation in the glass is dominated by small, mobile regions of high-energy cis/trans carbonate conformer defects, the volume fraction of which vary with thermal history and strain, their intrinsic mobility remaining largely unaffected.

The organization of the paper is as follows. The experiment and sample preparation methods are described first. Next, the results and analysis using the KWW form are presented. Finally, an interpretation and discussion are given in terms of a distribution of defect-rich and defect-poor regions of the glass.

(II.) EXPERIMENT

We have used samples of bisphenol-A polycarbonate cut from a large sheet which was carefully press polished between glass plates under vacuum at 170 °C for five hours in an air autoclave and then slowly cooled in the autoclave to room temperature over 18 hours. This material (which we designate "press-polished" (PP)) had a residual birefringence of ~ 1x10^{-7} and a residual thermal stress of 0.1 psi. Sections were cut from this sheet and then thermally quenched from 148 °C to 0 °C. Subsections from these quenched materials were annealed at 125 °C for 36 hours. The recovery experiment is illustrated in Figure 1, and is similar to one described by Kohlrausch [15]. The method consists of imposing a fixed uniaxial strain , ε_h , for a hold time , t_h , while monitoring the force. At the end of the hold time t_h , the sample is completely unloaded and both the residual strain and birefringence are measured as functions of recovery time. The strain-optical coefficients for samples of all thermal histories are the same up to strains of .035 with an estimated error of less than .5%. Multiple tests at the same strain and for the same hold times on these materials indicated an experimental variance ≤ 5%.

(III.) RESULTS AND FITTING TO KWW

As discussed below, it was thought that polycarbonate would be a good example of a defect-dominated glass (owing to the simple, sharp backbone conformational distribution), so that the strain and birefringence recovery results were analyzed using the Kohlrausch form;

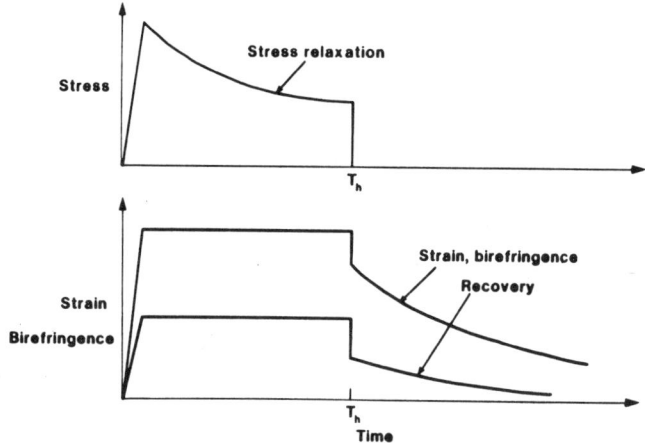

Figure 1. Illustrating stress/time, strain/time, and birefringence/time sequences in the recovery experiment. At time t=0, the strain is taken from zero to some specified hold value. The birefringence follows a parallel path to a constant value. During the hold period, stress relaxation takes place as shown at the top of the figure. At the end of the hold period, T_h, the load is released, the stress falls to zero and the decay of the strain and birefringence is monitored.

TABLE I

Values of the various parameters

Material	History	Strain	Hold Time (sec)	α	τ (sec)	$E_u - E_r$ (GP)	E_r (GP)
BPAC	Press Pol.	.01	5000	.15	400	.208	2.30
		.02	5000	.15	400	.264	1.96
		.03	5000	.15	400	.365	1.64
	Quenched	.01	5000	.15	400	.704	1.98
		.02	5000	.15	400	.864	1.67
		.03	5000	.15	400	1.441	1.24

$$x_{ne}(t) = x_0 \exp[-(\tfrac{t}{\tau})^\alpha] + x_p \tag{2}$$

where t is the recovery time since load release. x_{ne} is the non-elastic component of x, that is, the part of x which does not return instantly to zero when the load is removed. x_p is the "persistent" part of x which recovers (if at all) on a longer time scale than x_0.

The kinetic parameters α and τ evaluated from both strain and birefringence recovery were found to be the same to within experimental error, and are presented in Table 1. α and τ were also found to be independent of hold time, t_h, strain and thermal history. Figures 2 and 3 show typical recovery data from which α and τ were found.

The relaxation modulus during the strain hold (see Figure 1) was taken to have the form;

$$E(t) = E_r + (E_u - E_r) \exp[-(\tfrac{t}{\tau})^\alpha] \tag{3}$$

Values of $E_u - E_r$ and E_r were evaluated by substituting the kinetic parameters α and τ obtained from the recovery experiments into the rhs of eq 3, and setting this equal to the measured modulus at different times. These results are also presented in Table 1.

Figures 4 and 5 show fits of eq 3 to ε = 2% stress relaxation data for a press-polished and a quenched sample of polycarbonate. The figures illustrate the quality of the KWW fitting as well as the influence of thermal history on $E_u - E_r$ and E_r, and hence on the steepness of the relaxation curves. α and τ in Figures 3 and 4 are identical, as also seen in Table 1.

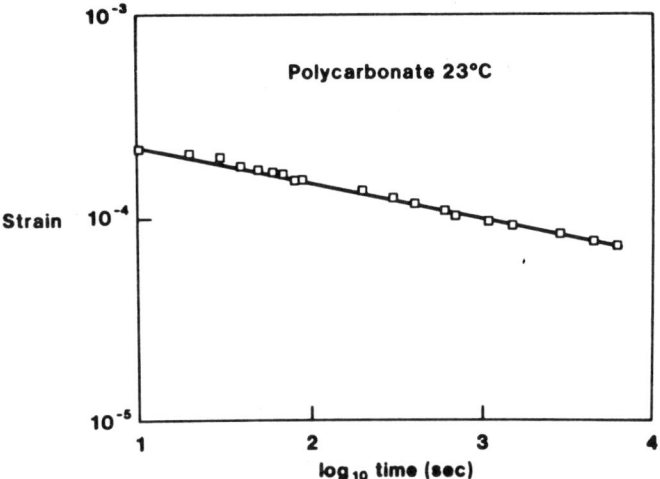

Figure 2. Time dependence of strain recovery of polycarbonate following load release.

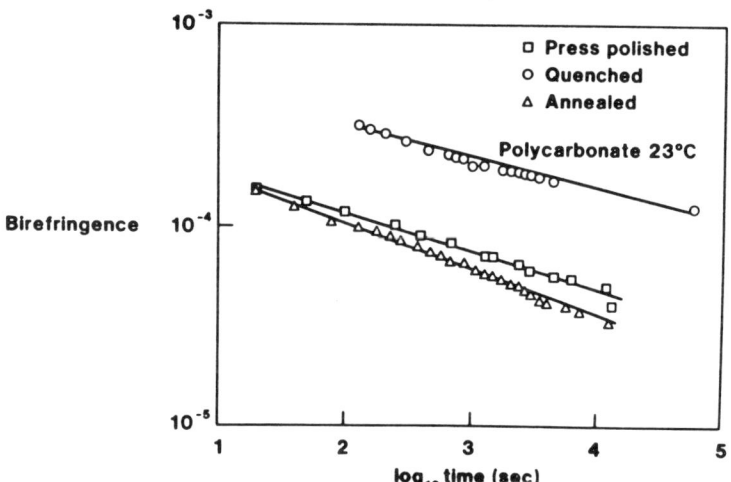

Figure 3. Time dependence of birefringence recovery of polycarbonate following load release for three different thermal histories.

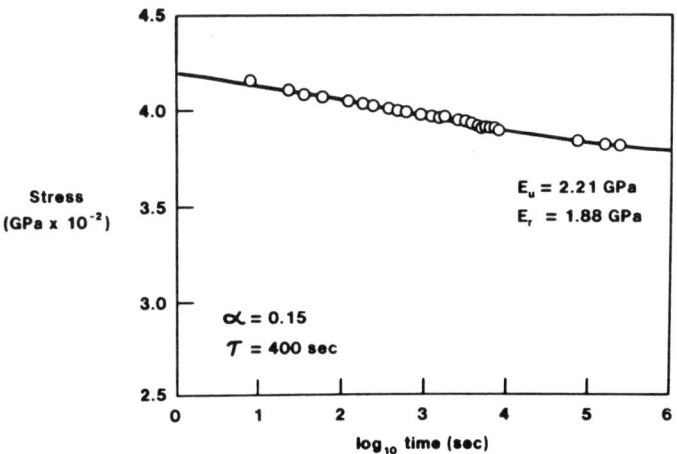

Figure 4. Stress relaxation of press-polished polycarbonate at 23 °C.

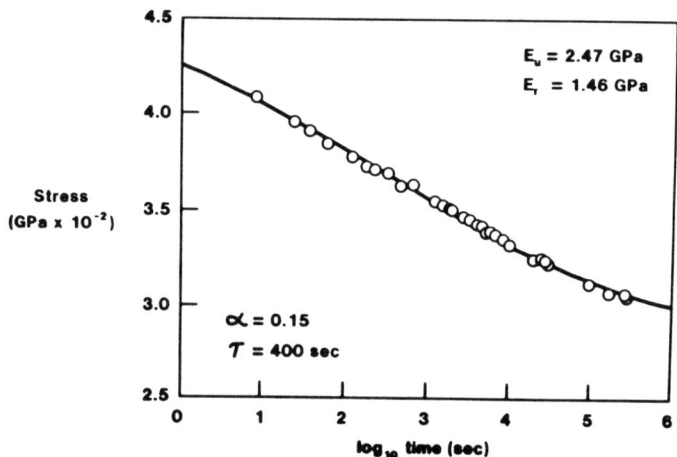

Figure 5. Stress relaxation of quenched (149-0 °C) polycarbonate at 23 °C.

(IV.) DISCUSSION

The defect diffusion model of the KWW decay law [8,9,17] is an attractive one in the case of polycarbonate since there is computational evidence [18] and NMR, dielectric and mechanical relaxation data [19] indicating that a single conformational defect, a high-energy cis/trans carbonate linkage, persists in the solid and leads to the prominent low-temperature loss peak at -100 °C. In addition, Jones has shown that the cis-defect may diffuse along the polymer chain with little backbone distortion [19]. It is therefore gratifying, but not unexpected, to find that all three sets of data (i.e., stress relaxation, birefringence and strain recovery) are described by KWW, with the same α and τ. It is worthwhile to emphasize that the exponent α reported in this work is nearly identical to that seen in the NMR and mechanical loss [19], though the τs vary by six orders of magnitude.

Of course, the novel feature of the results presented in Table 1 is the insensitivity of the kinetic parameters to either thermal history or strain. The large differences seen in the course of quenching and straining to different levels appears to be caused by strong shifts in the relaxed modulus. We interpret this behavior as coming about through changes in the population of cis defects brought about by quenching or strain. At high temperature in the melt, the cis defect ought to be rather plentiful since it has only 1,100 calorie/mole greater energy than the trans form. Upon cooling through Tg, most of the cis conformers are converted to trans though a few may be trapped and lead to isolated "hot-spots" or soft regions in the solid. When the solid is strained, these soft regions are able to undergo much larger distortion than the all-trans "hard" regions, and we propose that the anelastic response is chiefly due to the extra distortion of the defect-rich regions. The volume fraction of defects in the glass should change with the strength of the quench, since a fast quench will trap a larger number. If the soft regions are also able to grow in the presence of a strain bias, this would account for the strong non-linearity seen in Table 1. We are presently undertaking additional experiments to confirm and test these ideas, and quantify the effects of cooling rate and strain.

In summary, we find that the defect diffusion picture of glass relaxation provides a plausible (and predictive) explanation for our mechanical and optical relaxation experiments on polycarbonate. It accounts for the KWW form of the relaxation kinetics, the variability of the relaxation intensity with sample preparation history and strain level, and is in accord with NMR, dielectric and mechanical spectroscopy results. The defect proposed is one which ab initio Hartree Fock calculations suggest as most likely. Additional experiments are underway to examine the influence of temperature, molecular weight and strain rate. Also, simulations of strain-rate jumps, yielding, and creep are in progress.

REFERENCES

1. For example, see " Structure and Mobility in Molecular and Atomic Glasses", J.M. O'Reilly and M. Goldstein, Eds., Ann. N.Y. Acad. Sci. 371, 1981.

2. R. Kohlrausch, Pogg. Ann. Phys. 91, 198 (1854).

3. G. Williams and D.C. Watts, Trans. Faraday Soc. 66, 80 (1971).

4. C.T. Moynihan and A.V. Lesikar, Ann. N.Y. Acad. Sci. 371, 151 (1981) and

references therein.

5. L.C.E. Struik, Physical Aging In Amorphous Polymers and Other Materials, Elsevier Scientific Publishing Company, Amsterdam, 1978.

6. G.D. Patterson, Adv. Polymer Science 48, 127 (1983).

7. K.L. Ngai, Comments Solid State Physics 9, 127-147 (1979).

8. M.F. Shlesinger and E.W. Montroll, Proceedings National Academy Sci., (U.S.A.) 81, 1280 (1984).

9. J.T. Bendler and M.F. Shlesinger, Macromolecules 18, 591 (1985).

10. R.G. Palmer, D.L. Stein, E. Abrahams, and P.W. Anderson, Phys. Rev. Lett. 53, 958 (1984).

11. A.K. Jonscher, J. Mater. Sci. 16, 2037 (1981); L.A. Dissado and R.M. Hill, Proc. R. Soc. Lond. A 390, 131 (1983); R.M. Hill and L.A. Dissado, J. Mater. Sci. 19, 1576 (1983).

12. J.R. Macdonald, J. Appl. Phys. 58, 1955 (1985) and references therein.

13. J. Klafter and M.F. Shlesinger, Proc. Natl. Acad. Sci. (USA), 83, 848 (1986).

14. J.T. Bendler and M.F. Shlesinger, Bull. Amer. Phys. Soc. 30, 583 (1985).

15. F. Kohlrausch, Pogg. Ann. Phys. 119, 352 (1863). Also as Described in J.F. Bell, "The Experimental Foundations of Solid Mechanics", Handbuck der Physik, VIa/1 (Springer-Verlag, Berlin, 1973), pp. 74-82.

16. J.T. Bendler, W.V. Olszewski, and D.G. LeGrand, Polymer Preprints 26, No. 2, 90 (1985).

17. J.T. Bendler and M.F. Shlesinger, "The Wonderful World of Stochastics," Eds., M.F. Shlesinger and G.H. Weiss, (North Holland, 1985) Chapter 1.

18. J.T. Bendler, Ann. N.Y. Acad. Sci., 371, 299 (1981).

19. A.A. Jones, Macromolecules 18, 902 (1985).

ON THE ORIGIN OF NON-EXPONENTIAL DECAY PROCESSES IN AMORPHOUS SYSTEMS WITH APPLICATION TO POLYMERS

Edmund A. Di Marzio and Isaac C. Sanchez
Institute for Materials Science and Engineering
National Bureau of Standards
Gaithersburg, MD. 20899

ABSTRACT

The KWW and hyperbolic decay laws are viewed as arising from a phase particle escaping potential energy wells. Each well gives exponential decay, $p(t,E)=(1/\tau) \exp(-t/\tau)$ with $\tau=\tau_0 \exp(E/kT)$, where E is well depth, but because of dispersion in well depths the resultant decay, $p(t)$, is non-exponential. The formula is: $p(t)=\int Q(E)p(t,E)dE/\int Q(E)dE$, where $Q(E)$ is the distribution of particles in wells. The formula is general since it arises simply from the ergodic theorem and the principle of detailed balance. When $\int Q(E)dE$ is the equilibrium partition function $p(t)$ is the after-effect function which is much like the KWW relation. When $Q(E)$ is exponential $p(t)$ has the hyperbolic form. This latter form which has as its domain of attraction the KWW form is expected in non-equilibrium, and additionally is consistent with minimum information. Application to polymer glasses is made. As the system is cooled the allowed phase points, as measured by the configurational entropy, become fewer in number and it becomes more difficult to move from one allowed point in phase space to another. A phase transition is viewed as a migration of phase points to a specific region of phase space, the migration being effected by a biasing potential. (In one dimension think of a random potential superimposed on a linear bias.) The biasing potential can be sufficiently strong to flush out the points from the potential minima, in which case the transition occurs, or the potential minima can increase in depth faster than the biasing potential as a function of supercooling so that there is trapping in the potential wells rather than flushing. In the first case, the larger the supercooling the faster the system moves towards equilibrium; in the second case the larger the supercooling the more tightly the system is locked into the wells. This latter feature results in "phase locking" into a non-equilibrium phase generally and for polymers can result in gel formation or glass formation. The ideas are applied to polymer glasses via a minimal model that incorporates both chain stiffness and accessible volume ideas.

1. INTRODUCTION

Polymeric materials display non-exponential long time relaxation behavior. It is the purpose of this paper to understand why this happens. Now the point should be immediately made that many non polymeric materials also display non-exponential long time relaxation. It would seem that a proper procedure would be to first understand the phenomenon in some general way so that the manifestation in polymers would be understood as a simple application of some general principles. Unfortunately there is no unanimity in understanding. At least 7 different explanations exist[1] as to why the stretched exponential form, $\exp(-(t/\tau)^{\alpha})$, describes long time relaxation. Before approaching the polymer problem logic requires that we decide which if any of the existing explanations are to be used.....or devise an alternate explanation. Actually what we shall try to show is that a very simple idea advanced previously[2] is the basic reason behind non exponential long time relaxation.

The crux of the idea is that the principle of detailed balance
determines the residence time of a given state of the system (the
persistence time of a phase point in a potential well in phase space).
This, plus the fact that the distribution function for the residence time
of a phase point in a well in phase space is exponential in time, and the
fact that there is a distribution of well depths in phase space,
necessarily results in non exponential decay[2]. Certain reasonable choices
in the well depth distribution result in distributions similar to the
Kohlrausch Williams Watts (KWW) distribution. This idea is an obvious
variant of the trapping models[3], the only difference being that the
particles in trapping models are conventional ones like electrons or
molecules while our particles are phase points.

Because of the large amount of recent literature the remainder of
this Introduction will serve to place the paper in a proper context.

In 1970 Williams and Watts settled on the fractional exponential as
an accurate representation of dielectric response[4].

$$\frac{\epsilon(\omega)-\epsilon(\infty)}{\epsilon(0)-\epsilon(\infty)} = - \int_0^\infty \exp(-i\omega t)\frac{d\phi(t)}{dt} dt \qquad (1)$$

where the autocorrelation function $\phi(t)$ is given by

$$\phi(t) = \exp(-(t/\tau)^\alpha), \qquad 0 < \alpha < 1 \qquad (2)$$

Values of the index α between .2 and .8 generally give good fits with
experimental data. For some materials Eq. (2) results in good fits to
$\epsilon(\omega)$ over 5 orders of magnitude,

But for other materials the stretched exponential is not a good fit at all. This is seen immediately for composite materials since $\phi(t)$ does not form a group under composition. We can clarify this statement by a specific example. Suppose our composite materials were constructed so that the total response ϕ were a sum

$$\phi = (\sum \phi_i v_i)/(\sum v_i) \tag{3}$$

where v_i is the amount of material of kind i. Then it is obvious that ϕ is not a stretched exponential if ϕ_i is, unless of course each ϕ_i had the same α dependence. ϕ would be a stretched exponential only if the ϕ_i combined as a product but there is no composite material that combines responses in this way.

For a homogeneous, amorphous polymer above the glass transition the KWW form can be a good fit to data. However, there is at least one case[5] where the sum of 2 KWW forms is needed to fit data close to the glass transition even though only one KWW form was needed farther from the glass transition.

In 1863 F. Kohlrausch proposed the identical form for mechanical relaxation in stretched fibers of various materials[6]. The same form had been used earlier (1847) by R. Kohlrausch for the residual charge in a capacitor[7]. We shall use the symbols KWW to describe the form of Eq. (2) no matter what the kind of response. The question immediately arises as to whether all response functions are of the KWW form.

This question has an immediate answer for the case of the complex modulus G and the complex viscosity η. Since[8] $G = i\omega\eta$ we have the result that if G is of the KWW form then η is its time derivative which is not of the KWW form.

Another relationship between response functions is

$$[\varepsilon(\omega)-\varepsilon(\infty)]/[\varepsilon(0)-\varepsilon(\infty)] = 1/(1-i\omega\zeta(\omega)) \quad , \quad \zeta = \eta(\omega)K \quad (4)$$

where ζ is the friction coefficient for a spherical bead in which a dipole is imbedded. The above equation is exactly like the Debye equation[9] except that the zero frequency viscosity is replaced by the complex viscosity. This equation was derived in 1974[10]. The point is that within the assumptions of the model there is a unique relationship between the normalized dielectric constant and the viscosity and if one obeys KWW then the other does not. Of course Eq. (4) implies that the reciprocal of the normalized dielectric constant and the modulus ($G=i\omega\eta$) do have the same KWW behavior. The model leading to Eq. (4) is that of a dipole fixed to a sphere which can rotate in a viscous fluid. Generalizing the model will result in a connection between $\varepsilon(\omega)$ and $\eta(\omega)$ other than Eq. (4) but will still force $\varepsilon(\omega)$ to be non-KWW if $\eta(\omega)$ is KWW. The general rule is that one does not have KWW behavior for all relaxation functions. However, algebraic response functions (susceptibilities) may well form a group under operations like Eq. (4).

It is possible for more than one relaxation function to obey the KWW form. For example, the photon correlation function obeys the KWW form for six orders of magnitude[11]. This is related to the fact that the dielectric relaxation function also obeys the KWW form. As Lindsey and Patterson observe the photon correlation function $C(t)$ is essentially the square of the dielectric relaxation function[12]. ($C(t) = 1 + \phi^2(t)$). It is because the relaxation functions are not in general multiples of each other that they can not all have the KWW form.

Other arguments against assigning a deep significance to the KWW distribution are that (1) because the curves are generally plotted on a log-log scale or at least on a semi-log scale there is a compression of data. (2) there is so little structure to long time relaxation curves that many other 3 parameter functions would also fit the data equally well.

The above arguments do not preclude the KWW form in all cases and since we know from experiment that it does occur we now inquire how this might be. The KWW (stretched exponential) form has an important place in probability theory as the transform of a stable distribution and in statistics as the Weibull distribution. Since the discovery of the applicability of the KWW form to physics predates the mathematical discoveries by about 75 years it is proper to ask if an explanation in terms of some general mathematical law is possible. The suspicion is that the same kind of statistics of compound events leading to the stretched exponential in probability theory also lead to the stretched exponential in physical applications.

Although little used in the natural sciences the Weibull distribution is widely used in statistics with over 1,400 literature references[13] as of July 1984. It provides a remarkable correlation of data in widely divergent areas such as size distribution of fly ash, and coal dust, yield strength of steel, fiber strength of cotton fibers, statures for adult males in the British Isles, length of crytoideae, etc.; to paraphrase Weibull[14]. Now the only justification advanced to account for the Weibull distribution is the "a chain is only as strong as its weakest link" argument. A chain is said to fail under a load x when any one of its links fails under the load. If P is the probability of failure of a link and P_n the probability of failure of the chain then

$$1-P_n = (1-P)^n, \text{ or } P_n = 1-\exp(-n\psi(x)) \text{ if } P = +\psi(x) = \int_0^x \psi(x)dx \qquad (5)$$

Whenever the probability of survival is a product of probabilities the above relation holds. Cumulative damage, electrical insulation breakdown where breakdown of any one of many parallel paths constitutes breakdown of the whole, the death of an organism whose survival is dependent on the survival of its many essential parts and functions, are other examples. The choice of ψ to be algebraic in x is made to fit the data. Notice that P_n need not be small even though P is small. This means that the asymptotic properties of P for long time (if time were our variable of interest) may well determine the properties of P_n in the non asymptotic range. This insight may enable one to avoid an objection raised by Zwanzig[15] to a paper by Palmer et al.[16]. Zwanzig correctly observed that many of the asymptotic arguments for the KWW form are valid only when the times are so large that the KWW form is reduced to less than 1% of its t=0 value, but the experimental observations show that the KWW form is a good fit to data in the 90% to 1% range. The above weakest link argument obviates the objection, but does not relieve one from the obligation of proposing an actual molecular model.

Another point about the Weibull distribution concerns its wide applicability. Perhaps the reason no one has found a theoretical justifications for it is simply that there isn't any. After all, the data is expected to be monotonic and is not highly structured, and the Weibull distribution has three adjustable parameters. With three adjustable parameters the general but not precise weakest link argument may have sufficient physical basis to allow a reasonable fit to the data.

Another, perhaps more pregnant, rationale for the applicability of the KWW form is provided by probability theory. Now the stretched exponential is the transform of a stable distribution (but only for $0<\alpha<2$). A stable distribution is one which when folded into itself gives the self same distribution[17] (simple definition). For example, the Gaussian distribution with index $\alpha = 2$, the Cauchy distribution with index $\alpha = 1$ and the log normal distribution with index $\alpha \to 0$ are all stable distributions. Now a non-stable distribution with finite variance will, when folded into itself many times, result in a Gaussian distribution for the sum of the independent variables ($S = \sum x_i$). The non-stable distribution is said to belong to the domain of attraction of the Gaussian distribution which is a stable distribution of index 2. This is a restatement of the central limit theorem. A valid generalization of the central limit theorem is that a given probability distribution many be within the domain of attraction of a stable distribution with index α. When such a distribution is folded into itself many times, it approaches a stable distribution of index α. The totality of indices α from 0 to 2 does not, however, label all domains of attraction. There exist domains of attraction that are not centered on the stable distributions[17].

The suggestion is that just as the Gaussian distribution can be the result of a sequence of physically meaningful independent random variables so too the KWW distribution is the result of a sequence of physically meaningful random variables. The only difference being that in one case the variance is finite while in the other it is infinite. On the face of it, it would seem that since there are physically meaningful situations where the Gaussian distribution appears there should also be physically meaningful situations where the more general KWW distribution appears. A transcription of the mathematics of Levy functions to physics would imply the following (1) The variable conjugate to time is the relevant independent variable. (2) The probability distribution in this variable is the relevant probability distribution. (3) The existence of random variables that are sums of random variables. These 3 implications are realized by (1) using the energy as the random variable conjugate to time. (2) Using the Boltzmann distribution as the probability distribution. (3) The way total energy is distributed within the system.

We conclude that since there are cases for which Eq. (2) works well we should inquire into the bases for its functional form, although we can not really believe, in view of the totality of the above comments that there exists a universally valid equation. The existence of non-exponential long time relaxation is beyond question but the KWW form is not universally valid.

2. THEORY

The main point of this paper is that there is a very simple view of why long time tails occur in materials and that under very general conditions all materials will display long time tails for some of their properties. The specific KWW form, however, is not ubiquitous: the algebraic (hyperbolic) form and the after effect function each are appropriate forms under certain conditions. Elementary statistical mechanical arguments can be used to justify these statements as follows.

If we integrate over momentum variables then the part of the classical partition function remaining can be discussed in terms of the phase space diagram (more accurately, configuration space diagram) of Fig. (1). Horizontal slices represent all of the position coordinates while the vertical axis gives the value of the potential energy of a system. If the potential energy of the system has a given value the phase particle moves in a horizontal plane but when kinetic energy feeds into the potential energy the phase point can move with a vertical component. Also, when the system energy changes through coupling to the radiation field or to particles of the heat bath then the system can change energy and the phase particle can move vertically. An important observation in this picture is that a phase point can be in a potential well, as is illustrated in Fig. (1), and that for these particles the process of flow involves the escape from the well. Because of the potential energy barrier horizontal motion is vibratory and does not result in flow except at the top of the barrier. This classical picture also serves to introduce the quantum mechanical problem since all that is required for Schrödinger's equation is a potential energy as a function of the particle coordinates. We use both languages interchangeably.

The ergotic theorem gives us the fraction of time, f_i, that a system spends in state i.

$$f_i = \exp(-E_i/kT) \tag{6}$$

where E_i is the energy of state i and T the temperature. However Eq. (6) tells us nothing about how fast the system is moving from state to state, or what is the residence time, or what is the renewal time for state i. We can extract some information on these questions from the principle of detailed balance.

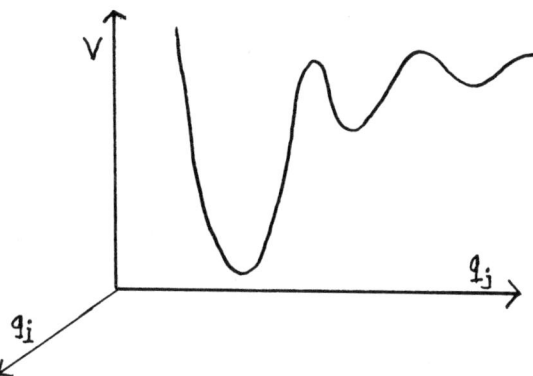

Fig (1). System potential energy vs. the configuration variables q_i. The number of potential minima decreases sharply at lower V. At low temperatures a point in configuration space spends most of its time vibrating in these deep minima. The exchange between potential and kinetic energy and coupling to the radiation field allows for vertical motion. Flow corresponds to motion of the point from well to well. Calculation of viscosity is equivalent to a determination of the geometry of the configuration potential; both as to the number of wells at level V and to their connectedness. The non-exponential decay of our system arises from a distribution in well depths. Depending on how these wells are sampled one can obtain the algebraic, the stretched exponential or the after-effect distribution laws, as well as others.

$$\alpha_{ij}/\alpha_{ji} = \exp(-[E_j-E_i]/kT) \qquad (7)$$

where α_{ij} is the probability of going from state i to j.

We now must discuss the α_{ij}. The first problem has to do with the apportioning of the energy difference between states, E_j-E_i, into the forward and backward reactions. We shall assume that the energy appears only in that direction for which it is a barrier. This is very natural to a picture where particles are in wells and the top of the well corresponds to an activated state. Thus

$$\alpha_{ij} = b(i,j) \qquad (8)$$

$$\alpha_{ij} = b(i,j)\exp(-[E_j-E_i]) \quad E_j > E_i \qquad (9)$$

The b(i,j) do depend on i,j and are generally non-zero when the corresponding wave functions overlap.

We are now in a position to derive the long time relaxation behavior. A particle in state i that surmounts a barrier to find itself in state j will do so in a characteristic time τ given by $1/\tau=\alpha_{ij}$. The normalized probability density distribution, $p(t,\tau)$, for this process is

$$p(t,\tau) = (1/\tau)\exp(-t/\tau), \qquad (10)$$

$$1/\tau = b(i,j)\exp(-[E_j-E_i]/kT) \qquad (11)$$

Equation (5) assumes exponential decay. We have used this form because a deep square well and a ramp potential[18] both give exponential decay, and because our attempts to pick a deep potential that has anything but exponential decay has always ended in failure. We believe that as long as the (deep) potential is monotonically increasing as one leaves the deepest part the probability for escape is exponential in time. A recent paper by deGennes bears on this question[19]. Equation (9) is a conditional probability. To obtain the total probability density distribution for a particle in state i going to state j we must multiply $p(t,\tau)$ by a weighting factor Q. Since Eq. (10) is normalized this probability is just the probability that state i is occupied. Let us now consider that the state j is an activated state, i.e. that it allows flow. Our picture is that a phase point in a well corresponds to no flow, but only vibrations. Once the particle escapes the well and is in state j at the top of the well, then there is a flow. We shall assume that the j state is one of the lowest energy states that is still at the top of the well. Even higher energy states will also be good states for flow but we shall argue that because of their higher energy they are not sampled often enough by the system to be important. The probability that particles in wells escape to state j and result in system flow is then

$$p(t,j) = \int Qp(t,\tau)di / \int Qdi \qquad (12)$$

The integral in the denominator serves to normalize $p(t,j)$. In order for the above formula to apply we imagine that our system has settled down sufficiently so that for a particle, every which way is up. This means that the temperature is sufficiently small that a phase point is virtually always in a well. The system flows only during the short periods of time that a phase point has escaped a well and is in the activated state j. After a short time the phase particle finds another well and settles in for another long time (compared to its flow time). The above picture will not hold at high temperatures when the phase particle spends a large portion of its time out of the wells.

Our observation is that for sensible values of Q, $p(t,j)$ shows algebraic behavior or KWW like behavior.

To obtain KWW like behavior we identify $\int Qdi$ with the partition function and Q with the Boltzmann exponential (weighted by the degeneracy factor). It is reasonable to suppose that the state i is visited according to the Boltzmann law. Our assumption is obviously equivalent to the instantaneous entropy assumption of fluctuation theory[20]. This means we are not, at this stage, allowing our system to fall out of equilibrium as certain other theories do[21]. We can expand Q about the minimum in free energy so that Q becomes a Gaussian function of the i variables. Symbolically, if i is a vector and A a matrix then

$$Q = \exp(-i^\dagger A i) \tag{13}$$

We can transform from the n variables of the i type to a set which contains E and n-1 other non energy variables. Under certain conditions on the Jacobian of the transformation we can suppress the other variables to obtain,

$$p(t,j) = \int \exp(-aE^2 - E/kT - bt\exp(-E/kT)) \, dE / \int Q \, dE \tag{14}$$

One recognizes this to be the after effect function tabulated by Jahnke and Emde[22]. As Fig. (2) shows this function behaves very much like the KWW form $\exp(-at^\alpha)$.

If $Q(E) = \exp(-\alpha E/kT)$ we obtain

$$p(t,i) = 1/t^{1+\alpha} \tag{15}$$

an expression which has been used many times for the pausing time distribution in the continuous time random walk model. An interesting observation is that the exponential form for $Q(E)$ can be obtained from information theory. If we retreat a bit in our thinking and accept that the only certainties are the fact that each system is in some state i and that the average energy of the systems is known, that is to say

$$\sum n_i = n \quad , \quad \sum n_i E_i = E \quad , \tag{16}$$

Then the exponential form is obtained by a simple application of information theory (maximum entropy principle). The information theory approach is justified only when other information is entirely lacking and

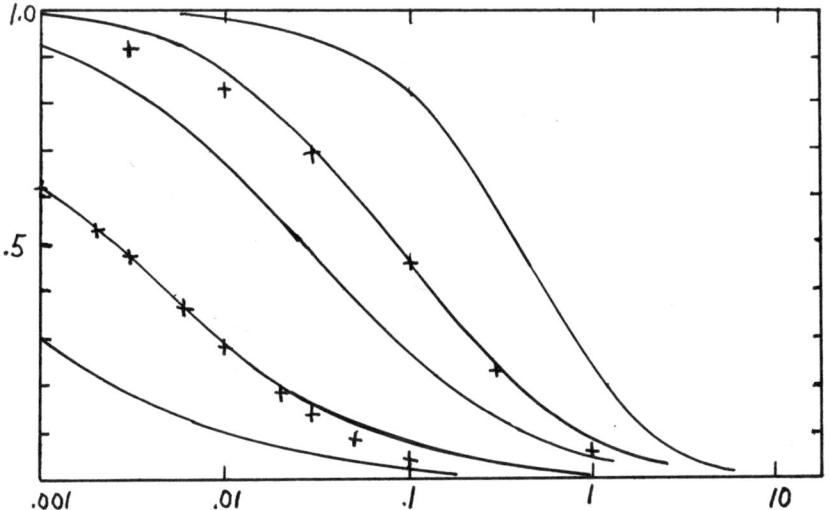

Fig. (2). The after-effect function (solid curves) and the KWW function (crosses). Each curve has the asymptotic values of one and zero. Also, we can make the curves as well as their derivatives coincide at any point. The similarity of the two functions is apparent from the figures.

in fact we can make a worst case scenario that this is so. The view that transition probabilities α_{ij} describe instantaneous jumps even for systems that flow infinitely slower than mollases may be wrong. And unless a reexamination of perturbation theory proves it valid for such slow processes then perhaps we should assume the least amount of information as in Eq. (16) above. It can also be argued that a physical law should not be assumed to hold to an infinite number of significant figures but only to as many as can be verified by experiment. We have no guarantee that the laws of statistical mechanics and in particular the Boltzmann law can predict probabilities so close to zero as to be unmeasurable.

Now the view of flow that we are discussing here is a variant of the hopping model. It is expected therefore that it will display long time tails just as the hopping model does. The hopping model and the continuous time random walk model are equivalent in the sense that the crucial equations are the same[23]; however the models are distinctly different and are definitely not equivalent! Many different physical models can result in one equation. In the same way our variant of the hopping model is not equivalent to the hopping models of solid state physics. First, our particle is a point in phase space rather than, for example, an electron in real space. The use of a reaction coordinate or a collective coordinate is a suitable way to generalize hopping models. Second, the particle need not continue in existence indefinitely. In the hopping model of electron transport the electron is made to traverse the whole sample, but in our model the particle needs to persist only long enough for the concept of surmounting a barrier to be meaningful. Since the interaction potential is necessarily not harmonic there can and will

be transitions to other states which means that the particles are
annihilated and created. Hopping models with annihilation and creation
are easily constructed.

A form of the hopping model which contains the ideas behind Eqs. (6)
to (12) in a perspicuous manner is to write

$$\dot{n}_i = \sum b(i,j)(n_j - n_i \exp((E_i - E_j)/kT)) \qquad (17)$$

which has as the obvious equilibrium solution

$$n_j/n_i = \exp(-[E_j - E_i]/kT) \qquad (18)$$

A recent paper by DeDominicis et. al. uses the master equation and
detailed balance to derive a KWW law[24].

3. APPLICATION TO POLYMERS

So far our discussion has been general and presumably of wide
applicability. We now apply our ideas to amorphous liquids and glasses,
particularly polymeric materials.

The glass transition in polymers has been studied extensively for many
years and more is known about polymer glasses than about glasses formed
from simple molecules. There are both experimental and theoretical
reasons for this.

On the experimental side, it is easier to form glasses from polymers.
Simple liquids tend to crystallize in spite of heroic efforts to keep them
amorphous while there are many polymers that can never crystallize and
always form glasses. Also there are more variables in polymeric systems.
Molecular weight, degree of branching or of cross-linking, copolymer

composition and amount of plasticizer do not have obvious analogues in monomeric systems. The experimental studies are thus much richer in polymeric systems.

On the theoretical side, it is known that the lattice model treated by mean field methods is highly successful in predicting the thermodynamic properties of polymers. Polymer solutions, polymer mixtures, liquid crystals, block copolymers and amorphous polymers have all been successfully dealt with. For amorphous polymers configurational entropy, which can easily be calculated from the lattice model, has been shown to be heavily involved in determining the properties of glasses[25]. The observation that the glass temperature occurs when the configurational entropy reaches a critical value as we cool has allowed us to successfully predict the glass temperature of polymers as a function of 1) molecular weight, 2) plasticizer content, 3) copolymer composition, 4) pressure, 5) stretch ratio for a rubber, 6) cross-link content for a rubber, and 7) blend composition. In addition the lattice model predicts the thermodynamic properties with reasonable accuracy. For example, the specific heat discontinuity can be predicted. The predictions are no-parameter predictions!

Because of the excellent experimental confirmation in the above 7 classes of systems we shall take the view that the glass temperature in polymers occurs when the configurational entropy decreases to some critical value. The qualitative explanation is as follows. The number of configurations (distinct arrangements of the system of molecules) available to a system decreases as we lower the temperature, for 2 reasons. First, according to the isomeric state model of a single chain, lower temperatures favor the lower energy shapes; in the limit of zero

temperature only the lowest energy conformation will occur. Second, in
the context of the lattice model, the number of distinct arrangments due
to the permutations of polymers and empty lattice sites (and solvent
molecules) decreases as the number of holes or solvent molecules
decreases. Thus as we lower the temperature the number of allowed points
in phase space gets smaller and smaller and the separation between them
gets farther and farther apart. Flow, which measures the rate at which
the phase particles move from one allowed point to another, becomes more
and more difficult. The viscosity then bears an inverse relation to the
configurational entropy. The above statements are made in the context of
the microcanonical ensemble. In the context of the canonical ensemble
all phase points are allowed but those of high energy are rarely visited.
This adds to the above picture the idea that the regions between allowed
points are of higher energy so that each allowed point lies in a well from
which it must escape in order to have flow. This picture of glasses is in
complete accord with the description of the motion of phase points given
in the previous parts of this paper (see Fig. (1)). We therefore conclude
on qualitative grounds alone that glasses show non-exponential long time
tails. The remainder of this paper is an attempt to quantify the above
simple pictorial description.

Some recent work by Stillinger on simple liquids bears directly on
the question of the geometry of the potential surface[26]. By means of
molecular dynamics he was able to show that the energy of the system as a
function of the configurational part of phase space consists of hills and
valleys with many local minima. The density of minima in the Energy vs.
configuration plot looks like "a portuguese man of war in configurational
sea", thus justifying as an approximation the quadratic density profile

leading to Eq. (14). Thus at low temperatures simple fluids behave in much the same way as described above. There is an added insight in the observation that the motion of the phase point corresponds in most cases to motion of one atom and its nearest neighbors; all other atoms being held fixed. An observation of Helfand and Weber on polymers is obviously related to this[27]. They show that a coordinated motion of 2 successive monomer units in the chain can result in chain motion without any significant motion of any other units. Thus the fundamental flow process in polymers may also involve only local motions rather than a cooperative motion of many monomers. Adam and Gibbs postulated that the fundamental unit of flow is some small number of spatially contiguous monomer units but that the size of the flow unit is determined by its configurational entropy[28]. Finally the original Turnbull-Cohen picture of flow also involves motion in a localized region[29]. These 4 separate investigations then suggest that flow can be understood as a process occurring in a small part of the system, and is therefore ammenable to treatments of several variables only.

Let us now reexamine our estimate of the number of deep wells as a function of energy. Although each allowed configuration represents, in the lattice model, a local minimum it can by no means be said to be a deep minimum since an adjacent configuration obtained by moving one bead into an adjacent vacant site will not result in a very different energy. The number of deep minima is but a small fraction of the total for each energy E, and in absence of contrary evidence we shall assume this fraction is independent of E. Because Q appears in both the numerator and denominator of Eq. (9) the proportionality constant does not appear in the final expressions.

However we can go further. Temperature allows us to determine which of the wells are deep. At high temperatures all wells are readily accessible so that the ones of lower energy are masked. As we lower the temperature only the lower energy wells are accessible, and near the glass temperature all the accessible wells are deep wells. This means that $Q(E)$ itself is the proper weighting function.

Our zeroth order theory then is very simple. We imagine a part of the system to be imbedded in the remainder which we take as a constant pressure- constant temperature reservoir. We simply use for the distribution function Q

$$Q = \frac{\exp(-(E-\langle E \rangle)^2 / 2\langle(E-\langle E \rangle)^2\rangle)}{(2\pi \langle(E-\langle E \rangle)^2\rangle)^{1/2}} \tag{19}$$

where[20]

$$\langle(E-\langle E \rangle)^2\rangle = kT^2 N \Delta c_p - kT^2 PV\Delta\alpha + kTP^2 V\Delta k_T \tag{20}$$

The reason for the deltas in Eq. (20) is that we are concerned only with the energies involved with the configurational part of the partition function. Accordingly we subtract out, as base line, the vibrational contributions.

The problem really requires a more comprehensive treatment of the geometry of the potential in configuration space than we have given here. A useful first step in the context of the lattice model might be to consider the well structure in terms of f and n_o, the fraction of bonds flexed and the number of empty lattice sites[25]. This is left for future work.

Notice that the size of our system enters linearly in the denominator of the argument of the exponential of Eq. (19). This is consistent with the idea of a characteristic size required for the flow process. The larger the size the smaller the coefficient a in Eq. (14) which means the farther the deviation from simple exponential behavior.

So far we have not reproduced the KWW form but only 2 other candidates, Eqs. (14) and (15). However the hyperbolic distribution function of Eq. (15) has been shown by Shlesinger and Montroll to lead to the KWW relation for dielectric response if reasonable physical assumptions are made[30]. Also there are many cases where the hyperbolic distribution is as good a fit to the data as the KWW distribution. Finally, as figure 2 shows, the after effect function is very close in its behavior to the KWW distribution.

Note should be made of a barrier model of Bendler and Shlesinger which obtains long time tails due to dispersion of barrier heights[31].

4. CALCULATION OF VISCOSITY

We have then in a general way explained non-exponential decay by following the progress of a phase point in configuraiton space. We must now, in order to proceed further, describe how the glass responds to the various experimental probes. This is the same problem as that of calculating the appropriate susceptibility or response function. Since the number of different response functions is large, being equal to the product of the number of generalized forces and the number of generalized responses, we will discuss only the one we believe to be the most basic, which is viscosity. It is basic because it measures the response to an applied shear field and also because other responses such as dielectric constant and photon correlation function depend on it. The nature of the relationship between ε and η has been discussed in the introduction and indeed we can view the Debye relation as the minimal model for the connection between ε and η. Similarly we can view the Navier-Stokes equations as a minimal model for the connection between the light scattering response of a fluid and the mechanical response. In both of these classes of theories one obtains a legitimate generalization by use of a frequency dependent, or time dependent, viscosity.

Thus we consider the calculation of the complex viscosity, $\eta(\omega)$ or alternatively $\eta(t)$. This is a very difficult problem, not even the molecular weight dependence of $\eta(t)$ is known from theory, or to say it more accurately, there is not a consensus in the polymer community that it is known; so we will content ourselves with a limited achievement. Since the viscosity and the configurational entropy are intimately related and since (in the lattice model) the number of holes and the number of flexes determine the entropy, we expect the number of holes and the number of flexes to be important determinants of the viscosity. Theories of the viscosity of simple liquids where there is no chain stiffness have been adapted to polymers, but the adaptations do not discuss the effects arising from the stiffness of the chains. We will limit ourselves in the remainder of this paper to describing the effect of chain stiffness on the viscosity of polymer glasses.

Probably the most simple model arises by adapting the Helfand-Weber observation to the Cohen-Turnbull model. We simply replace the total number of atoms by the total number of adjacent pairs of flexed bonds, which is proportional to f^2.

$$N \to Nf^2 \quad , \quad f = \exp(-\Delta\varepsilon/kT)/(1 + \exp(-\Delta\varepsilon/kT)) \qquad (21)$$

This modification results in an equation similar in form to the Macedo-Litovitz equation[32] which fits data for a wider class of materials than does the Cohen-Turnbull equation. It would be interesting to see if the values of ε that fit the viscosity data are close to the values of ε obtained from fitting to the glass transition. Another aspect of our modification is that the viscosity will show a change in slope at the second-order transition since according to the entropy theory of glasses[25] the volume and the number of flexes do.

References

(1) The contribution by A. K. Rajagopal and K. L. Ngai in Relaxation in Complex Systems, K. L. Ngai and G. B. Wright Editors, Naval Research Laboratory, Washington, D.C. 20375-5000, lists closer to 20 and this does not count works published after the fall of 1984.

(2) E. A. DiMarzio and I. C. Sanchez, Bull. Amer. Phys. Soc., 30, 583 (1985).

(3) G. Pfister and H. Scher, Advances in Physics, 27, 747 (1978).

(4) G. Williams and D. C. Watts, Trans. Faraday Soc., 66, 80 (1970).

(5) R. Ferguson, M. Lee, A. M. Jamieson and R. Simha, Bull. Amer. Phys. Soc., 30, 583 (1985).

(6) F. Kohlrausch, Pogg. Ann. Physik, 119, 352 (1863).

(7) R. Kohlrausch, Pogg. Ann. Physik, 91, 198 (1854).

(8) J. D. Ferry, Viscoelastic Properties of Polymers, 3rd Ed. (John Wiley and Sons, New York, 1980).

(9) P. Debye, Polar Molecules (Dover Publications Inc., 1929).

(10) E. A. DiMarzio and M. Bishop, J. Chem. Phys., 60, 3802 (1974).

(11) G. D. Patterson, C. P. Lindsey and J. R. Stevens, J. Chem. Phys., 70, 643 (1979).

(12) C. P. Lindsey and G. D. Patterson, J. Chem. Phys., 73, 3348 (1980).

(13) S. Eggwertz and N. C. Lind, Eds., Probabilistic Methods in the Mechanics of Solids and Structures (Springer-Verlag, N.Y. 1984)., page XVI.

(14) W. Weibull, J. Appl. Mechanics, 18, 293 (1951).

(15) R. Zwanzig, Phys. Rev. Lett., 54, 364 (1985).

(16) R. G. Palmer, D. L. Stein, E. Abrahams and P. W. Anderson, Phys. Rev. Lett., 53, 958 (1984).

(17) W. Feller, An Introduction to Probability Theory and Its Applications Vol. II, 2nd Ed. (John Wiley and Sons, N.Y. 1971).

(18) The methods of a recent paper by R. J. Rubin and G. H. Weiss, J. Chem. Phys. 78, 2039 (1983) is easily adapted to solve the problem of a particle in a ramp potential.

(19) P. G. deGennes, J. Statistical Phys., 12, 463 (1975).

(20) H. B Callen, Thermodynamics (John Wiley and Sons, N.Y. 1962).

(21) For example see G. H. Fredrickson and H. C. Anderson, Phys. Rev. Lett., 53, 1244 (1984).

(22) E. Jahnke and F. Emde, Tables of Functions, 4th Edition (Dover Publications, N.Y. 1945).

(23) F. W. Schmidlin, Phys. Rev. B16, 2362 (1977); Philos. Mag. B41, 535 (1980).

(24) C. deDominicis, H. Orland and F. Lainee, J. Physique Lett., 46, L463 (1985).

(25) J. H. Gibbs, J. Chem. Phys, 25, 185 (1956); E. A. DiMarzio, Annals. N.Y. Acad. Sci., 371, 1 (1981); E. A. DiMarzio, Relaxations in Complex Systems, K. L. Ngai and G. B. Wright Eds., (Naval Research Lab., Washington, D.C. 20375-5000 (1984)).

(26) F. H. Stillinger, Phys. Rev. B, 32, 3134 (1985).

(27) E. Helfand, Science, 226, 647 (1984).

(28) G. Adam and J. H. Gibbs, J. Chem. Phys., 43, 139 (1965).

(29) M. H. Cohen and D. Turnbull, 31, 1164 (1959); D. Turnbull and M. H. Cohen, Ibid. 34, 120 (1961); 52, 3038 (1970).

(30) M. F. Shlesinger and E. Montroll, Proc. Natl. Acad. Sci. U.S.A., 81, 1280 (1984).

(31) J. T. Bendler and M. F. Shlesinger, Macromolecules, 18, 591 (1985).

(32) P. B. Macedo and T. A. Litovitz, J. Chem. PHys., 42, 245 (1965).

CLASSICAL AND QUANTUM ANTS
IN A LABYRINTH

Takashi Odagaki
Department of Physics
Brandeis University
Waltham, Massachusetts 02254

ABSTRACT

The dynamic percolation in systems with static disorder is discussed for both classical and quantum processes. Differences in threshold, critical exponents and other properties are examined. The quantum percolation with tunneling effects is also studied.

1. Introduction

The concept of percolation was first introduced in 1957 by Broadbent and Hammersley.[1] The basic idea of the percolation approach is to understand the qualitative change in various physical properties of disordered materials by observing the underlying geometrical structure. In each application, a definition for the connection between some objects is assumed and the question is asked if there exists an infinitely connected channel in the system. Various physical properties are supposed to change their nature at the critical value of a parameter on one side of which an infinite channel of the object appears. The percolation approach has been successful in various applications to such as the dilute magnetism, gelation etc. The percolation phenomenon itself has also been studied extensively on the analogy of phase transitions.[2]

The underlying geometrical structure, however, may not be a sufficient information to determine certain properties. For example, the frequency-dependent (dynamic) response of the percolating object can be obtained only by studying its dynamics. Furthermore, the percolation threshold and the critical indices might depend on the character of the percolating object, because the connected channel for one object may not be a channel for other objects. In addition, one can imagine the case where the connections depend on time, that is, a system which involves a time-dependent (dynamic) disorder.[3] In order to study these problems, the dynamics of the percolating objects must be examined.

Recently, there has been a considerable effort to investigate the percolation process as a dynamic process of the percolating object. De Gennes first discussed the dynamics of percolating object as diffusion of ants in a labyrinth.[4] He suggested that the diffusion constant D should behave like $(p-p_c)^t$ near the percolation threshold p_c where $1-p$ is the fraction of blocked area in the system. Subsequently, the percolation problem is formulated in terms of a random walk equation and the critical behavior of the frequency-dependent (dynamic) diffusion constant has been examined.[5-7]

The percolation process of quantum mechanical objects has also been studied recently, where the quantum localization[8] is incorporated in the percolation phenomena.[17]

In this paper, we present an overview of the dynamic percolation process in the system with static disorder, emphasizing on a comparison of classical and quantum percolation. In Sec. 2, we discuss the classical dynamic percolation where the motion of percolating objects is governed by a random walk or a diffusion equation. After describing various definitions of the percolation threshold, we summarize the critical behavior of the dynamic diffusion constant. The percolation of quantum mechanical particles is discussed in Sec. 3, where the localization and tunneling effects are studied. An explicit comparison of a critical exponent for a one-dimensional percolation problem is given and it is shown that the classical and quantum processes give rise to incompatible critical exponents. We also discuss possible phase diagrams for percolating system with quantum effects.

2. Classical Dynamic Percolation

2.1 A Classification

To describe the motion of a classical percolating object (hereafter referred to as a classical (C-) ant), we employ a microscopic random walk equation or a diffusion equation. For discrete processes, we assume that a C-ant can occupy a set of discrete points $\{\vec{s}\}$ and its motion obeys the random walk master equation

$$\frac{\partial}{\partial t}P(\vec{s},t|\vec{s}_0,0) = \sum_{\vec{s}'\neq\vec{s}} \{w_{\vec{s}\vec{s}'} P(\vec{s}',t|\vec{s}_0,0) - w_{\vec{s}'\vec{s}} P(\vec{s},t|\vec{s}_0,0)\}, \qquad (2.1)$$

where $P(\vec{s},t|\vec{s}_0,0)$ denotes the conditional probability of a C-ant being at site $\vec{s}$ at time t when it started site $\vec{s}_0$ at time $t = 0$, and $w_{\vec{s}\vec{s}'}$ is the jump rate from site $\vec{s}'$ to $\vec{s}$. According to the distribution of sites $\{\vec{s}\}$ the percolation in lattices and in spatially disordered systems are distinguished. The discrete lattice problem with nearest neighbor interactions can usually be classified into two cases: Bond problems, in which the jump rate $w_{\vec{s}\vec{s}'}$, ($\vec{s}$ and $\vec{s}'$ are nearest neighbors) obeys the probability distribution

$$P(w_{\vec{s}\vec{s}'}) = p\delta(w_{\vec{s}\vec{s}'} - w_0) + (1-p)\delta(w_{\vec{s}\vec{s}'}) \qquad (2.2)$$

with $0 \leq p \leq 1$ and $w_0 > 0$, and site problems, in which the jump rate $w_{\vec{s}\vec{s}'}$ is written as $w_{\vec{s}\vec{s}'} = w_0 \, n_{\vec{s}} n_{\vec{s}'}$ and the parameter $n_{\vec{s}}$ obeys the probability distribution

$$P(n_{\vec{s}}) = x\delta(n_{\vec{s}} - 1) + (1-x)\delta(n_{\vec{s}}) \qquad (2.3)$$

with $0 \leq x \leq 1$. It is straightforward to extend these definitions to more general processes such as the systems with further neighbor jump or directed bonds.

For the continuum percolation, a C-ant is assumed to be anywhere in the space and obey the diffusion equation

$$\frac{\partial}{\partial t} P(\vec{r},t) = \vec{\nabla}\{w(\vec{r})\vec{\nabla}P(\vec{r},t)\} \qquad (2.4)$$

where the local diffusion constant $w(\vec{r})$ is determined by the geometrical structure of blocks, and $P(\vec{r},t)$ denotes the probability density that one finds the C-ant at $\vec{r}$ at time t.

The foregoing definitions are summarized in Table I.

2.2 Percolation threshold

We can conceive various definitions of the percolation threshold, depending on the physical quantity used in the definition. The dynamic diffusion constant for the discrete problems is given by[6]

$$D(\omega) = -\frac{\omega^2}{2d} \sum_{\vec{s}} \langle (\vec{s}-\vec{s}_0)^2 \, \tilde{P}(\vec{s},i\omega|\vec{s}_0) \rangle_{\vec{s}_0} \qquad (2.5)$$

where d is the dimensionality and $\tilde{P}(\vec{s},u|\vec{s}_0)$ is the Laplace transform of $P(\vec{s},t|\vec{s}_0,0)$

$$\tilde{P}(\vec{s},u|\vec{s}_0) = \int_0^\infty e^{-ut} P(\vec{s},t|\vec{s}_0,0)dt. \qquad (2.6)$$

Table I. A classification of various dynamic percolation processes with static disorder

Dynamic Percolation			Govorning Equation/Hamiltonian	
			Classical	Quantum
Discrete	Lattice	Site	Master Equation Eq. (2.1)	Tight Binding Hamiltonian Eq. (3.1)
		Bond		
	Spatial Disorder			
Continuum			Diffusion Equation Eq. (2.4)	One Particle Hamiltonian with a random potential Eq. (3.4)

The average $\langle \rangle_{\vec{s}_0}$ in Eq. (2.5) is taken over the distribution of $\vec{s}_0$. Using the static diffusion constant $D(0)$ we define a (diffusion) threshold p_D by

$$p_D = \text{Sup}\{p|\ \text{Re}D(0) = 0\}. \qquad (2.7)$$

We define the strong $(L^S(\vec{s}))$ and weak $(L^W(\vec{s}))$ localization functions of site $\vec{s}$ by[18]

$$L^S(\vec{s}) = \lim_{t \to \infty} P(\vec{s},t|\vec{s},0) = \lim_{u \to 0} u\tilde{P}(\vec{s},u|\vec{s}) \qquad (2.8)$$

and

$$L^W(\vec{s}) = \int_0^\infty P(\vec{s},t|\vec{s},0)dt = \lim_{u \to 0} \tilde{P}(\vec{s},u|\vec{s}). \qquad (2.9)$$

The percolation threshold in the strong sense (p_S) and in the weak sense (p_W) are defined by

$$p_S = \text{Sup}\{p|\ L^S(\vec{s}) \neq 0 \text{ for all } \vec{s}\} \qquad (2.10)$$

and

$$p_W = \text{Sup}\{p|\ L^W(\vec{s}) = \infty \text{ for all } \vec{s}\}. \qquad (2.11)$$

These definitions can be easily modified for the continuous percolation.

It has been rigorously proven that the following relations among the different definitions of the threshold:[18]

$p_D > p_c$, $p_S = p_c$,

and

$p_W = p_c$ for $d > 2$, $p_W = 1$ for $d < 2$,

hold, where p_c is the percolation threshold defined by the usual geometrical connection alone. Although some approximate treatments suggest that $p_D = p_c$ also holds, no rigorous proof has been given yet.

2.3 Critical behavior of the dynamic diffusion constant

As mentioned in the Introduction, it is known that the static diffusion constant near the percolation threshold behaves as $(p-p_D)^t$ for $p > p_D$ with $t \sim 1.1$ for two dimensions and $t \sim 1.7$ for three dimensions.[19] The ac parts of the diffusion constant also show a critical behavior at the percolation threshold. For example, the ac parts, $D(\omega) - D(0)$ of the diffusion constant determined by the coherent medium approximation[6,20] can be written as

$$D(\omega) - D(0) = w_0 a^2 \left[A(p) \left(\frac{\omega}{w_0}\right)^\sigma i + B(p) \left(\frac{\omega}{w_0}\right)^\tau \right] \tag{2.12}$$

near the static limit $\omega = 0$ in three-dimensions. The exponent σ is equal to $1/2$ when $p = p_D$, otherwise unity, and τ is equal to $3/2$ when $p > p_D$, $1/2$ when $p = p_D$, and 2 when $p < p_D$. The coefficients $A(p)$ and $B(p)$ are critical at $p = p_D$ as shown in Fig. 1.

Similar behavior of the ac parts of the diffusion constant is seen in other dimensions too. In particular, for a one-dimensional chain the exact results $\sigma = 1$, $\tau = 2$ and

$$A(p) = \frac{p}{2(1-p)^2} , \quad B(p) = \frac{p(1+p)^2}{4(1-p)^4} , \tag{2.13}$$

are obtained.[5] For a one-dimensional classical percolation process, the continuum model leads to the same critical exponents as in Eq. (2.13). As an example, let us consider a one-dimensional chain in which blocks of length b are distributed randomly. A C-ant is always confined in a finite segment between two blocks, whose width, ℓ, obeys the distribution $\frac{\ell(1-p)^2}{b^2 p} e^{-\ell(1-p)/bp}$, p being the fraction of the unblocked part in the chain. It can be shown[21] that the leading term of the ac part of the diffusion constant in the limit of $\omega = 0$ is given by

$$D(\omega) - D(0) = \frac{b^2 p^3}{2(1-p)^2} \omega i. \tag{2.14}$$

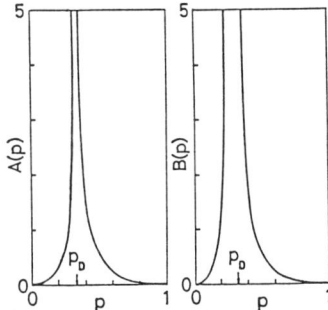

Figure 1. The critical behavior of the real and imaginary parts of the frequency dependent diffusion constant $D(\omega)$. The ac part of $D(\omega)$ is written as $D(\omega)-D(0)/a^2 w_0 \simeq A(p)(\omega/w_0)^\sigma$ i + $B(p)(\omega/w_0)^\tau$ near the static limit, and the coefficients $A(p)$ and $B(p)$ are plotted against p, which are determined by the coherent medium approximation.[6] In this approximation, $p_D = 1/3$ for the simple cubic lattice.

3. Quantum Effects in Percolation

3.1 A classification and models

It is known that a quantum mechanical particle subject to a random potential may be localized when the degree of randomness exceeds a critical value.[8] Moreover, a quantum particle can escape (tunnel) through a potential barrier which may prohibit traveling of a classical particle. Apparently, the localization will increase the percolation threshold and the tunneling will reduce it. These two competing effects determine the actual percolation threshold.

First, we consider the case where tunneling is prohibited. The percolation process of a quantum mechanical object (referred to as a quantum (Q-)ant) can be classified in the same way as for the classical problem [see Table I]. For the discrete problems, the Q-ant is described by a tight binding Hamiltonian

$$H = \sum_{\vec{s}} |\vec{s}\rangle \epsilon_{\vec{s}} \langle\vec{s}| + \sum_{\vec{s}'\neq\vec{s}} |\vec{s}\rangle v_{\vec{s}\vec{s}'} \langle\vec{s}'|. \tag{3.1}$$

For lattice problems, state $|\vec{s}\rangle$ is associated with the lattice site $\vec{s}$. The range of transfer energy $v_{\vec{s}\vec{s}'}$ depends on the problem. For simple problems, $v_{\vec{s}\vec{s}'}$ is assumed to be non-zero only when $\vec{s}$ and $\vec{s}'$ are the nearest neighbors. The site problem is characterized by the distribution of the site energy $\epsilon_{\vec{s}}$

$$P(\varepsilon_{\vec{s}}) = x\delta(\varepsilon_{\vec{s}}) + (1 - x)\delta(\varepsilon_{\vec{s}} - \varepsilon_1) \tag{3.2}$$

with $\varepsilon_1 = \infty$ and $v_{\vec{s}\vec{s}'} = v_0$ (constant); the bond problem is characterized by the distribution of $v_{\vec{s}\vec{s}'}$,

$$P(v_{\vec{s}\vec{s}'}) = p\delta(v_{\vec{s}\vec{s}'} - v_0) + (1 - p)\delta(v_{\vec{s}\vec{s}'} - v_1) \tag{3.3}$$

with $v_1 = 0$, v_0 being a constant and $\varepsilon_{\vec{s}} = 0$ for all $\vec{s}$. For spatially disordered systems $\{|\vec{s}\rangle\}$ are distributed randomly in space, and $v_{\vec{s}\vec{s}'}$ is nonzero when the distance $|\vec{s} - \vec{s}'|$ is less than a specified length.

For continuum problems, a Q-ant is described by a Hamiltonian

$$H = -\frac{\hbar^2 \nabla^2}{2m} + V(\vec{r}) \tag{3.4}$$

where $V(\vec{r})$ takes the value of either V_0 or V_1 ($= \infty$) randomly.

The transition probability of a $\vec{Q}$-ant that it is at $\vec{s}$ at time t when it started from $\vec{s}_0$ at t = 0 is given by

$$P_Q(\vec{s}, t | \vec{s}_0, 0) = |\langle \vec{s} | e^{-iHt/\hbar} | \vec{s}_0 \rangle|^2. \tag{3.5}$$

The quantum percolation threshold is defined in exactly the same way as in Sec. 2.2, using $P_Q(\vec{s}, t | \vec{s}_0, 0)$ in place of $P(\vec{s}, t | \vec{s}_0, 0)$. Therefore, we can define the diffusion threshold, the thresholds in the strong sense and in the weak sense. Apparently, these thresholds are larger than or equal to the one defined by the geometrical connection, and $p_D \geqslant p_S \geqslant p_c$, $\geqslant p_W \geqslant p_S \geqslant p_c$ should hold. However, the explicit relation among these thresholds has not been obtained yet.

The effects of tunneling can also be studied using Hamiltonian (3.1) or (3.4). Namely, we consider that $\varepsilon_1 = \infty$ or $v_1 = 0$ or $V_1 = \infty$ is caused by infinitely high barriers. When the barrier is reduced so that tunneling is allowed, we assign a finite value to ε_1 or V_1 and a nonzero value to v_1. These values assigned to ε_1, V_1 or v_1 measure the strength of tunneling.

3.2 One-dimensional system - A comparison of a critical exponent

We consider the quantum percolation without tunneling in one-dimensional chain. Obviously, the static diffusion constant vanishes and no percolation occurs if any disruption exists. However, the frequency-dependent response such as the dynamic diffusion constant may not vanish even when disruptions exist.

As an example, we obtain the frequency-dependent polarization of an electron, which is related to the imaginary part of the ac conductivity or diffusion constant. First we consider the bond problem in which the electron is confined within a finite segment, (say, N sites segment). Within the linear response, the polarizability $\alpha_N(\omega)$ near the static limit is given by

$$\alpha_N(\omega) \propto A_N + C_N \hbar^2 \omega^2 / v_0^2 \tag{3.6}$$

where A_N and C_N are given vigorously by[21]

$$A_N = \gamma_N [15\gamma_N^2/\sigma_N^2 + 13 - (N+1)^2]/24\sigma_N^2$$

$$C_N = \gamma_N [\frac{21\gamma_N^2}{8\sigma_N^4} + \frac{7\gamma_N^2}{2\sigma_N^2} + \frac{113}{120} + \frac{(N+1)^2}{48} - \frac{(N+1)^4}{40}]/8\sigma_N^4 \tag{3.7}$$

with $\sigma_N = \sin \frac{\pi}{N+1}$ and $\gamma_N = \cos \frac{\pi}{N+1}$. Averaging $\alpha_N(\omega)$ over all possible lengths of the segment, we obtain the polarizability $\alpha(\omega)$ near the static limit in the vicinity of the percolation point $p = p_c \equiv 1$ as

$$\alpha(\omega) \propto A_Q(p) + C_Q(p) \hbar^2 \omega^2 / v_0^2 \tag{3.8}$$

and

$$A_Q(p) = \frac{5(15-\pi^2)}{\pi^4} (p_c-p)^{-4} \; ; \; C_Q(p) = \frac{1134(105-\pi^4)}{\pi^8} (p_c-p)^{-8}. \tag{3.9}$$

The critical exponent -4 for $A_Q(p)$ should be compared with that of $A(p)$ in Eq. (2.13). Thus, the classical and quantum models for the one-dimensional bond percolation problem give rise to different critical exponents.

This is also seen for the continuum model, whose classical counterpart was discussed in Sec. 2.3. In the linear response regime, the polarizability of an electron confined in a well of width ℓ is given by[21]

$$\alpha_\ell^Q(\omega) \propto \frac{15-\pi^2}{12\pi^4} \ell^4 + \frac{105-\pi^4}{40\pi^8} \ell^8 \left(\frac{m\omega}{\hbar}\right)^2 \tag{3.10}$$

near the static limit. Thus the static polarizability behaves as

$$\alpha^c \propto (p_c-p)^{-4} \tag{3.11}$$

near $p = p_c(\equiv 1)$. The exponent is different from the classical counterpart given in Eq. (2.14). These differences in the critical exponent can be easily understood from a scaling argument. The polarizability α has the dimension [Charge^2 Length^2]/[Energy]. For the quantum problem, the relevant scales for the length and energy are ℓ and $\hbar^2/m\ell^2$ which lead to $\alpha \propto \ell^4$. The corresponding scales for the classical problem are ℓ and kT which lead to $\alpha \propto \ell^2$.

3.3 Discrete quantum percolation

First, note that the distributions Eqs. (3.2) and (3.3) do not have any moments of $\varepsilon_s^{\rightarrow}$ or $\log|v_{\underset{ss}{\rightarrow\rightarrow},}|$. Therefore, the distribution involved in quantum percolation process should be considered as a class different from the standard Anderson's model in which the second moment of the distribution for $\varepsilon_{\rightarrow}$ or $\log|v_{\underset{ss}{\rightarrow\rightarrow},}|$ serves as the measure of randomness and when the degree of randomness exceeds a critical value the absense of diffusion takes place. This is the case even when the tunneling is allowed as described in Sec. 3.1, because while the variance of the distribution is always maximum at $p = 0.5$ the localization transition will be expected to occur not necessarily at $p = 0.5$ as we shall see later.

There have been considerable efforts to locate the quantum percolation threshold. While some of the earlier works[9,10] based on computer simulation concluded that the difference between the classical and quantum thresholds would be small, recent works showed a significant difference between both thresholds.[12-17] In particular, the quantum percolation threshold for two-dimensional lattices is believed to be unity. This seems to be consistent with the result for the Anderson's problem, based on the scaling argument[22] that all states are localized in any two dimensional random system. However, it should be remarked that (1) there exist various definitions of percolation threshold; (2) the scaling theory concerns with the conductance so that the related threshold is the diffusion threshold; 3) none of the estimations calculate directly the threshold defined before. A further study is needed to elucidate the relation among various definitions of percolation threshold.

3.4 Effects of tunneling

As an example, we consider the bond percolation model with tunneling, i.e., the tight binding Hamiltonian subjected to the probability distribution (3.3) with $v_1 \neq 0$. The limit $v_1 = 0$ corresponds to the usual quantum percolation and nonvanishing v_1 represents the tunneling through the barrier. First, we employ the homomorphic cluster coherent potential approximation[23] to obtain the density of states. Figure 2 shows the density of states as a function of the energy E and the parameter p when $v_1/v_0 = 0.1$, where the semi-elliptic density of states $(36v_0^2 - E^2)^{1/2}/18\pi v_0^2$ is used for $p = 1$ as a model three dimensional system. Note that spectral gaps appear near $E = \pm v_0$ when $p \approx 0$.

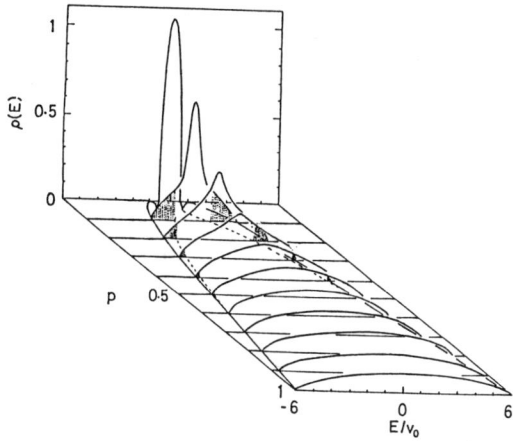

Figure 2. The density of states for the quantum bond percolation model with tunneling. The density of states is obtained by the homomorphic cluster CPA for $v_1/v_0 = 0.1$ and the semi-elliptic density of states is assumed for $p = 1$. The states in the shaded part of the density of states are localized according to the $F^*(E)$ criterion.

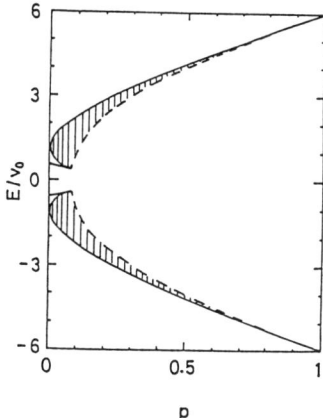

Figure 3. The band edges and mobility edges for the quantum bond percolation model with tunneling, $v_1/v_0 = 0.1$. The mobility edges are determined by the $F^*(E)$ criterion and states in the shaded area are localized.

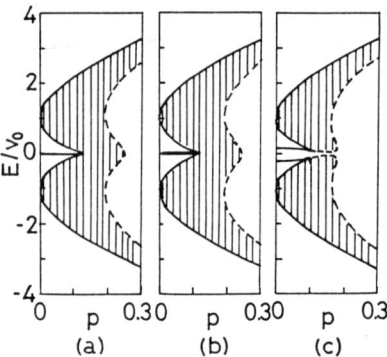

Figure 4. Similar to Fig. 3 for various values of v_1: (a) $v_1/v_0 = 0.000001$, (b) $v_1/v_0 = 0.001$, (c) $v_1/v_0 = 0.03$.

It is clearly seen in Figure 3 in which the band edges are shown. In fact, the impurity levels at $p = 0$ due to a single impurity bond of v_1 appears at

$$E^i_\pm = \pm \frac{1}{2} \{v_1 - 7v_0 - [v_1^2 + 16v_0 v_1 + 91v_0^2 + 216v_0^3/(v_1 - v_0)]\}^{1/2}, \quad (3.12)$$

whence the spectral gaps appear when $6|v_0| < |E^i_\pm|$. The states near the band edges are supposed to be localized. To locate the mobility edges, we employ a modified F(E) function[24] defined by

$$F^*(E) = | 6\sigma_{od} / (E - \sigma_d) | \quad (3.13)$$

where σ_d and σ_{od} are the diagonal and off-diagonal elements of the coherent Hamiltonian determined by the HCPA. According to the F(E) criterion, the states which satisfy $F^*(E) < 1$ are localized. The states in the shaded area in the density of states in Fig. 2 are localized, and in Fig. 3 the mobility edges are shown with the shaded area for localized states.

When the "strength of tunneling" v_1 is reduced, the mobility edges move inward [see Fig. 4]. They merge together at about $v_1 \simeq 0.3v_0 (\equiv v_1^*)$ and for $v_1 < v_1^*$ all states become localized for a limited range of p which is far from the range at which the variance of the distribution is largest. It is, therefore, tempting to regard the p^* at which the state at $E = 0$ becomes localized when p is reduced for a given value of v_1 as the percolation thresh-

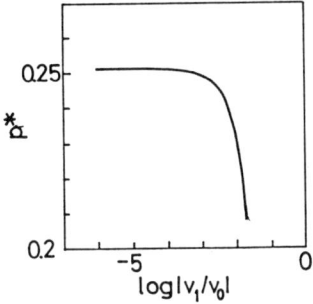

Figure 5. The critical probability p* vs log $|v_1/v_0|$ where the limit of p* at log $|v_1/v_0|$ = $-\infty$ is considered to be the usual quantum bond percolation threshold. p* exists only when $|v_1| < 0.03 |v_0|$.

old for the quantum percolation with tunneling. The critical concentration p* exists, thus, only for $|v_1| \leq |v_1^*|$ and is a decreasing function of $|v_1|$. We show in Fig. 5 the v_1 dependence of p*. The limiting value of p* ~ 0.252 at v_1 = 0 will be regarded as the quantum percolation threshold for the process without tunneling.

The site problem also shows a similar behavior.[25,26] Namely, one can define a critical probability x* for a given ε_1 at which all states associated with the site energy ε_0 become localized. The critical probability is an increasing function of ε_1, approaching the quantum percolation threshold in the limit ε_1 = ∞. Figure 6 shows the ε_1 dependence of the critical probability determined by the L(E) criterion. Detailed studies will be presented elsewhere.[26]

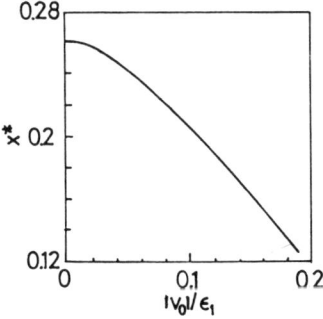

Figure 6. The critical probability x* vs $1/\varepsilon_1$, where the limit of x* at $1/\varepsilon_1$ = 0 can be regarded as the usual quantum site percolation threshold.

3.5 Summary

In the quantum percolation regime, one can have various possible phases instead of just two phases (percolated or nonpercolated) in the classical percolation regime. The present consensus that the classical and quantum percolation threshold are significantly different immediately implies the existence of three phases: classically localized, qauntum localized and extended, as schematically depicted in Fig. 7(a). (There will be more phases if the diffusion threshold, strong and weak percolation thresholds are different.)

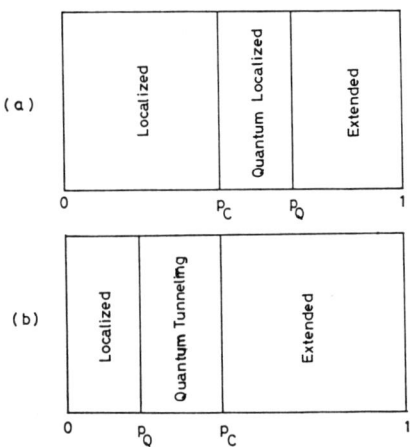

Figure 7. A schematic figure of examples of possible phases in quantum percolation picture: (a) without tunneling; (b) with sufficient tunneling.

When tunneling is allowed, we can have completely different phases. If the strength of tunneling is weak, we expect to have similar phases to those given in Fig. 7(a). When the tunneling is strong enough, the threshold can be smaller than the classical one and the possible three phases are; localized, quantum tunneling and extended as depicted in Fig. 7(b). Some experiments showing more than two phases near the classical percolation point may be explained in terms of the quantum percolation.

4. Concluding Remark

In this paper the notion of the dynamic percolation in systems with static disorder was explained for both classical and quantum mechanical objects. The classical and quantum percolations give rise to incompatible results for percolation threshold, critical exponents and other properties. In particular, the quantum percolation picture with or without tunneling suggests the possibility of having various phases instead of just two phases in the geometrical percolation, and hence enhances tremendously the possible range of application of the percolation theory.

Acknowledgments: I would like to thank Professor M. Lax of CCNY for discussions and encouragement, and K. C. Chang and D. Nguyen for their assistance. This work was supported in part by a grant from Research Corporation.

References

1. S.R. Broadbent and J.M. Hammersley, Proc. Camb. Phil. Soc. 53, 629 (1957).
2. For review see V.K.S. Shante and S. Kirkpatrick, Adv. Phys. 20, 325 (1971); D. Stauffer, Phys. Rep. 54, 1 (1979); See also, Annals of the Israel Physical Society, Vol. 5. "Percolation Structures and Processes", eds (G. Deutsher, R. Zallen and J. Alder, (Israel Physical Society, 1983).
3. A.K. Harrison and R. Zwanzig, Phys. Rev. A32, 1072 (1985).
4. P.G. de Gennes, La Recherche 7, 919 (1976).
5. T. Odagaki and M. Lax, Phys. Rev. Lett. 45, 847 (1980).
6. T. Odagaki and M. Lax, Phys. Rev. B24, 5284 (1981).
7. I. Webman, Phys. Rev. Lett. 47, 1496 (1981); J. Nieuwoudt and S. Mukamel, J. Stat. Phys. 36, 677 (1984).
8. P.W. Anderson, Phys. Rev. 109, 1492 (1958).
9. S. Kirkpatrick and T.P. Eggarter, Phys. Rev. B6, 3598 (1972).
10. T. Odagaki, N. Ogita and H. Matsuda, J. Phys. C13, 189 (1980).
11. R. Raghavan and D.C. Mattis, Phys. Rev. B23, 4791 (1981); R. Raghavan, Phys. Rev. B29, 748 (1984).
12. Y. Shapir, A. Aharony and A.B. Harris, Phys. Rev. Lett. 49, 486 (1982).
13. V. Srivastava and D. Weaire, Phys. Rev. B18, 6635 (1978); V. Srivastava and M. Chaturvedi, Phys. Status Solidi B120, 421 (1983).
14. S.N. Evangelou, Phys. Rev. B27, 1397 (1983).
15. A.B. Harris, Phys. Rev. Lett. 49, 296 (1982).
16. T. Odagaki and K.C. Chang, Phys. Rev. B30, 1612 (1984).
17. S. Mukamel, in this proceeding.
18. T. Odagaki and M. Lax, Phys. Rev. B26, 6480 (1982).
19. J.P. Straley, Phys. Rev. B15, 5733 (1977).
20. T. Odagaki, M. Lax and A. Puri, Phys. Rev. B28, 2755 (1983).
21. D. Nguyen and T. Odagaki, to be published.
22. E. Abrahams, P.W. Anderson, D.C. Licciardello and T.V. Ramakrishnam, Phys. Rev. Lett. 42, 673 (1979).
23. T. Odagaki and F. Yonezawa, J. Phys. Soc. Japan 47, 379 (1979).
24. E.N. Economou and M.H. Cohen, Phys. Rev. B5, 2931 (1972).
25. E.N. Economou, S. Kirkpatrick, M.H. Cohen and T.P. Eggarter, Phys. Rev. Lett. 25, 520 (1970).
26. K.C. Chang and T. Odagaki, in preparation.

Self-Consistent Mode-Coupling Theory of
Excitation Transport in Disordered Media:
Application to Anderson Localization and
Incoherent Transport

Roger F. Loring and Shaul Mukamel[†]
Department of Chemistry
University of Rochester
Rochester, New York 14627

ABSTRACT

We present a novel mode-coupling procedure for constructing self-consistent equations for excitation transport in disordered systems. The present approach allows incoherent hopping models and fully quantum mechanical models, including those expected to display Anderson localization, to be treated in a unified way. This procedure applies both to short-range (e.g. nearest neighbor) and to long-range (e.g. dipole-dipole) interactions. Our formulation is based on self-consistent equations for $P_o(\epsilon)$ and $D(\vec{k},\epsilon)$. $P_o(\epsilon)$ is the Laplace transform of $P_o(t)$, the conditional probability that an excitation that occupies a given spatial position at a given time will be found at that position after a time t has elapsed. $D(\vec{k},\epsilon)$ is the frequency and wavevector dependent diffusion coefficient. The mode-coupling procedure is applied to an incoherent hopping model and to quantum percolation. For the hopping problem, our procedure correctly reproduces the long-time power law decay of $P_o(t)$. For quantum percolation, we derive an equation that is similar in structure to the equation obtained by Vollhardt and Wölfle for a Fermi gas in a random potential. The frequency dependence of the ac conductivity and the critical exponents determined from this equation agree with the predictions of scaling theory.

[†]Camille and Henry Dreyfus Teacher-Scholar

I. Introduction

A wide variety of theoretical techniques has been applied to the study of incoherent hopping transport of excitations in disordered media, including the continuous time random walk formalism[1-4], self-consistent calculations of the Green function[5-9], and effective medium approximations[10,11]. By incoherent transport, we mean a process whose dynamics are described by a Pauli master equation. In this case, the quantum mechanical phase of the wave function does not play a significant role. A great deal of attention is also currently focussed on fully quantum mechanical models of transport in disordered media, in which the dynamics are governed by the Schrödinger equation rather than a master equation. The problem of Anderson localization is of particular interest[12-20] and has been studied in the contexts of electrical conduction and excitation transport. The theoretical procedures that have been applied to this class of problems are very different from those used to investigate incoherent hopping models. Most treatments of the quantum problem center on the wave function and the character of the transport is determined from the degree to which the wave function can be considered to be localized or extended.

In this work, we present a new approach to performing self-consistent calculations of transport properties for any model involving the dynamics of a single excitation in a disordered system whose ensemble average is translationally invariant. This approach can thus be applied both to incoherent hopping models and to fully quantum mechanical models. The two major dynamical quantities in this approach are $D(\vec{k},\varepsilon)$, the generalized diffusion kernel in Fourier-Laplace space (see Eq.(2)) and $P_o(t)$, the probability that an excitation that occupies a particular position at time τ can be found at that position at a later time $t+\tau$. Our approach starts with an exact relation between $P_o(\varepsilon)$, the Laplace transform of $P_o(t)$, and the diffusion kernel (Eq.(3)). We then develop a hierarchy of approximations in which the diffusion kernel is expressed as a functional of $P_o(\varepsilon)$, resulting in a second relation between these two quantities. When the second relation is substituted into the first, we obtain a closed self-consistent equation for $P_o(\varepsilon)$. This equation contains an integration over the wavevector $\vec{k}$ that is conjugate to the position variable. This type of equation is characteristic of mode-coupling theories of nonlinear hydrodynamics, in which self-consistent equations are derived that contain similar integrations over wavevector[21-23]. An important feature of the mode-coupling theories is their capacity to correctly predict the long-time asymptotic power-law decays of hydrodynamic variables ("long-time tails"). In a similar fashion, the approach presented here can be used to generate approximations to $P_o(t)$ that reproduce the correct long-time power-law behavior of that quantity, namely $P_o(t) \sim t^{-d/2}$ in d dimensions[1,3,4,7]. The major advantage of the present formulation is its capacity to treat both quantum and incoherent transport with short-range or long-range interactions within the same theoretical framework.

In Sec. II, we present a general formulation of our mode-coupling approach. In Sec. III, this methodology is applied to the incoherent hopping problem on a substitutionally disordered lattice. In Sec. IV, we apply the mode-coupling theory to quantum percolation. We derive a self-consistent equation for the diffusion kernel that is similar in structure to the Vollhardt-Wölfle equation for a Fermi gas in a random potential[18,19]. The results of Sec. III for incoherent transport and the results of Sec. IV for quantum percolation are compared. Section V contains a comparison to previous work.

II. The Mode-Coupling Procedure

In this section, we present the self-consistent procedure for calculating transport properties of dilute excitations (e.g., electrons or Frenkel excitons) on disordered lattices. We shall consider a general class of models involving the motion of a single excitation in a disordered system whose ensemble average is translationally invariant. Our goal is the calculation of $P(\vec{r},t)$, the ensemble-averaged probability that an excitation undergoes a displacement $\vec{r}$ during a time t. $P(\vec{k},\varepsilon)$, the Fourier-Laplace transform of $P(\vec{r},t)$, is defined by

$$P(\vec{k},\varepsilon) = \sum_{\vec{r}} \exp(i\vec{k}\cdot\vec{r}) \int_0^\infty dt \exp(-\varepsilon t) \; P(\vec{r},t). \qquad (1)$$

$D(\vec{k},\varepsilon)$, the diffusion kernel, is defined by

$$P(\vec{k},\varepsilon) = [\varepsilon + k^2 \, D \, (\vec{k},\varepsilon)]^{-1}. \qquad (2)$$

If $D(\vec{k},\varepsilon)$ approaches a limit $D(0,0)$ as $k\to 0$ and $\varepsilon\to 0$, then for small k and ε, $P(\vec{k},\varepsilon)$ approaches the Green function of an ordinary diffusion equation with diffusion constant $D(0,0)$. $P_o(t)$ is defined to be the probability that an excitation that resides at a particular site at time τ will be located at that site at time $t+\tau$. ($P_o(t) \equiv P(\vec{r}=0,t)$.) $P_o(\varepsilon)$, the Laplace transform of $P_o(t)$, is related to the diffusion kernel by

$$P_o(\varepsilon) = \Omega^{-1} \int d\vec{k} \, [\varepsilon + k^2 \, D \, (\vec{k},\varepsilon)]^{-1}. \qquad (3)$$

The integration in Eq.(3) is carried out over the first Brillouin zone, whose "volume" is Ω. ($\Omega = (2\pi)^d/\Omega_o$, where Ω_o is the volume of the unit cell.)

Equation (3) provides an exact relation between $P_o(\varepsilon)$ and $D(\vec{k},\varepsilon)$. In order to obtain a closed set of self-consistent equations, we need another independent relation between these two quantities. To that end, we introduce the ansatz

$$D(\vec{k},\varepsilon) = \tilde{D}(\vec{k}, \, P_o(\varepsilon)). \qquad (4)$$

According to this ansatz, D is expressed as a functional of P_o, which depends implicitly on ε through the ε dependence of P_o. In the Appendix we present a systematic procedure by which $\tilde{D}$ can be determined for a general class of models of a disordered lattice. In these models, which will be considered in detail in the subsequent sections, a fraction c of the lattice sites participates in the excitation transport, and the remaining sites are blocked. In the Appendix, we demonstrate that $\tilde{D}$ can be obtained order-by-order in c from the naive density expansions of D and P_o, which are the only microscopic input to the theory[9]. (The naive (unrenormalized) density expansions of these quantities are defined in Eqs.(A2) and (A3).) Substitution of Eq.(4) into Eq.(3) yields

$$P_o(\varepsilon) = \Omega^{-1} \int d\vec{k} \, [\varepsilon + k^2 \, \tilde{D} \, (\vec{k}, P_o)]^{-1}. \qquad (5)$$

We propose the following self-consistent procedure for calculating $P(\vec{k},\varepsilon)$. (i) Approximate the functional $\tilde{D}(\vec{k},P_o)$. (ii) Substitute this expression into the right side of Eq.(5), and solve Eq.(5) for $P_o(\varepsilon)$. (iii) Substitute this expression for $P_o(\varepsilon)$ into the approximate functional $\tilde{D}(\vec{k},P_o)$ to generate $D(\vec{k},\varepsilon)$, (Eq.(4)). This expression for $D(\vec{k},\varepsilon)$, when substituted into the right side of Eq.(2), yields $P(\vec{k},\varepsilon)$.

III. Application to Incoherent Transport: The Master Equation

We consider a lattice with a fraction c of the sites occupied by active particles that are capable of retaining and transferring an excitation. These particles are assumed to be randomly distributed among the lattice sites. The unoccupied sites do not participate in the transport. The dynamics of a particular realization of the random system are governed by a Pauli master equation.

$$\frac{d\vec{p}}{dt} = \underset{\approx}{W} \cdot \vec{p} \qquad (6)$$

$\vec{p}$ is an N component vector whose i'th element is the probability that an excitation resides on the particle labelled i. N is the number of active particles. W is an NxN matrix whose elements are given by

$$W_{ij} = w_{ij} - \delta_{ij} \sum_m w_{mj}. \tag{7}$$

w_{ij} is the rate of excitation transfer between particles labelled i and j. $P(\vec{r},t)$, the configuration-averaged probability that an excitation undergoes a displacement $\vec{r}$ in a time t is given by

$$P(\vec{r},t) = (N-1) \langle [\exp(-Wt)]_{21} \, \delta_{\vec{r},\vec{r}_{12}} \rangle$$
$$+ \delta_{\vec{r},\vec{0}} \, \langle [\exp(-Wt)]_{11} \rangle \tag{8}$$

$\vec{r}_{12} = \vec{r}_1 - \vec{r}_2$, where $\vec{r}_n$ is the position of the particle labelled n. The angular brackets in Eq.(8) denote a configuration average.

In order to apply the mode-coupling procedure described in the previous section to this model, we need to calculate the functional $\tilde{D}(\vec{k},P_0)$. In the Appendix, we expand $\tilde{D}$ in a power series in c:

$$\tilde{D}(\vec{k},P_0) = \sum_{j=1}^{\infty} c^j \, \tilde{D}_j(\vec{k},P_0). \tag{9}$$

The expansion of $\tilde{D}$ in Eq.(9) should be understood as follows. $\tilde{D}$ depends on c explicitly, in addition to depending implicitly on c through the dependence of P_0 on c. Equation (9) is an expansion of $\tilde{D}$ with respect to its explicit dependence on c. The coefficients $\tilde{D}_j$ depend on c only through the dependence on c of P_0. Each term in the expansion depends on c to infinite order. Equations (4) and (9) can, therefore, be regarded as a partially resummed density expansion of $D(\vec{k},\epsilon)$. In the Appendix, we prove that $\tilde{D}_n$ can be uniquely determined from the coefficients of c^m with $m \leq n$ in the density expansions of D and P_0 (Eqs.(A2) and (A3)). The expansion in Eq.(9) can be truncated to obtain an approximation $\tilde{D}^{(n)}$, that is exact to order n in c.

$$\tilde{D}^{(n)}(\vec{k},P_o) = \sum_{j=1}^{n} c^j \tilde{D}_j(\vec{k},P_o) \tag{10}$$

Equation (10) can be substituted into Eq.(5) to yield a self-consistent equation for P_o. We can improve upon approximations of the form given in Eq.(10) by noting that at c=1 (no disorder), D can be calculated exactly, since the system is translationally invariant. In this case, D is independent of ε, and hence $\tilde{D}(\vec{k},P_o)$ at c=1 does not depend on $P_o(\varepsilon)$. Denoting $\tilde{D}$ at c=1 by $\tilde{D}^*$, we have

$$\tilde{D}^*(\vec{k},P_o) = k^{-2} \sum_{\vec{r}} w(\vec{r}) [1-\exp(i\vec{k}\cdot\vec{r})]. \tag{11}$$

We now have two pieces of information regarding $\tilde{D}$: a truncated expansion in c (Eq.(10)) that will be accurate for c < 1 and the exact form at c=1 (Eq.(11)). By constructing an approximation to $\tilde{D}$ that interpolates between Eqs.(10) and (11), we guarantee that the self-consistent equation will give exact results at c=1. A hierarchy of such approximations can be derived by constructing Padé approximants in the variable $\gamma \equiv c/(1-c)$:

$$\tilde{D}(\vec{k},P_o) = \frac{a_1\gamma + a_2\gamma^2 + \cdots a_j\gamma^j}{1 + b_1\gamma + b_2\gamma^2 + \cdots b_j\gamma^j}. \tag{12}$$

The coefficients a_n and b_n in Eq.(12) are obtained by requiring that Eq.(12) reduce to Eq.(10) in the limit c→0 and to Eq.(11) at c=1. The first member of the hierarchy is

$$\tilde{D}(k,P_o) = \frac{\gamma \tilde{D}_1(\vec{k},P_o)}{1 + \gamma \tilde{D}_1(\vec{k},P_o)/\tilde{D}^*(\vec{k},P_o)} \tag{13}$$

According to Eq.(A8) of the Appendix, $\tilde{D}_1(\vec{k},P_o)$ equals $D_1(\vec{k},P_o)$, where $D_1(\vec{k},\varepsilon^{-1})$ is the first term in an expansion in c of the diffusion kernel. $D_1(\vec{k},\varepsilon^{-1})$ can be obtained from the exact solution of the master equation for two particles[5].

$$\tilde{D}_1(\vec{k},P_o) = k^{-2} \sum_{\vec{r}} (1-\exp(i\vec{k}\cdot\vec{r})) \frac{w(\vec{r})}{1+2w(\vec{r})P_o} \qquad (14)$$

The summation in Eq.(14) runs over all displacements in the lattice. The first member of our hierarchy of approximations to $\tilde{D}$ is given by Eq.(13), together with Eqs.(11) and (14). According to the mode-coupling procedure described in the previous section, Eq.(13) can then be substituted into the right side of Eq.(5) to obtain the self-consistent equation for P_o.

An important feature of the present calculation is that $P(\vec{k},\varepsilon)$ can be obtained from Eqs.(5) and (13) for any transfer rate $w(\vec{r})$. In the case of a long-range transfer rate, such as that arising from dipole-dipole interactions ($w(\vec{r}) \sim r^{-6}$), the self-consistent equation must be solved numerically. The results of such calculations will be presented elsewhere. Here we will consider a transfer rate that is finite only between sites that are nearest neighbors on the lattice. In this case, the long-time behavior of the transport properties may be obtained analytically. In addition, the case of a transfer rate of finite range represents a stringent test of the theory, since a percolation transition is expected[24-26]. Below the critical concentration for site percolation, the excitations will be confined to clusters of finite size, and above this threshold the excitations can move throughout the system. For a nearest-neighbor transfer rate and a hypercubic lattice in d dimensions, the self-consistent equations (Eqs.(5) and (13)) assume the form:

$$P_o(\varepsilon) = \Omega^{-1} \int d\vec{k} [\varepsilon + k^2 \tilde{D}(\vec{k}, P_o)]^{-1} \qquad (15a)$$

$$\tilde{D}(\vec{k}, P_o) = \frac{k^{-2} 2cw(d - \sum_{j=1}^{d} \cos(\vec{k} \cdot \vec{a}_j))}{1 + 2(1-c) P_o w} . \qquad (15b)$$

w is the magnitude of the transfer rate between nearest-neighbor sites on the lattice. $\vec{a}_j$ is the lattice vector in the jth direction. Equations (15) constitute our final self-consistent equations for P_o and $\tilde{D}$ for incoherent transport with nearest-neighbor interactions. They may be solved iteratively. The long-time behavior can be obtained analytically by the following procedure. We postulate that $P_o(\varepsilon) \sim A \varepsilon^{\alpha-1}$ as $\varepsilon \to 0$, where A is a function of c. If this form is substituted into Eqs.(15), A and α can be determined analytically. The Tauberian theorem for Laplace transforms can then be used to infer that $P_o(t) \sim A t^{-\alpha}$ for $t \to \infty$[27]. With this procedure, we obtain the following results. For a cubic lattice (d=3) there exists a critical concentration c*, such that for $c^* < c \leq 1$, transport becomes diffusive in the limit of long times and large distances, and $P_o(t)$ decays as $t^{-3/2}$. This power law decay of $P_o(t)$ agrees with scaling arguments, which predict that $P_o(t)$ in d dimensions should decay as $t^{-d/2}$ at long times[3,4,7]. c* is predicted to be ~ 0.325, which agrees well with the currently accepted numerical estimate of 0.310 for the site percolation threshold on a simple cubic lattice[28]. The diffusion constant D(0,0), which is defined to be the $\vec{k} \to 0$ and $\varepsilon \to 0$ limit of $D(\vec{k}, \varepsilon)$ vanishes as c approaches c* from above as $(c-c^*)^s$ with s=1. This "mean-field" value for the critical exponent s has also been obtained in effective medium theories of dynamical percolation[10,11]. This exponent has been estimated from Monte Carlo calculations to be 1.75 ± 0.1[29]. For c<c*, $P_o(t)$ decays to a finite value, which indicates that the excitations are localized on clusters of finite size.

$$P_o(t) \sim (c^*-c)^2, \quad t\to\infty \tag{16a}$$

$$D(0,\varepsilon) \sim (c^*-c)^{-2}\varepsilon, \quad \varepsilon\to 0 \tag{16b}$$

For a square lattice (d=2), Eqs.(15) predict $c^*=1$. $P_o(t)$ decays to a finite value for all $c<1$.

$$P_o(t) \sim \begin{cases} \exp[-2\pi c/(c^*-c)] & c<c^* \\ t^{-1} & c=c^*. \end{cases} , t\to\infty \quad \begin{matrix}(17a)\\(17b)\end{matrix}$$

$$D(0,\varepsilon) \sim \begin{cases} \{\exp[2\pi c/(c^*-c)]\}\varepsilon & c<c^* \\ W & c=c^*, \end{cases} , \varepsilon\to 0 \quad \begin{matrix}(17c)\\(17d)\end{matrix}$$

The exact threshold for site percolation on a square lattice occurs at $c^*\approx 0.59$[26]. For a linear chain (d=1), the theory again predicts $c^*=1$, which is exact in this case.

$$P_o(t) \sim \begin{cases} (c^*-c)^2 & c<c^* \\ t^{-1/2} & c=c^* \end{cases} , t\to\infty \quad \begin{matrix}(18a)\\(18b)\end{matrix}$$

$$D(0,\varepsilon) \sim \begin{cases} (c^*-c)^{-2}\varepsilon & c<c^* \\ W & c=c^* \end{cases} , \varepsilon\to 0 \quad \begin{matrix}(18c)\\(18d)\end{matrix}$$

IV. Application to Quantum Transport: Anderson Localization

In Section III, we considered a model in which the dynamics of excitation transport are governed by a Pauli master equation (Eq.(6)). The master equation description is valid when the quantum coherences do not play a significant role in the dynamics. In this section we apply the ideas of Sec. II to a fully quantum mechanical model in which the dynamics are described by the Schrödinger (or quantum Liouville) equation. We consider a lattice in d dimensions with a fraction c of the sites occupied by randomly distributed active particles. For a given configuration of the random system, the dynamics are governed by the tight-binding Hamiltonian

$$H = \sum_{i<j} \hbar J_{ij}(\vec{r}_{ij})[|i\rangle\langle j| + |j\rangle\langle i|]. \tag{19}$$

The indices i and j in Eq.(19) label the active particles. $P(\vec{r},t)$, the configuration-averaged probability that an excitation propagates a distance $\vec{r}$ in time t, is given by

$$P(\vec{r},t) = (N-1) \langle [\exp(-iLt)]_{22,11} \, \delta_{\vec{r},\vec{r}_{12}} \rangle$$
$$+ \delta_{\vec{r},\vec{0}} \langle [\exp(-iLt)]_{11,11} \rangle. \tag{20}$$

Equation (20) is the fully quantum mechanical analog of Eq.(8), which defines $P(\vec{r},t)$ for the master equation. L is the Liouville operator corresponding to the Hamiltonian of Eq.(19). (The action of L on an operator M is defined by LM = [H,M].) We have adopted tetradic notation[30], in which $[\exp(-iLt)]_{\ell\ell,mm}$ is the conditional probability that an excitation propagates from a particle at position $\vec{r}_m$ to a particle at position $\vec{r}_\ell$ in time t. The diffusion kernel is related to $P(\vec{r},t)$ by Eq.(2).

In order to apply the mode-coupling equation (Eq.(5)) to this model, we need to calculate the functional $\tilde{D}$. To that end, we expand $\tilde{D}(\vec{k},P_0)$ in a power series in c, as in Eq.(9). The coefficients $\tilde{D}_n(\vec{k},P_0)$ can be calculated according to the procedure set forth in the Appendix. (See Eq.(A7).) In our treatment of incoherent transport in Sec. III, we proposed a hierarchy of approximations to $\tilde{D}$ that are obtained by truncating Eq.(9) to yield Eq.(10) and then using a Padé approximant to interpolate between this expression and the exact result at c=1 (Eq.(11)). In the incoherent hopping model, the diffusion kernel assumes a simple, frequency-independent form at c=1, as shown in Eq.(11). Since the diffusion kernel is independent of ε, it is also independent of $P_0(\varepsilon)$, and we can immediately identify

$\tilde{D}(\vec{k}, P_o)$ at c=1. For the present model, the diffusion kernel can also be calculated exactly at c=1, and it is found to depend on ε[31]. In principle, since $P_o(\varepsilon)$ can also be calculated exactly at c=1, we can express the diffusion kernel in terms of $P_o(\varepsilon)$ to obtain $\tilde{D}(\vec{k}, P_o)$ at c=1. Thus, Padé approximants of the type shown in Eq.(12) could also be constructed for the present model to yield approximations to $\tilde{D}$ that are exact at c=1. In practice, determining $\tilde{D}(\vec{k}, P_o)$ at c=1 for the present model is quite tedious, and we will restrict ourselves to approximations of the type given in Eq.(10), which do not yield an exact result at c=1. The lowest order approximation of this type is obtained by equating $\tilde{D}$ to the first term in its density expansion:

$$\tilde{D}(\vec{k}, P_o) = c\, \tilde{D}_1(\vec{k}, P_o). \qquad (21)$$

$\tilde{D}_1(\vec{k}, P_o)$ can be calculated from the solution of the Schrödinger equation for two particles with the Hamiltonian of Eq.(19), according to the procedure described in the Appendix. The final result is

$$\tilde{D}_1(\vec{k}, P_o) = k^{-2} \sum_{\vec{r}} (1-\exp(i\,\vec{k}\cdot\vec{r})) \frac{2[J(\vec{r})]^2\, P_o}{1 + 4\,[J(\vec{r})\, P_o]^2}. \qquad (22)$$

We shall now specialize to the quantum percolation model in which the transfer matrix element $J(\vec{r})$ is nonzero only between sites that are nearest neighbors on the lattice. The quantum percolation problem[32-35] is a special case of the more general problem of Anderson localization[12-20]. In the master equation model of Sec. III, with an excitation transfer rate of finite range, the excitation is confined to finite clusters below the percolation threshold and can move infinitely far from its starting point above the threshold. The following behavior is expected in quantum percolation. Below the percolation threshold, the quantum excitation is necessarily localized on clusters of finite size. However, above the threshold, the quantum excitation may still be localized. An intuitive picture of this situation is that a cluster may have infinite extent but may be sufficiently ramified (i.e., contains many holes) that it cannot be completely explored by a quantum excitation. In fact, it is found that an excitation in the quantum percolation model undergoes a transition from being localized at long time to being delocalized at long time at a critical concentration that lies above the usual percolation threshold[32-35]. We will refer to this concentration as the "quantum percolation threshold", although it should be noted that the usual percolation threshold is a static property of the lattice, while the quantum percolation threshold is a dynamical property of the model. The relationship between the quantum threshold and the static percolation threshold is particularly interesting in two dimensions. According to the scaling theories of Anderson localization, the quantum excitation should be localized in two dimensions in the presence of a finite amount of disorder[15-17]. The quantum percolation threshold, c^*, should be unity. In contrast, it is well known that the static percolation threshold for two-dimensional lattices is less than unity[26].

Substituting a transfer matrix element $J(\vec{r})$ that has the value J for nearest neighbors on the lattice and is zero otherwise into Eqs.(21) and (22) and making use of Eq.(5), we obtain a self-consistent equation for quantum percolation with nearest neighbor interactions:

$$P_o(\varepsilon) = \Omega^{-1} \int d\vec{k}[\varepsilon + k^2 \tilde{D}(\vec{k}, P_o)]^{-1} \tag{23a}$$

$$\tilde{D}(\vec{k}, P_o) = \frac{4 k^{-2} c J^2 P_o}{1 + 4(JP_o)^2} [d - \sum_{j=1}^{d} \cos(\vec{k} \cdot \vec{a}_j)]. \tag{23b}$$

$\vec{a}_j$ is the lattice vector in the jth direction. $D(\vec{k}, \varepsilon)$ can now be calculated self-consistently for the quantum transport problem from

Eqs.(4) and (23). Before doing that, however, we shall modify
Eq.(23a) as follows. We shall separate $P_o(\varepsilon)$ into two parts: a
contribution from integration over small values of k in Eq.(23a),
which is assumed to contain the critical behavior, and $\bar{P}_o(\varepsilon)$, the
contribution from integration over large values of k in Eq.(23a).
$\bar{P}_o(\varepsilon)$ is assumed to be well-behaved in the vicinity of a critical
point. Equation (23a) becomes

$$P_o(\varepsilon) = (a/\pi)^d \int_0^{k_o} k^{d-1} \, dk \, [\varepsilon + k^2 \tilde{D}(0,P_o)]^{-1} + \bar{P}_o(\varepsilon) \qquad (24)$$

The division of a dynamical variable into contributions from small and
large wavevectors is common in mode-coupling theories of critical
dynamics[21,22]. The consequences of partitioning Eq.(23a), as shown
in Eq.(24), will be discussed later. In Eq.(24), $\tilde{D}(0,P_o)$ is the $k \to 0$
limit of $\tilde{D}(\vec{k},P_o)$. a is the lattice spacing. We have introduced a cut-
off wavevector k_o in Eq.(24). The behavior of $P_o(\varepsilon)$ near a critical
point is assumed to be dominated by the contribution to Eq.(23a) with
$k < k_o$. To lowest order in c, $\tilde{D}(0,P_o)$ can be obtained from Eq.(23b).

$$\tilde{D}(0,P_o) = 2J^2 a^2 c P_o / [1 + 4J^2 (P_o)^2]. \qquad (25)$$

For $c < c^*$, the critical concentration, $P_o(\varepsilon)$ will diverge as ε^{-1} for
small ε. For $c > c^*$, the $\varepsilon \to 0$ limit of P_o is finite, but diverges as c^*
is approached. Therefore for small ε, in the vicinity of the critical
point (above and below), $JP_o \gg 1$ in Eq.(25). In this limit, Eq.(25)
becomes

$$\tilde{D}(0,P_o) = ca^2/(2P_o). \qquad (26)$$

We next substitute Eq.(24) for $P_o(\varepsilon)$ into the denominator of the right
side of Eq.(26) to obtain a self-consistent equation for the diffusion
kernel that is valid for small frequencies and wavevectors and for
concentrations near the critical point.

$$\frac{D(0,\varepsilon)}{D_o} = 1 - \left(\frac{2\ a^{d-2}}{c\ \pi^d}\right) \int_0^{k_o} \frac{k^{d-1}\, dk}{k^2 + \varepsilon/D(0,\varepsilon)} \qquad (27)$$

$$D_o = c\ a^2/[2\bar{P}_o(0)]. \qquad (28)$$

In Eq.(27), the $\varepsilon \to 0$ limit of $\bar{P}_o(\varepsilon)$, the non-critical contribution to $P_o(\varepsilon)$, is assumed to be finite in the vicinity of a critical point. We treat D_o as an additional input to the calculation. It can be estimated with a theory that is valid at concentrations greater than and far from the critical point.

Equation (27) is our final self-consistent equation for quantum percolation with nearest-neighbor interactions. It is similar in structure to the equation derived by Vollhardt and Wölfle (VW) for a gas of noninteracting fermions at T = 0 in the presence of randomly placed static scatterers[18-20]. The VW equation can be obtained from Eq.(27) by replacing the factor $2a^{d-2}/(c\pi^d)$ with $d\lambda k_F^{2-d}$, where k_F is the Fermi wavevector, and λ is a dimensionless parameter specifying the degree of disorder. Since the blocked sites in the quantum percolation model can be regarded as randomly placed static scatterers, the types of disorder in the two models are very similar. However, there are fundamental differences between the two approaches. First, in the present work we consider the dynamics of a single quantum particle, while the resummation procedure used by VW to derive their self-consistent equation is closely connected to the many-body properties of the Fermi gas. The present treatment can be applied to a variety of physical problems that are not described by the model of VW, such as the motion of dilute electronic or vibrational excitations[36]. Second, the starting point for the VW resummation procedure is a perturbation expansion in scattering events, in which the zeroth order problem is the <u>delocalized</u> Fermi gas. Our resummation procedure is based on an expansion in c, the concentration of conducting sites, and hence, the zeroth order problem in the present treatment is a <u>localized</u> electron. Finally, in the VW treatment, the two relevant dynamical quantities are $D(0,\varepsilon)$, the frequency-dependent diffusion coefficient, and $M(\vec{k},\varepsilon)$ the current relaxation kernel. In the present treatment, we focus on D and P_o. The fact that our treatment leads to an equation that is similar in structure to the VW equation suggests that the critical dynamics of these two different models may be related.

The primary predictions of Eq.(27) are as follows[18-20]. In one and two dimensions, as $\varepsilon \to 0$,

$$D(0,\varepsilon) \sim \varepsilon, \tag{29}$$

for all values of c. Therefore, the excitation is always localized, and no transition from localized to extended states occurs. In three dimensions, Eq.(27) takes the form

$$\frac{D(0,\varepsilon)}{D_o} = 1 - \left(\frac{c^*}{c}\right) + \left(\frac{\pi c^*}{2 k_o}\right)\left(\frac{\varepsilon}{D(0,\varepsilon)}\right)^{1/2}. \tag{30}$$

There exists a critical concentration $c^* = 2k_o a/\pi^3$, such that for $c<c^*$,

$$D(0,\varepsilon) \sim \varepsilon(c^*-c)^{-2\nu}, \tag{31}$$

and for $c>c^*$,

$$D(0,0) \sim (c-c^*)^s. \tag{32}$$

The critical exponents are $\nu=1$ and $s=1$. These exponents obey the scaling relation $s = (d-2)\nu$, derived by Wegner[37]. At the critical point ($c = c^*$), $D(0,\varepsilon) \sim \varepsilon^{1/3}$, and hence $\langle r^2(t) \rangle$, the mean-squared-displacement of an excitation, is proportional to $t^{2/3}$. On the insulating side of the transition, the ac conductivity (which is proportional to Re $D(0,-i\omega)$) goes as ω^2 for small ω, and as $\omega^{1/3}$ at higher frequencies. The crossover frequency approaches zero as $c \to c^*$.

The frequency dependence of the ac conductivity calculated from Eq.(27) agrees with the predictions of scaling theories, as discussed by Shapiro[20]. The present mode-coupling calculation, which is based on Eq.(24), can thus be used to calculate critical exponents. It does not yield numerical values for the critical concentration c^*. This is not surprising, since c^* depends on the lattice geometry as well as the dimensionality, unlike the critical exponents which are universal and are expected to depend only on the dimensionality. In Eq.(24) we focus on long wavelength properties which will be sensitive to the dimensionality, but not the lattice type.

We shall now comment on the significance of the partitioning introduced in Eq.(24). A self-consistent equation for $P_o(\epsilon)$, which does not involve the division of P_o into contributions from long and short wavelengths, is given by Eqs.(23), which yield the following results. In one and two dimensions, Eq.(29) holds, and the excitation is always localized. In three dimensions, a transition occurs at $c^* \approx 0.48$. For $c<c^*$, Eq.(31) holds with $\nu=1$. However, for $c>c^*$, $s=1/2$ in Eq.(32). According to the scaling theories, s should equal 1 if $\nu=1$. It is, therefore, clear that the modified self-consistent calculation based on Eqs.(24) and (25) is preferable to the calculation based on Eqs.(23). Both calculations are based on an approximation to $\tilde{D}$ (Eq.(23b)) that is valid for small values of c. There is no reason to expect such an approximation to work well at concentrations above the critical point. By introducing the cut-off wavevector k_o in Eq.(24), we introduce a scaling ansatz that holds in the vicinity of the critical point, both from above and from below.

In this work we have used the mode-coupling ideas of Sec. II to derive approximate self-consistent equations for incoherent transport (Eqs.(15)) and for fully quantum mechanical transport (Eq.(27)). Because of the complexity of the quantum problem, Eq.(27) was derived under more restrictive conditions than Eqs.(15). Equations (15) can be used to calculate $P(\vec{r},t)$ for all concentrations, times, and displacements in the incoherent hopping problem. Equations (5), (11), (13), and (14) can be used to perform calculations for any excitation transfer rate $w(\vec{r})$. Equation (27) gives the long-time and long-wavelength transport properties for the quantum problem with interactions that are finite only between nearest neighbors on the lattice at concentrations near (above or below) the critical point. In order to explore the relationship between the critical dynamics in the incoherent hopping model and in the quantum model, we will derive a self-consistent equation for the long wavelength diffusion kernel $D(0,\varepsilon)$ in the incoherent model from the results of Sec. III. This equation is a special case of Eqs.(15), which are valid for all ε and k. Taking the $k \to 0$ limit of Eq.(15b) yields

$$\tilde{D}(0,P_o) = \frac{c\,w\,a^2}{1+2(1-c)P_o w} \qquad (33)$$

w is the magnitude of the excitation transfer rate between nearest neighbors on the lattice. Equation (33) is the analog for the master equation of Eq.(25). Substitution of Eq.(15a) for $P_o(\varepsilon)$ into the denominator of the right side of Eq.(33) yields a self-consistent equation for $D(0,\varepsilon)$ in d dimensions.

$$\frac{D(0,\varepsilon)}{D_o} = 1 - \left(\frac{1-c}{c}\right) I_d[a^2\varepsilon/(2D(0,\varepsilon))] \qquad (34)$$

$$D_o \equiv c\,w \qquad (35)$$

$$I_d(x) \equiv \pi^{-d} \int_0^\pi dq_1 \cdots \int_0^\pi dq_d \, [x+d- \sum_{j=1}^{d} \cos(q_j)]^{-1} \qquad (36)$$

In the vicinity of the critical point (either above or below), $a^2\varepsilon/D(0,\varepsilon)$ approaches zero for small ε. Above the critical point, $D(0,\varepsilon)$ approaches a constant value as $\varepsilon \to 0$. Below the critical point, $D(0,\varepsilon)$ is proportional to ε, with a constant of proportionality that diverges as the critical concentration is approached. In either case, we are interested in small values of $a^2\varepsilon/D(0,\varepsilon)$. The limiting behavior of $I_d(x)$ for small x is given below for $d=1,2,$ and 3.

$$I_d(x) \approx \begin{cases} (2x)^{-1/2} & d=1 \\ (2\pi)^{-1}\ln(x^{-1}) & d=2 \\ I_3(0) - (4/\pi)(2x)^{1/2} & d=3. \end{cases} \quad (37)$$

The constant $I_3(0)$ is 0.4822 to four digits[38]. In order to facilitate comparison between Eq.(27) (for the quantum problem) and Eq.(34) (for the master equation), we consider the limiting behavior of the wavevector integral on the right side of Eq.(27) for small values of $a^2\varepsilon/D(0,\varepsilon)$.

$$K_d(x) \equiv \int_0^{q_0} \frac{q^{d-1} \, dq}{q^2 + x} \, . \quad (38)$$

For small values of x,

$$K_d(x) = \begin{cases} (\pi/2)x^{-1/2} & d=1 \\ [\ln(x^{-1})]/2 & d=2 \\ q_0 - (\pi/2)x^{1/2} & d=3 \, . \end{cases} \quad (39)$$

For small values of x, $K_d(x)$ has the same x dependence as $I_d(x)$ in Eq.(37). Substitution of Eq.(39) into Eq.(27) and of Eq.(37) into Eq.(34) shows that the self-consistent equations for $D(0,\varepsilon)$ for the quantum and incoherent transport problems have the identical structure. For example, consider Eq.(34) in three dimensions. From Eq.(37), we have

$$\frac{D(0,\varepsilon)}{D_0} = 1 - \left(\frac{1-c^*}{1-c}\right)\left(\frac{c^*}{c}\right) + \left(\frac{(1-c)}{c}\right)\left(\frac{4}{\pi}\right)\left(\frac{\varepsilon}{D(0,\varepsilon)}\right)^{1/2} , \quad (40)$$

where $c^* = I_3(0)/(1+I_3(0)) \approx 0.325$. Equation (40) should be compared with Eq.(30). The two equations have the same structure and yield the same critical exponents.

Within the approximations that were made in Sec. III for the incoherent hopping problem and that were made in this section for the quantum problem, the mode-coupling approach yields the same critical dynamics for the two models. We expect that higher order approximations to the functional $\tilde{D}$ will be capable of elucidating the differences in the critical dynamics of these two models, and hence, of providing insight into the influence of quantum coherences on the transport properties of quantum systems.

V. Comparison to Previous Work

In this section, we compare the present self-consistent procedure to previous treatments of incoherent excitation transport (Pauli master equation) and quantum transport (Schrödinger equation). The problem of incoherent motion on disordered lattices with transfer rates of finite range has been successfully treated with the coherent potential approximation (CPA), which in this context is also denoted the coherent medium approximation or the effective medium approximation[10,11]. This technique was first applied to lattices with bond disorder by Odagaki and Lax[10] and to lattices with site disorder by Korzeniewski, Friesner, and Silbey[11]. The lowest order CPA predicts percolation thresholds at $c^*=2/z$ for either site or bond disorder on a lattice with coordination number z. The CPA correctly predicts $c^*<1$ in two dimensions and yields the correct asymptotic behavior of $P_o(t)$, $P_o(t) \sim t^{-d/2}$. For the fully quantum mechanical problem (Anderson localization), the CPA has been used to calculate the density of states[39]. Velicky has developed a generalized CPA that can be used to calculate transport properties such as electrical conductivity[40]. His results, however, cannot be used to predict critical points and exponents. A disadvantage of the CPA is that, although it is useful for models with short-range (e.g., nearest-neighbor) interactions, it cannot be applied in a straightforward manner to long-range (e.g., dipole-dipole) interactions.

Another self-consistent resummation procedure for the master equation was developed by Gochanour, Andersen, and Fayer, who treated incoherent excitation transport among particles randomly distributed in a continuum[5]. Their approach was generalized to substitutionally disordered lattices by Loring, Andersen, and Fayer[8]. This formulation is based on a self-consistent equation for $P_o(\epsilon)$ that was derived diagrammatically and has the capacity to treat both short-range and long-range interactions. Within this approach the predicted $D(\vec{k},\epsilon)$ is not self-consistent with the predicted $P_o(\epsilon)$, in that these quantities do not obey Eq.(3). A consequence of this feature is that the $P_o(t)$ predicted with this formulation decays exponentially at long times ~ $\exp(-\Gamma t)$, rather than as ~ $t^{-d/2}$ as predicted by scaling theory.[6,9,41]

The treatment of Anderson localization by Vollhardt and Wölfle is based on the derivation of self-consistent equations for $D(0,\epsilon)$ and $M(\vec{k},\epsilon)$, where $M(\vec{k},\epsilon)$ is the current relaxation kernel[18]. They consider a Fermi gas in a random potential, and their method cannot be easily generalized to incoherent or quantum excitation transport arising from interactions of arbitrary distance dependence.

The present approach, like other mode-coupling procedures, can be formulated in the time domain rather than in the frequency domain, as in Eq.(5)[23,42]. A comparison of these procedures will be presented in a future work[43]. The present mode-coupling procedure is the only approach that has the capacity to treat both the Anderson localization and the incoherent transport problems with short-range or long-range interactions.

ACKNOWLEDGMENTS

The support of the National Science Foundation, the Office of Naval Research, the Army Research Office, and the Petroleum Research Fund, administered by the American Chemical Society, is gratefully acknowledged.

APPENDIX

In this work, we have made the ansatz that the diffusion kernel could be expressed as a functional of P_o. For the models of Sections III and IV, this ansatz (Eq.(4)) takes the form

$$D(\vec{k},\varepsilon,c) = \tilde{D}(\vec{k}, P_o(\varepsilon,c),c). \tag{A1}$$

In this Appendix we write P_o as $P_o(\varepsilon,c)$, D as $D(\vec{k},\varepsilon,c)$, and $\tilde{D}$ as $\tilde{D}(\vec{k},P_o,c)$ to emphasize the dependence of these quantities on concentration. Equation (A1) is a definition of the functional $\tilde{D}$. In this Appendix, we present a systematic prescription for evaluating $\tilde{D}$. We begin by expanding D and P_o in powers of c.

$$D(\vec{k},\varepsilon,c) = \sum_{j=1}^{\infty} c^j D_j(\vec{k},\varepsilon^{-1}) \tag{A2}$$

$$P_o(\varepsilon,c) = \varepsilon^{-1} + \sum_{j=1}^{\infty} c^j Q_j(\varepsilon^{-1}). \tag{A3}$$

D_j and Q_j can be calculated from the exact solution of the transport problem for j or fewer particles. We shall take $\{D_j\}$ and $\{Q_j\}$ as given and use them to construct an expression for $\tilde{D}$. $\tilde{D}(\vec{k},P_o,c)$ depends explicitly on c, as well as depending implicitly on c through the c dependence of P_o. We expand $\tilde{D}$ in a power series with respect to its explicit c dependence.

$$\tilde{D}(\vec{k}, P_o, c) = \sum_{j=1}^{\infty} c^j \tilde{D}_j(\vec{k}, P_o). \tag{A4}$$

The function $D_n(\vec{k}, \varepsilon^{-1})$ in Eq.(A2) is given by

$$D_n(\vec{k}, \varepsilon^{-1}) = \frac{1}{n!} \frac{\partial^n}{\partial c^n} D(\vec{k}, \varepsilon, c) \bigg|_{c=0}$$

$$= \frac{1}{n!} \frac{\partial^n}{\partial c^n} \tilde{D}(\vec{k}, P_o, c) \bigg|_{c=0}. \tag{A5}$$

The second equality in Eq.(A5) follows from the definition of $\tilde{D}$ in Eq.(A1). Substitution of Eq.(A4) into Eq.(A5) and application of Leibnitz' Rule for the nth derivative of a product yields

$$D_n(\vec{k}, \varepsilon^{-1}) = \sum_{j=1}^{n} \frac{1}{(n-j)!} \frac{\partial^{n-j}}{\partial c^{n-j}} \tilde{D}_j(\vec{k}, P_o) \bigg|_{c=0}. \tag{A6}$$

An expression for $\tilde{D}_n(\vec{k}, P_o)$ is obtained by solving Eq.(A6) for $\tilde{D}_n(\vec{k}, \varepsilon^{-1})$ and replacing ε^{-1} with P_o.

$$\tilde{D}_n(\vec{k}, P_o) = D_n(\vec{k}, P_o) - \sum_{j=1}^{n-1} \frac{1}{(n-j)!} \frac{\partial^{n-j}}{\partial c^{n-j}} \tilde{D}_j(\vec{k}, P_o(x,c)) \bigg|_{\substack{c=0 \\ x=P_o}}$$

$$\tag{A7}$$

Equation (A7) relates $\tilde{D}_n$ to D_n and to $\{\tilde{D}_j\}$ with $j<n$. Using Eqs.(A2), (A3), and (A7), we can relate the unknown function $\tilde{D}_n$ to the known functions D_j and Q_j with $j<n$. In this manner, we may calculate each term in the series for $\tilde{D}$ in Eq.(A4). For illustrative purposes, we will calculate $\tilde{D}_n(\vec{k}, P_o)$ explicitly for n=1 and n=2. For n=1, Eq.(A7) becomes

$$\tilde{D}_1(\vec{k}, P_o) = D_1(\vec{k}, P_o). \tag{A8}$$

For n=2, we have

$$\tilde{D}_2(\vec{k},P_o) = D_2(\vec{k},P_o) - \frac{\partial}{\partial c}\left. \tilde{D}_1(\vec{k},P_o(x,c))\right|_{\substack{c=0 \\ x=P_o}}$$

$$= D_2(\vec{k},P_o) - \left.\frac{\partial \tilde{D}_1(\vec{k},y)}{\partial y}\right|_{y=P_o} \left.\frac{\partial P_o(x,c)}{\partial c}\right|_{\substack{c=0 \\ x=P_o}}$$

$$= D_2(\vec{k},P_o) - Q_1(P_o) \left.\frac{\partial D_1(\vec{k},y)}{\partial y}\right|_{y=P_o}. \quad (A9)$$

REFERENCES

1. E. W. Montroll and G. H. Weiss, J. Math. Phys. **6**, 167 (1965); G. H. Weiss and R. J. Rubin, Adv. Chem. Phys. **52**, 363 (1983).

2. H. Scher and M. Lax, Phys. Rev. B **7**, 4491 (1973); **7**, 4502 (1973).

3. M. F. Shlesinger, J. Stat. Phys. **10**, 421 (1974); M. F. Shlesinger and E. W. Montroll in Studies in Statistical Mechanics, Vol. **11**, p.1, J. L. Lebowitz, ed., North Holland (1984).

4. J. Klafter and R. Silbey, J. Chem. Phys. **72**, 849 (1980); **72**, 843 (1980).

5. C. R. Gochanour, H. C. Andersen, and M. D. Fayer, J. Chem. Phys. **70**, 4254 (1979).

6. B. Movaghar and G. W. Sauer, J. Phys. C**13**, 4933 (1980).

7. S. Alexander, J. Bernasconi, W. R. Schneider, and R. Orbach, Rev. Mod. Phys. **53**, 175 (1981).

8. R. F. Loring, H. C. Andersen, and M. D. Fayer, J. Chem. Phys. **80**, 5731 (1984).

9. J. Nieuwoudt and S. Mukamel, Phys. Rev. B **30**, 4426 (1984); J. Stat. Phys. **36**, 677 (1984).

10. T. Odagaki and M. Lax, Phys. Rev. B **25**, 2307 (1982); **26**, 6480 (1983).

11. G. Korzeniewski, R. Friesner, and R. Silbey, J. Stat. Phys. **31**, 451 (1983); G. Korzeniewski and D. Calef, J. Phys. C **16**, 4599 (1983).

12. P. W. Anderson, Phys. Rev. **109**, 1492 (1958).

13. N. F. Mott and E. A. Davis, Electronic Processes in Non-crystalline Materials, (Clarendon, Oxford, 1979).

14. E. N. Economou and M. H. Cohen, Phys. Rev. B **5**, 2931 (1972).

15. D. J. Thouless, in Ill-Condensed Matter, eds. R. Balian, R. Maynard, and G. Toulouse (North-Holland, Amsterdam, 1979).

16. Anderson Localization, eds. Y. Nagaoka and H. Fukuyama (Springer, Berlin, 1982)

17. E. Abrahams, P. W. Anderson, D. C. Licciardello, and T. V. Ramakrishnan, Phys. Rev. Lett. **42**, 673 (1979).

18. D. Vollhardt and P. Wölfle, Phys. Rev. Lett. **45**, 842 (1980); Phys. Rev. B **22**, 4666 (1980); Phys. Rev. Lett. **48**, 649 (1982).

19. S. Mukamel, in Random Walks and Their Applications in the Physical and Biological Sciences, eds. M. F. Shlesinger and B. J. West, AIP Conference Proceedings **109**, 1984, p. 205.

20. B. Shapiro, Phys. Rev. B **25**, 4266 (1982).

21. K. Kawasaki, in Phase Transitions and Critical Phenomena, eds. C. Domb and M. S. Green (Academic, New York, 1971) Vol. **5A**, p. 166.

22. T. Keyes, in Statistical Mechanics, ed. B. J. Berne (Plenum, New York, 1977), Vol. **6B**.

23. S. Mukamel, Phys. Rev. B **25**, 830 (1982).

24. S. R. Broadbent and J. M. Hammersley, Proc. Camb. Phil. Soc. **53**, 629 (1957).

25. V. K. S. Shante and S. Kirkpatrick, Adv. Phys. **20**, 325 (1971).

26. D. Stauffer, Phys. Rep. **54**, 1 (1979).

27. G. H. Hardy, Divergent Series (Oxford University Press, Oxford, 1949).

28. M. F. Sykes, D. S. Gaunt, and M. Glen, J. Phys. A **9**, 1705 (1976).

29. J. Straley, Phys. Rev. B **15**, 5733 (1977).

30. S. Mukamel, Phys. Rep. **93**, 1 (1982).

31. V. M. Kenkre, in Exciton Dynamics in Molecular Crystals and Aggregates, (Springer, Berlin, 1982).

32. S. Kirkpatrick and T. P. Eggarter, Phys. Rev. B **6**, 3598 (1972).

33. R. Raghavan and D. C. Mattis, Phys. Rev. B **23**, 4791 (1981).

34. Y. Shapir, A. Aharony, and A. B. Harris, Phys. Rev. Lett. **49**, 486 (1982).

35. T. Odagaki and K. C. Chang, Phys. Rev. B **30**, 1612 (1984).

36. R. Kopelman, in Modern Problems in Condensed Matter Sciences, eds. V. M. Agranovich and A. A. Maradudin (North-Holland, Amsterdam, 1983), Vol. **4**.

37. F. J. Wegner, Z. Phys. B **25**, 327, (1976).

38. S. Katsura, S. Inawashiro, and Y. Abe, J. Math. Phys., **12**, 895 (1971).

39. F. Yonezawa and K. Morigaki, Suppl. Prog. Theor. Phys. **53**, 1 (1973).

40. B. Velicky, Phys. Rev. **184**, 16 (1969).

41. S. G. Fedorenko and A. I. Burshtein, Chem. Phys. **98**, 314 (1985).

42. S. Mukamel, J. Stat. Phys. **27**, 317 (1982); **30**, 179 (1983).

43. R. F. Loring and S. Mukamel, to be published.

CARRIER TRANSPORT IN DYNAMICALLY DISORDERED SYSTEMS

Stephen D. Druger
Department of Chemistry and Materials Research Center
Northwestern University, Evanston, IL 60201 USA

ABSTRACT

Existing transport theories must be modified when the host medium undergoes microscopic structural changes on a timescale comparable to that of the carrier motion, corresponding thereby to dynamic disorder. Earlier relations giving average transport behavior for dynamically disordered systems in terms of that for statically disordered systems are summarized. A new, generalized, and far more widely applicable relation of this kind is shown to follow from the renewal properties of the models considered and is applied to several special cases. The formal relation between the dynamic-disorder renewal process of fluctuations in hopping rates and the renewal process of the hopping itself in a generalized CTRW model is examined, and the relation to other recent work on dynamic disorder is also briefly discussed.

I. Introduction

Charge or mass transport often occurs in a host medium that undergoes microscopic structural changes on a timescale comparable to that of the carrier motion itself; examples of such systems include biomembranes,[1] solid protonic conductors,[2] and polymer electrolytes.[3]

In certain biomembranes such as gramicidin-A, ions move through molecular channels along which they encounter potential barriers that fluctuate in time, corresponding to a structural host evolution that controls the carrier transport. Similarly, the Grotthus mechanism of protonic conduction allows site-to-site carrier motion only between those neighboring H_2O or NH_3 groups that have the proper relative orientation; thermally activated rotation of these groups is the structural host process interacting with the carrier motion. The polymer electrolytes (e.g., polyethylene oxide complexed nonstoichiometrically with a salt such as NaSCN) have been of primary interest in the present work. These systems are almost always studied above the glass transition temperature, where the carriers move between well-defined solvation sites in a host environment that undergoes a great variety of large-scale non-vibrational motions, so that favored pathways are continually changing into unfavored pathways, and vice versa.

In the usual models[4] applied to statically disordered systems, the transfer rates $\{W_{ij}\}$ are assigned permanently between pairs of sites in some random way, and an ensemble average is performed over various $\{W_{ij}\}$ configurations. The systems of present interest are characterized in addition by dynamic disorder, which means that as the carrier motion within each configuration occurs, the W_{ij} rates between sites are concurrently randomly reassigned at some representative rate λ characterizing the microscopic structural evolution of the host. A special case of this dynamic disorder hopping (DDH) model is the dynamic bond percolation (DBP) model, in which hops are allowed only between nearest-neighbor sites on a periodic lattice, and the corresponding W_{ij} hopping rates are restricted to the binary choice of values w (with probability f) and 0 (with probability 1-f). Many of our previously-published[5,6] results already apply to the more general DDH model based on a possibly-continuous distribution of W_{ij} values. Included are expressions giving the average transport behavior for the dynamically disordered systems in terms of average transport behavior for a corresponding (and usually more easily and widely studied) statically disordered system; these relations play a pivotal role in applications to experiment, but have previously been derived and studied only for the special case in which the probability for a reassignment event in the time interval (t,t+dt] is λdt independent of any other reassignment (so that the probability distribution of waiting times is exponential).

Two essential properties of the random $\{W_{ij}\}$ reassignment events we consider are (1) that each one reestablishes the same statistical state existing immediately after any other reassignment (i.e., with the carrier at its particular location surrounded by completely random $\{W_{ij}\}$ configurations), and (2) that the temporal probability density $\psi(t)$ for the next reassignment to occur at time t later (i.e., the "waiting time distribution") immediately begins evolving anew from t=0 upon reassignment. The theme of this presentation is the way in which these two "renewal" properties imply both the previously published relation and other far more general relations giving average dynamic disorder transport in terms of average static disorder transport behavior.

The next section reviews (in a somewhat generalized form) some previous results that are relevant to the present paper (but which are still based on restrictive assumptions such as the exponential waiting-time distribution) and briefly describes their preliminary application to experiment. In Sec. III, the precise form taken by the renewal properties for our dynamic-disorder models is stated, is shown to be a generalization of the property previously derived for a specific sequence of reassignment events, and is shown to imply very general relations giving dynamic-disorder transport in terms of static-disorder transport. Under additional special assumptions, more satisfactory and less restrictive results are generated for some of the cases previously considered. Section IV examines the formal relationship between our DDH models and the CTRW models describing the hopping process itself, and finally comments on the relation of the results presented here to other recent work on dynamic disorder.

II. Previous Dynamic Disorder Results

For the transport on a periodic lattice of sites, the behavior described by the master equation

$$\dot{P}_i(t) = \sum_j W_{ij} P_j - \left[\sum_j W_{ji}\right] P_i \tag{1}$$

(for $P_i(t)$ the probability of occupancy of site i at time t) is equivalent to a random walk with W_{ij} the probability per unit time for an instantaneous hop from site j to site i; there are, however, interesting random walks[7] not described by the simple master equation (1). The behavior of interest is an average over configurations of randomly assigned $\{W_{ij}\}$ values and, in the special case of bond percolation, with an average fraction f of nearest neighbor $W_{ij} = w$ being the only non-zero transfer rates. Dynamic disorder (in both the general and bond percolation cases) has been studied in our earlier work by assuming an exponential waiting-time distribution for reassignment events, in which the W_{ij} configuration is randomly reassigned independently of the carrier location[5]. (The term "renewal" has been used previously by us[5,6] to mean simply "W_{ij} reassignment," but the reassignments considered in this present paper also constitute a special case of a "renewal process" in the sense used in probability theory.)

Transport coefficients can be obtained for these models by considering such quantities as the derivative $\frac{d}{dt}\langle\vec{r}\rangle(t)$ of the mean displacement from initial position, or alternatively, by considering the behavior of

$$\langle\delta r^2\rangle(t) \equiv \langle|\vec{r}(t) - \vec{r}_0|^2\rangle - |\langle\vec{r}\rangle(t)|^2 \tag{2}$$

and the corresponding generalized diffusion coefficient[8]

$$D(\omega) = n_d(-i\omega)^2 \lim_{\epsilon\to 0^+} \int_0^\infty e^{-\epsilon t} e^{-i\omega t} \langle\delta r^2\rangle(t) dt . \tag{3}$$

Equations (2) and (3) at zero applied field, together with the Einstein relation

$$\sigma(\omega) = \frac{ne^2}{k_B T} D(\omega) , \tag{4}$$

give the conductivity, with $n_d = 1/(2d)$ for dimensionality d. Of course, at zero field $\langle\vec{r}\rangle(t)$ is zero so that $\langle\delta r^2\rangle(t)$ is simply $\langle r^2\rangle(t)$, but Eqs. (2) and (3) are used here in a form that allows extension to the strong-field case.

Consider first the relation between dynamic-disorder and static-disorder transport for a specific sequence of times $(t_1, t_2, \ldots, t_N)$ at which $\{W_{ij}\}$ reassignments occur, with no other reassignments in the intervals $(t_0 = 0, t_1)$ and (t_N, T), and with the process of random reassignment and random hopping having been ongoing from $t = -\infty$ up to t_0. For the average behavior without further reassignment, we allow the mean carrier displacements (such as $\langle\vec{r}\rangle_0(t)$ or $\langle\delta r^2\rangle_0(t)$) to each have a functional form that can differ depending on whether $t = 0$ is a random starting time during the ongoing process or is the time of the last previous reassignment. In this and other respects, the behavior of $\langle\delta r^2\rangle(t)$ shown here to result from the reassignment sequence (and illustrated schematically in Fig. 1) is more general than demonstrated previously.[5]

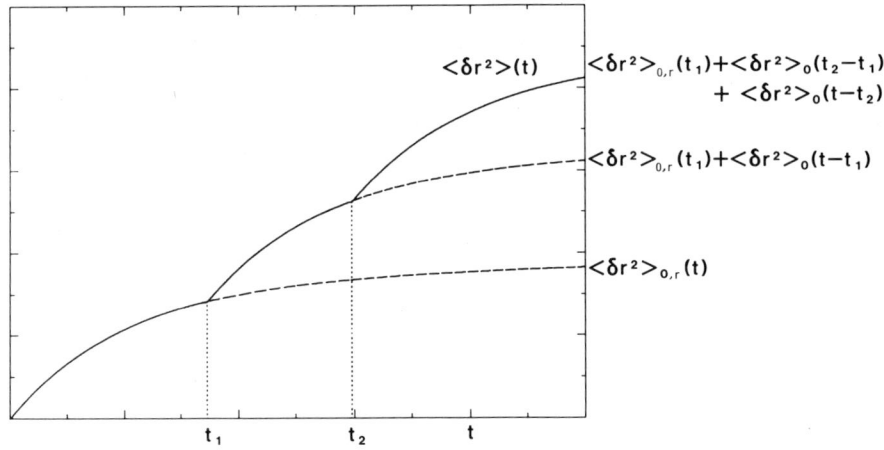

Fig. 1. Mean-square displacement for a specified sequence of reassignment events.

Let $p_N(\vec{s}, t_N + \tau)$ be the probability at time $t_N + \tau$ for displacement of the carrier by s from its initial position (at $t = 0$). (The subscript "M" in "p_M" or "$\langle\delta r^2\rangle_M$", etc., denotes "after reassignment at t_M".) In terms of summation over lattice sites $\{\vec{s}\}$

$$\langle r^2\rangle_N(t_N + \tau) = \sum_s s^2 \, p_N(\vec{s}, t_N + \tau). \tag{5}$$

But because each reassignment reestablishes the condition of a completely random $\{W_{s,s'}\}$ assignment independent of the carrier location, $p_N(\vec{s}, t_N + \tau)$ is given by

$$p_N(\vec{s}, t_N + \tau) = \sum_{s'} p_{N-1}(\vec{s'}, t_N) p_0(\vec{s}-\vec{s'}, \tau) \tag{6}$$

where $p_0(\vec{s}, \tau)$ is the probability for carrier displacement $\vec{s}$ <u>given</u> that there is a renewal at $\tau = 0$ and no subsequent renewal. Then the corresponding mean-square displacement at $T = t_N + \tau$, found by expanding $s^2 = [(\vec{s}-\vec{s}') + \vec{s}']^2$ in (5), and then using (6) together with the normalization condition $\int p(s,t) = 1$ and lattice periodicity, becomes

$$\langle r^2 \rangle_N(t_N + \tau) = \langle r^2 \rangle_0(\tau) + \langle r^2 \rangle_{N-1}(t_N) + 2\langle \vec{r} \rangle_0(\tau) \cdot \langle \vec{r} \rangle_{N-1}(t_N). \tag{7}$$

Similar evaluation of $\langle \vec{r} \rangle_N(t_N + \tau)$ gives

$$\langle \vec{r} \rangle_N(t_N + \tau) = \langle \vec{r} \rangle_0(\tau) + \langle \vec{r} \rangle_{N-1}(t_N). \tag{8}$$

Subtracting the square of (8) from (7) leads to

$$\langle \delta r^2 \rangle_N(t_N + \tau) \equiv \langle r^2 \rangle_N(t_N + \tau) - \left| \langle \vec{r} \rangle_N(t_N + \tau) \right|^2$$

$$= \langle \delta r^2 \rangle_{N-1}(t_N) + \langle \delta r^2 \rangle_0(\tau) \tag{9}$$

where the subscript "0" consistently denotes "starting from reassignment at $t = 0$ without further reassignment." Then (9) (with τ replaced by $t_N - t_{N-1}$) can be used to express $\langle \delta r^2 \rangle_{N-1}(t_N)$ in terms of $\langle \delta r^2 \rangle_{N-2}(t_{N-1})$ and the procedure repeated iteratively to yield

$$\langle \delta r^2 \rangle_N(t_N + \tau) = \langle \delta r^2 \rangle_{0,r}(\tau_0) + \sum_{i=1}^{N} \langle \delta r^2 \rangle_0(\tau_i). \tag{10}$$

Similar iterative use of (8) gives

$$\langle \vec{r} \rangle_N(t_N + \tau) = \langle \vec{r} \rangle_{0,r}(\tau_0) + \sum_{i=1}^{N} \langle \vec{r} \rangle_0(\tau_i) \tag{11}$$

where the subscript "0,r" denotes "starting from a random instant for the ongoing process and continuing without further reassignment," and where

$$\tau_i = \begin{cases} t_1 & i = 0 \\ t_{i+1} - t_i & 0 < i < N \\ T - t_N & i = N. \end{cases} \tag{12}$$

The resulting behavior, illustrated in Fig. 1, is always diffusion-like for observation times $T \gg \lambda^{-1}$ (where λ is a representative renewal rate). Similar behavior for $\langle x \rangle(t)$ follows immediately from (11), and for other displacements such as $\langle \delta x^2 \rangle(t)$ can be proven by similar methods.

The physical problem of interest, however, involves an average over all possible random reassignments. We have previously considered[6] the simplified case of reassignments equally likely to occur (at rate λ) without regard for the occurrence of any other reassignment, and have used a generating function technique and a somewhat more elaborate (though less general) version of Eq. (6) to show that the diffusion coefficient in the absence of an applied field is given by

$$D(\lambda, i\omega) = n_d(\lambda + i\omega)^2 \int_0^\infty e^{-(\lambda+i\omega)t} \langle r^2 \rangle_0(t) \, dt \qquad (13)$$

(cf., Eq.(3)). Then (13) implies

$$D(\lambda, \omega) = D(0, \omega - i\lambda) \qquad (14)$$

showing how the dynamic-disorder diffusion coefficient can be obtained from analytic expressions for the static-disorder diffusion coefficient through the formal substitution $i\omega \to \lambda + i\omega$. This result is not limited to a binary choice of W_{ij} values as w or 0, and is an exact relationship for the class of models considered. (Our previous derivation[6] of (13) does, however, employ Fourier summations requiring a periodic lattice of sites.) Results similar to (14) have been shown by Harrison and Zwanzig[9] to hold for a closely related dynamic bond percolation model under an effective medium approximation. The requirement that the right-hand sides of Eqs. (13) and (3) be equal for all ω also leads[6] to a relationship giving $\langle r^2 \rangle(t)$ (with renewal) in terms of $\langle r^2 \rangle_0(t)$ (without renewal), namely

$$\frac{d^2}{dt^2} \langle r^2 \rangle(t) = e^{-\lambda t} \frac{d^2}{dt^2} \langle r^2 \rangle_0(t) \qquad (15)$$

with initial conditions $\langle r^2 \rangle(0) = 0$, and

$$\left. \frac{d\langle r^2 \rangle}{dt} \right|_{t=0} = \left. \frac{d\langle r^2 \rangle_0}{dt} \right|_{t=0}. \qquad (16)$$

An analogous result holds for $\langle x^2 \rangle(t)$.

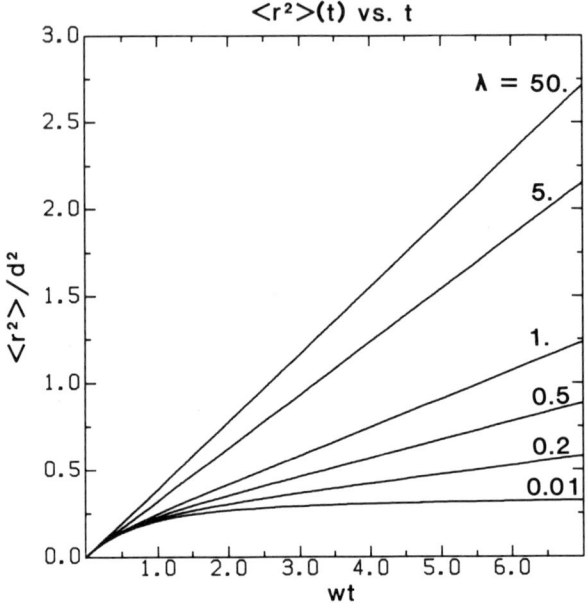

Fig. 2. Mean-square displacement in one-dimension for various reassignment rates, calculated for f = 0.2.

All of these relations allow the results of extensive work on static disorder hopping models and static percolation transport to be applied in studying dynamic disorder transport. An example is the application of Eqs. (13)-(16) to exact analytic solutions for $\langle r^2 \rangle_0(t)$ and $D(0,\omega)$ in the case of one-dimensional static bond percolation, converting them into solutions for dynamic percolation. As illustrated in Fig. 2, the resultant long-time slope of $\langle r^2 \rangle(t)$ vs. t (corresponding to the zero-frequency diffusion coefficient) increases approximately linearly with λ for small λ/w, corresponding to a renewal-controlled transport regime. For large λ/w, at least one renewal occurs before almost every hop, and because a $\{W_{ij}\}$ reassignment under present assumptions completely randomizes the $\{W_{ij}\}$ configuration and has no effect on the time distribution of other reassignments, further increase in the average number of renewals between hops has almost no effect. These limiting behaviors can also be predicted directly from Eqs. (13)-(16).

A further example of the utility of Eqs. (13)-(16) is their preliminary application[10,11] to frequency-dependent conductivity data in polymer electrolytes, based on a number of ad hoc assumptions. Specifically, we assume[11] (1) that the conductivity response of charge distributions bound to the polymer host and the response resulting from long-range carrier motion are separable and additive, (2) that the long-range motion can be described by dynamic bond percolation (DBP) below the percolation threshold, and (3) that $\langle r^2 \rangle_0(t)$ for the long-range motion is approximately described by simple exponential saturation behavior

$$\langle r^2 \rangle_0(t) \approx A_0 [1 - e^{-\alpha_0 t}] \tag{17}$$

while the corresponding behavior for the bound-carrier motion (to be used in the linear-response result for $\sigma(\omega)$) is approximately a sum of N terms of the form (17). We thereby obtain the conductivity expressed in the Debye form

$$\sigma(\omega) = \sigma(0) + \sum_{J=0}^{N} A_J \frac{i\omega\tau_J}{1 + i\omega\tau_J} \tag{18}$$

where τ_J is α_J^{-1} except for the long-range τ_0 contribution, which is $(\alpha_0 + \lambda)^{-1}$. Although the presence of a large number of fitting parameters $\{A_J, \tau_J\}$ not determined in terms of the DBP model itself is unsatisfactory, Eq. (18) provides some information about the general behavior of $\sigma(\omega)$ to be expected from a more detailed application. Specific conclusions obtained from a fit to experimental data for $(PEO)_8NH_3CO_3SF_3$ and $(PEO)_x \cdot NaSCN$ are discussed elsewhere.[10,11]

III. Generalized Dynamic Disorder Hopping Relations

Previous work on dynamic disorder hopping has been based on restrictive model assumptions such as those described in Sec. II. We now show how most of these restrictions can be relaxed.

A. The Generalized Result

For the sake of generality, consider some observable F whose average is F(t) under the influence of dynamic disorder, and which exhibits the behavior illustrated in Fig. 1. Specifically, we require that for any sequence of $\{W_{ij}\}$ reassignment times $(t_1, t_2, \cdots, t_N)$ during the period $(t_0=0, T)$, that $F(T)$ be given by

$$F(T) = g(\tau_0) + \sum_{i=1}^{N} f(\tau_i) \tag{19}$$

where $g(0) = f(0) = 0$. Here $f(\tau)$ gives the growth in $F(t)$ that would be observed without further reassignment starting from a previous reassignment at $\tau = 0$, while $g(\tau)$ is the corresponding growth observed without reassignment starting from the arbitrary time $t_0 = 0$ before which random reassignment had been an ongoing process. The $\{\tau_i\}$ are defined by (12). In the average over a distribution of $\{\tau_i\}$, we also allow the probability density $\phi(\tau)$ for the first reassignment after $t_0 = 0$ (corresponding to probability $\phi(\tau)d\tau$ for reassignment in the interval $(\tau, \tau + d\tau]$) to differ from the corresponding waiting-time probability density $\psi(\tau)$ from one reassignment to the next. From this, without any additional assumption about the nature of the hopping process that leads to the growth laws $g(\tau)$ and $f(\tau)$ for the statically disordered system, we express the Laplace transform of $F(t)$ for the dynamically disordered system in terms of Laplace transforms involving only the probability densities ϕ and ψ and the growth laws g and f describing static disorder behavior. The mean displacements $\langle \delta r^2 \rangle(t)$ and $\langle x \rangle(t)$ have been shown in Sec. II to satisfy the required conditions on $F(t)$, and judicious use of the relation linking F with f, g, ψ, and ϕ for these displacements then gives the desired expressions for dynamic disorder transport coefficients.

To evaluate $F(T)$ as the average of (19) over possible $\{\tau_i\}$, first take the time axis to be discretized into intervals of length Δt with the $\Delta t \to 0$ limit to be taken later. The summation over averages for zero renewals (given by S_0), for one renewal (S_1), and for $2 \leq N < \infty$ renewals (S_2) then yields

$$F(T) = S_0 + S_1 + S_2 \tag{20}$$

$$= \Phi(T)g(T) + \sum_{\tau_0} [\phi(\tau_0)\Delta t]\Psi(T - \tau_0)[g(\tau_0) + f(T - \tau_0)]$$

$$+ \sum_{N=2}^{\beta} \sum_{\{\tau_i\}} [\phi(\tau_0)\Delta t] [\prod_{i=1}^{N-1} \psi(\tau_i)\Delta t] \Psi(T - \sum_{i=0}^{N-1} \tau_i)]$$

$$\times [g(\tau_0) + \sum_{i=1}^{N} f(\tau_i)] \tag{21}$$

where

$$\Phi(t) = 1 - \int_t^\infty \phi(\tau) d\tau; \quad \Psi(t) = 1 - \int_t^\infty \psi(\tau) d\tau \qquad (22)$$

are the probabilities for no reassignment until time t starting, respectively, from an arbitary time at t = 0 or from a previous reassignment at t = 0. The summation over $\{\tau_i\}$ is for all allowable combinations of the discrete $\tau_0, \tau_1, \cdots$ such that (cf. Eq. (12))

$$\sum_{i=0}^N \tau_i = T. \qquad (23)$$

It proves useful to express S_0 as

$$S_0 = \int_0^\infty \delta(T - \tau)\Phi(\tau)g(\tau) d\tau$$

$$= \frac{1}{2\pi} \int_{-\infty}^\infty d\omega \, e^{i\omega T} [\int_0^\infty e^{-i\omega\tau} \Phi(\tau)g(\tau)d\tau] . \qquad (24)$$

With the $\Delta t \to 0$ limit taken, and with

$$\int_0^\infty \delta(T - \sum_{i=0}^N \tau_i) d\tau_N = \frac{1}{2\pi} \int_0^\infty d\tau_N \int_0^\infty d\omega \exp[i(\omega T - \sum_j \omega \tau_j)] \qquad (25)$$

introduced to take account of the constraint (23), we find after some rearrangements

$$S_2 = \sum_{N=2}^\infty \int_{-\infty}^\infty e^{i\omega T} \int_0^\infty e^{-i\omega\tau_0} \{\Phi(\tau_0)[\prod_{i=1}^{N-1} \int_0^\infty e^{-i\omega\tau_i} \psi(\tau_i)]$$

$$\times \int_0^\infty \Psi(\tau_N) e^{-i\omega\tau_N} [g(\tau_0) + \sum_{k=1}^N f(\tau_k)]\} \{d\tau_i\} d\omega . \qquad (26)$$

Then S_2 is the sum of three terms s_1, s_2, and s_3 arising, respectively, from $g(\tau_0)\Phi(\tau_0)$, from $f(\tau_i)\psi(\tau_i)$ for $1 \leq i < N$, and from $f(\tau_N)\Psi(\tau_N)$. The first term is, for example,

$$s_1 = \frac{1}{2\pi} \sum_{N=2}^\infty \int_{-\infty}^\infty d\omega e^{i\omega T} \int_0^\infty \cdots \int_0^\infty [g(\tau_0)e^{-i\omega\tau_0}][\prod_{j=1}^{N-1} \psi(\tau_j)e^{-i\omega\tau_j}]$$

$$\Psi(\tau_N)e^{-i\omega\tau_N} \{d\tau_i\}$$

$$= \frac{1}{2\pi} \sum_{N=2}^{\infty} \int_{-\infty}^{\infty} d\omega e^{i\omega T} \tilde{g}(i\omega)[\tilde{\psi}(i\omega)]^{N-1} \tilde{\Psi}(i\omega) \tag{27}$$

so that, with s_2 and s_3 similarly evaluated, we obtain

$$S_2 = s_1 + s_2 + s_3$$

$$= \frac{1}{2\pi} \sum_{N=2}^{\infty} \int_{-\infty}^{\infty} d\omega \, e^{i\omega T} [\tilde{\psi}^{N-1} \tilde{g} \tilde{\Psi} + (N-1)\tilde{\psi}^{N-2} \tilde{\phi} \tilde{f} \tilde{\Psi}$$

$$+ \tilde{\psi}^{N-1} \tilde{\Phi} \tilde{f}_N] \tag{28}$$

in terms of Laplace transforms involving $\psi(t)$, $\phi(t)$, $f(t)$, and $g(t)$ at Laplace argument $i\omega$. Specifically, for a function $h(t)$, define

$$L[h(t)] \equiv \int_0^{\infty} e^{-i\omega t} h(t) \, dt \tag{29}$$

as the desired Laplace-Fourier transform. For the probabilities in (28), the definition used is

$$\tilde{\psi} \equiv L[\psi(t)]; \quad \tilde{\phi} \equiv L[\phi(t)]; \quad \tilde{\Psi} \equiv L[\Psi(t)]; \quad \tilde{\Phi} \equiv L[\Phi(t)] \,. \tag{30}$$

The transforms of $f(t)$ and $g(t)$ employed here are, however, defined as

$$\tilde{f} \equiv L[\psi(t)f(t)]; \quad \tilde{f}_N \equiv L[\Psi(t)f(t)]; \quad \tilde{f}_0 \equiv L[f(t)], \tag{31}$$

$$\tilde{g} \equiv L[\phi(t)g(t)]; \quad \tilde{g}_N \equiv L[\Phi(t)g(t)]; \quad \tilde{g}_0 \equiv L[g(t)]. \tag{32}$$

Similar evaluation of S_1 shows that Eq. (28) for S_2 can be changed to $S_1 + S_2$ merely by starting the summation at $N = 1$ instead of $N = 2$, and the summations over N can then be evaluated using

$$\sum_{N=1}^{\infty} x^{N-1} = (1-x)^{-1}; \quad \sum_{N=1}^{\infty} (N-1)x^{N-2} = (1-x)^{-2}. \tag{33}$$

We thereby obtain

$$F(T) = S_1 + S_2 + S_3$$

$$= \frac{1}{2\pi} \int_{-\infty}^{\infty} \{ \tilde{g}_N + \frac{\tilde{g} \tilde{\Psi} + \tilde{\phi} \tilde{f}_N}{1 - \tilde{\psi}} + \frac{\tilde{\phi} \tilde{f} \tilde{\Psi}}{(1-\tilde{\psi})^2} \} e^{i\omega T} \, d\omega \,. \tag{34}$$

(A small imaginary part is implicitly included in ω.) The integral over ω is the Bromwich formula[12] for the inverse Laplace transform (for imaginary $u = i\omega$), so that if $\widetilde{F}(\omega) = L(F)$, then Eq. (34) implies

$$\widetilde{F} = \widetilde{g}_N + \frac{\widetilde{g}\,\widetilde{\Psi} + \widetilde{\phi}\,\widetilde{f}_N}{1 - \widetilde{\psi}} + \frac{\widetilde{\phi}\,\widetilde{f}\,\widetilde{\Psi}}{(1 - \widetilde{\psi})^2} \quad . \tag{35}$$

This can be simplified by noting that the definition of $\Phi(t)$ in terms of $\phi(t)$ implies a relation between Φ and ϕ, specifically $d\Phi/dt = -\phi(t)$ (with a corresponding relation between Ψ and ψ) so that a partial integration of

$$\widetilde{\Phi}(i\omega) = \int_0^\infty e^{-i\omega t} \Phi(t)\, dt \tag{36}$$

(with analogous treatment of Ψ) yields

$$i\omega\widetilde{\Phi}(i\omega) = 1 - \widetilde{\phi}(i\omega); \quad i\omega\widetilde{\Psi}(i\omega) = 1 - \widetilde{\psi}(i\omega). \tag{37}$$

Also, $\widetilde{f}(i\omega)$ is related to $\widetilde{f}_N(i\omega)$, as seen by considering the partial integration result

$$L[\tfrac{d}{dt}(\Phi f)] = e^{-i\omega t}\Phi(t)f(t)\Big|_0^\infty + i\omega L(\Phi f) \tag{38}$$

with $f(t=0) = 0$. We thereby obtain from (38)

$$\widetilde{f}(i\omega) = -i\omega\widetilde{f}_N(i\omega) + L\!\left(\Psi\tfrac{df}{dt}\right) \quad . \tag{39}$$

Similarly, for $g(i\omega)$ and $g_N(i\omega)$ we obtain

$$\widetilde{g}(i\omega) = -i\omega\widetilde{g}_N(i\omega) + L\!\left(\Phi\tfrac{dg}{dt}\right) \quad . \tag{40}$$

Finally, Eqs. (37)–(40) simplify Eq. (35) to

$$(i\omega)^2 \widetilde{F}(i\omega) = \frac{\widetilde{\phi}(i\omega)}{\widetilde{\Psi}(i\omega)} L(\Psi\tfrac{df}{dt}) + i\omega\, L(\Phi\tfrac{dg}{dt}) \tag{41a}$$

$$= (\widetilde{\phi}/\widetilde{\Psi})\,[\widetilde{f} + i\omega\widetilde{f}_N] + i\omega\,[\widetilde{g} + i\omega\widetilde{g}_N] \quad . \tag{41b}$$

Equation (41) is the desired relationship giving the behavior of some average observable $\widetilde{F}$ with renewal in terms of the corresponding average growth laws f(t) and g(t) without renewal. This result is a consequence of the renewal property described by Eq. (19) together with g(t=0) = f(t=0) = 0, and does not depend further on the physical origin of the growth laws f(t) and g(t). In accordance with intuition, the second term (which dominates at high frequency) probes the growth of F(t) (given predominantly by g(t)) just after t_0, while the first term dominates at $\omega \approx 0$ and probes the growth of F(t) over long periods of time, which is determined mainly by f(t).

Some relevant physical applications of Eq. (41) arise by setting $F(t) = n_d \langle \delta r^2 \rangle (t)$ (consistent with Eqs. (10) and (19)), so that

$$(i\omega)^2 \, F(i\omega) = n_d (i\omega)^2 \int_0^\infty \varepsilon^{-i\omega t} \langle \delta r^2 \rangle (t) \, dt = D(\omega) \tag{42}$$

is the frequency-dependent diffusion coefficient. Then with $\langle \delta r^2 \rangle_{0,r}(t)$ the mean-square displacement (defined by Eq. (9)) without reassignment during the initial interval and $\langle \delta r^2 \rangle_0 (t)$ the corresponding mean-square displacement starting from a previous reassignment, we have for the dynamic-disorder $D(\omega)$

$$D(\omega) = \frac{\tilde{\phi}(i\omega)}{\tilde{\Psi}(i\omega)} L(\Psi \frac{d}{dt} \langle \delta r^2 \rangle_0) + i\omega \, L(\Phi \frac{d}{dt} \langle \delta r^2 \rangle_{o,r}) \, . \tag{43}$$

Alternatively, the charge transport can be treated directly in terms of each given component (x) of the current density $\vec{j}$ using

$$\tilde{j}_x(\omega) = L(ne \frac{d\langle x \rangle}{dt}) = i\omega \tilde{F} \tag{44}$$

for $F(t) = ne \langle x \rangle (t)$, since $\langle x \rangle (t)$ also obeys Eq. (19).

Equations (41)-(44) are far more general than the previous relation (of Ref. 6) between dynamic-disorder and static-disorder transport in at least the following respects (some already discussed):

(1) The renewal probability can depend on when a previous renewal last occurred.

(2) The averaged behavior of the motion (given by g(t)) starting from an arbitrary time can differ from that starting from a last previous renewal (given by f(t)).

(3) The weakness of the restrictions on the growth laws f(t) and g(t) arising from static-disorder hopping make Eq. (41) applicable to a far broader class of models, including those in which:

(a) The hopping waiting time functions $\psi_{ij}^{(h)}(t)$ in a CTRW are randomly assigned between sites, and randomly reassigned upon renewal,

(b) or the available sites are not on a periodic lattice but nonetheless give rise to growth laws f(t) and g(t) in accordance with (19),

(c) or the statically disordered system is described by a site-percolation model, rather than bond percolation.

B. Special Cases:

1. If the probability per unit time for a reassignment is λ (for constant λ) regardless of any other reassignment, then $\phi(t) = \psi(t) = \lambda e^{-\lambda t}$. All of the Laplace transforms defined by Eqs. (30)-(32) are then easily evaluated. If also f(t) = g(t) at all t (as we previously assumed for zero applied field), then we obtain from Eq. (41)

$$(i\omega)^2 \widetilde{F} = (\lambda + i\omega)^2 \int_0^\infty e^{-(\lambda+i\omega)t} f(t) \, dt$$

$$= (\lambda + i\omega)^2 \, \widetilde{f}_0(\lambda + i\omega) \qquad (45)$$

(with $\widetilde{f}_0$ defined in Eq.(31)). For $f(t) = n_d \langle \delta r^2 \rangle_0(t)$, this becomes

$$D(\omega) = n_d \, (\lambda + i\omega)^2 \int_0^\infty e^{-(\lambda+i\omega)t} \langle \delta r^2 \rangle_0(t) \qquad (46)$$

in agreement with the previously-derived Eq. (13) except that now neither a periodic lattice nor zero-applied field has been assumed; also, the essential role of the reassignment property (19) is now more clearly displayed.

It often occurs that the condition $W_{ij} = W_{ji}$ is not satisfied, so that arguments based on detailed balance do not lead to the conclusion that $f(t) \equiv g(t)$. An example is the one-dimensional dynamic bond percolation model under a sufficiently intense field for $\lambda/w \ll 1$. At a random instant, most carriers are trapped at the end of a cluster of connected bonds, so little initial increase is expected in

$g(t) = \langle x \rangle_{0,r}(t)$ (further reassignment being forbidden in calculating $g(t)$), whereas if a reassignment has just occurred at $t = 0$, then $f(t) = \langle x \rangle_0(t)$ is expected to increase rapidly at small t.

When $f(t)$ and $g(t)$ differ for the constant reassignment rate case, we have

$$(i\omega)^2 \tilde{F}(i\omega) = (\lambda+i\omega)[\lambda \tilde{f}_0(\lambda+i\omega) + i\omega \tilde{g}_0(\lambda+i\omega)] \tag{47}$$

(with $\tilde{f}_0(i\omega)$ and $\tilde{g}_0(i\omega)$ defined by Eqs. (31) and (32)).

2. At zero frequency ($\omega \to 0$), partial integration and l'Hospital's rule must be used to handle the 0/0 form of the first term in Eq. (43) while the factor multiplying $i\omega$ in the second term is found to be finite. All dependence on $g(t) = \langle \delta r^2 \rangle_{0,r}(t)$ disappears, and we find

$$D(\omega \to 0) = n_d \frac{\int_0^\infty \psi(t) \langle \delta r^2 \rangle_0(t) \, dt}{\int_0^\infty t\psi(t) \, dt} \tag{48}$$

in agreement with the result we previously obtained[5,6], except that now the applied field need not be zero, no assumption about the hopping process itself is made beyond that required for Eq. (19), and the derivation is more satisfactory than before insofar as the observation time for all renewal sequences averaged over is constrained to be exactly T.

3. Finally, we note the precise relation between $\phi(t)$ and $\psi(t)$ when $\psi(t)$ is the <u>literal</u> waiting time function as employed by Tunaley[13] and commented upon by Haus and Kehr;[14] specifically $\Psi(t) = \lambda^{-1}\phi(t)$, and therefore

$$\psi(t) = -\lambda^{-1} \frac{d\phi(t)}{dt} \tag{49}$$

where λ^{-1} is the mean renewal time and λ therefore a representative renewal rate and, in particular, the rate of occurrence for the first observed renewal is seen to be $\phi(0) = \lambda$. In this case

$$D(\omega) = \lambda L(\Psi \frac{df}{dt}) + i\omega L(\Phi \frac{dg}{dt}). \tag{50}$$

IV. Relation to the CTRW

The hopping process itself for a continuous-time random walk (CTRW) on a periodic lattice is also described by a property similar to Eq. (19), inasmuch as a hopping event at elapsed time τ since the previous hop adds a quantity $\langle d^2 \rangle(\tau)$ onto the mean-square displacement $\langle \delta r^2 \rangle(t)$ and immediately restores the initial condition in which both $\langle d^2 \rangle(\tau)$ and the waiting time function $\psi_h(\tau)$ start evolving again from $t = 0$. (The "h" subscript will be used to distinguish all quantities defined for the hopping process in obvious analogy to their counterparts in the dynamic disorder renewal process.) Note that[8,15] the time-dependence in $\langle d^2 \rangle(\tau)$ yields a generalization of the usual CTRW.

For a given sequence of times $(t_1 > 0, t_2, \ldots, t_N)$ at which hops occur, randomly producing various displacements $\vec{d}$, let $p_h(\vec{d}, \tau)$ be the conditional probability that a <u>hop</u> given to occur an elapsed time τ after the previous hop will displace the carrier by $\vec{d}$. (The corresponding displacement probability for the first hop after the arbitrary initial time $t=0$ is denoted as $p_{h,r}(\vec{d}, \tau)$.) Also, let $P_t(\vec{s}, t)$ be the probability that a <u>total</u> displacement $\vec{s}$ (from the initial carrier location at $t=0$) has already been attained at time t. (Note that the use of τ_I to denote the length of the interval (t_{I-1}, t_I) is consistent with definition (12).) Then at any time $t_I + \Delta t$ during the interval (t_I, t_{I+1}), the mean-square displacement is

$$\langle r^2 \rangle(t_I + \Delta t) = \sum_s P_t(\vec{s}, t_I + \delta t) \sum_d p_h(\vec{d}, \tau_I) [\vec{d} + \vec{s}]^2$$

$$= \langle r^2 \rangle(t_{I-1} + \delta t) + \langle d^2 \rangle(\tau_I) + 2 \langle \vec{d} \rangle(\tau_I) \cdot \langle \vec{r} \rangle(t_{I-1} + \delta t) \quad (51)$$

where we have used the normalization of both $P_t(\vec{s}, t)$ and $p_h(\vec{s}, t)$ (upon summation over $\vec{s}$), where δt is a positive infinitesimal time, and where

$$\langle \vec{d} \rangle(\tau) = \sum_d p_h(\vec{d}, \tau) \vec{d} \quad (52)$$

(with a corresponding definition for $\langle r^2 \rangle(t)$). Similarly we find

$$\langle \vec{r} \rangle(t_I + \Delta t) = \langle \vec{r} \rangle(t_{I-1} + \delta t) + \langle \vec{d} \rangle(\tau_I). \quad (53)$$

Then (53) and (51) imply

$$\langle \delta r^2 \rangle(t_I + \delta t) = \langle \delta r^2 \rangle(t_{I-1} + \delta t) + \langle \delta d^2 \rangle(\tau_I) \quad (54)$$

where $\langle\delta\xi^2\rangle \equiv \langle\xi^2\rangle - \langle\xi\rangle^2$. Equation (54) verifies the statements at the beginning of this section. (A similar argument, but with $p_{h,r}(\vec{d},t)$ instead of $p_h(\vec{d},t)$, holds also for $t_1 + \delta t$.)

The version of (19) that now results from (54) for the CTRW differs in one essential respect from that for dynamic disorder reassignment; we take $g_h(t)$ to be the $\langle\delta d^2\rangle(t)$ that evolves starting from an arbitrarily chosen time and $f_h(t)$ that from a previous hopping event, but the last contribution to the new version of (19) is $f_h(t_N)$ with the $f_h(T-t_N)$ term omitted, since no hop and therefore no increase in $F_h(t)$ occurs in (t_N,T).

We do not give the detailed derivation here but merely quote the result that follows by applying the procedure of Sec. II:

$$(i\omega)^2 \tilde{F}_h(i\omega) = \frac{\tilde{\Phi}_h(i\omega)}{\tilde{\Psi}_h(i\omega)} \tilde{f}_h(i\omega) + i\omega \tilde{g}_h(i\omega) \qquad (55)$$

with $\tilde{F}_h(\omega)$ the Laplace transform of $\langle\delta r^2\rangle_h(t)$, so that for the (static disorder) CTRW

$$\frac{1}{n_d} D(\omega) = \frac{\tilde{\phi}_h(i\omega)}{\tilde{\Psi}_h(i\omega)} \langle\tilde{\delta r}^2\rangle_{o,h}(i\omega) + i\omega\langle\tilde{\delta r}^2\rangle_{o,h,r}(i\omega). \qquad (56)$$

A similar result has been derived also by Kumar and Heinrichs[15], but the very direct averaging over distributions of hopping events here leads to a probably more transparent derivation which does not require the joint temporal and lattice-periodic Fourier transforms and sums. We present the result here primarily for comparison with the formally similar results of the previous section. It is seen that (56) differs from (41b) by omission of the $\tilde{f}_N$ and $\tilde{g}_N$ contributions describing growth in the interval (t_N,T), consistent with intuitive expectations.

Under the Scher/Lax assumption that $\tilde{\phi} = \tilde{\psi}$ and $\tilde{f} = \tilde{g}$, Eq. (56) reduces immediately to

$$\frac{1}{n_d} D(\omega) = \frac{i\omega\langle\tilde{\delta r}^2\rangle_{0,h}}{1 - \tilde{\psi}_h}$$

in agreement with the results of Scher and Lax.

Finally, we point out the work of Harrison and Zwanzig[9] on a dynamic disorder model similar, but not identical, to our own. Their

theory considers dynamic bond percolation for a constant rate of
reassignment events that occur independently for each pair of neighboring sites, in contrast to our study of reassignment events for the
entire $\{W_{ij}\}$ configuration. Under an effective medium approximation,
they obtain essentially the same $i\omega \to \lambda + i\omega$ rule that we have also
derived and which applies exactly for our DDH models at constant
renewal rate; correction terms corresponding physically to the correlation between motion over several hops and the specific long-range
$\{W_{ij}\}$ configuration then describe the difference between the two models
in higher-order effective medium theory. While the HZ results are
derived specifically for a bimodal $\{W_{ij}\}$ distribution of a percolation
model, it seems plausible that Eq. (19), and therefore also the $\lambda + i\omega$
substitution rule, would apply in an effective medium scheme for a more
general distribution in their model also.

V. Summary

The previous work outlined here has included the following:
(1) The theory of transport in a random medium has been extended to
    include the possibility of pathway reassignment, corresponding to
    dynamic disorder.
(2) In the special case of dynamic bond percolation, predictions are
    obtained in agreement with intuitive notions of how
    renewal ($\{W_{ij}\}$ reassignment) participates in diffusion below
    percolation.
(3) Most importantly, we believe that the dynamic disorder models
    provide a framework in which to better understand carrier transport
    in systems that undergo microscopic structural changes on the same
    timescale as the carrier motion itself; this approach appears to
    offer considerable advantage for the polymer electrolytes, where
    the models usually employed at this time are quasithermodynamic and
    not based on actual carrier motion.[16]

Previously reported expressions[5,6,9] describing dynamic disorder
transport in terms of static disorder transport are shown here to be
special cases of a relationship implied generally by the renewal properties of a broad class of dynamic disorder models. Analysis of the
more general results should enhance the range of possible application

of dynamic disorder transport theory to specific physical problems, as well as providing additional insight into the relation between dynamic and static disorder for systems that satisfy the reassignment rule (Eq. (19)) only approximately.

Acknowledgments: I am very grateful to Mark A. Ratner for helpful discussions and suggestions, and to the AFOSR and NSF/MRL program (the latter through the Northwestern MRC) for partial support of this research.

REFERENCES

1. See, e.g., E. E. Polymeropoulos and J. Brickmann, Ann. Rev. Biophys. $\underline{14}$, 315 (1985); H. Schröder, J. Chem. Phys. $\underline{79}$, 1981, 1991 (1983).
2. See, e.g., S. Chandra, "Superionic Solids, Principles and Applications," (North-Holland, 1981).
3. For a recent review, see M. B. Armand, Sol. St. Ionics $\underline{10}$, 745 (1984).
4. E.g., G. H. Weiss, Adv. Chem. Phys. $\underline{52}$, 363 (1983).
5. S. D. Druger, A. Nitzan, and M. A. Ratner, J. Chem. Phys. $\underline{79}$, 3133 (1983).
6. S. D. Druger, M. A. Ratner, and A. Nitzan, Phys. Rev. $\underline{B3}$, 3939 (1985).
7. V. M. Kenkre, E. W. Montroll, and M. F. Shlesinger, J. Stat. Phys. $\underline{9}$, 45 (1973).
8. H. Scher and M. Lax, Phys. Rev. $\underline{B7}$, 4491 (1973); E. W. Montroll and G. H. Weiss, J. Math. Phys. $\underline{6}$, 167 (1965).
9. A. K. Harrison and R. Zwanzig, Phys. Rev. $\underline{A32}$, 1072 (1985).
10. S. M. Ansari, M. Brodwin, M. Stainer, S. D. Druger, M. A. Ratner, and D. F. Shriver, Sol. St. Ionics $\underline{17}$, 101 (1985).
11. S. D. Druger, M. A. Ratner, and A. Nitzan, Sol. St. Ionics (In Press).
12. E.g., M. Abramowitz and I. A. Stegun, Handbook of Mathematical Functions (U.S. Gov't Printing Office, Wash., D.C., 1964) p. 1019; P. M. Morse and H. Feshbach, Methods of Theoretical Physics (McGraw-Hill, New York, 1953), p. 468.
13. J. K. E. Tunaley, Phys. Rev. Lett. $\underline{33}$, 1037 (1974); with regard to the CTRW problem, see also Refs. 13-14 and M. Lax and H. Scher, Phys. Rev. Lett. $\underline{39}$, 781 (1977).
14. J. W. Haus and K. W. Kehr, Phys. Rev. $\underline{B28}$, 3573 (1983).
15. A. A. Kumar and J. Heinrichs, J. Phys. $\underline{C13}$, 5971 (1980).
16. S. D. Druger, M. A. Ratner, and A. Nitzan, Sol. St. Ionics $\underline{9/10}$, 1115 (1983).

EFFECTIVE MEDIA EQUATIONS FOR THE CONDUCTIVITY OF MACROSCOPIC MIXTURES

D S McLachlan

Department of Physics and Condensed Matter Physics Research Group
University of the Witwatersrand, Johannesburg

ABSTRACT

Two equations which describe the conductivity (resistivity) of a wide variety of binary macroscopic conducting mixtures as a function of the conductivity of the components, the volume fraction of each, the space dimension and a single morphology parameter are presented. These equations are interpolations between Bruggeman's symmetric and asymmetric effective media theories. The parameter is determined from a critical composition, for example in a metal-perfect insulator mixture, it is the metal-insulator transition point. Excellent agreement is found with a wide range of experimental data.

I. Introduction

The electrical properties (conductivities and dielectric constants) of macroscopic mixtures have been studied for over one hundred years. They are usually an extremely sensitive function of the relative concentrations of the different components. However, different mixtures show a variety of shapes and arrangements of the components, which also strongly influence the electrical properties of the medium. In the case of isotropic media there appear to be two naturally occurring limiting cases. The symmetric case is where the two (or more) components fill all space but are completely randomly distributed. The asymmetric case is where the surfaces of the particles of one component (the dispersion) are always completely covered by the other (the host). This may persist over many orders of magnitude of dispersion particle size.

It is the purpose of this paper to present two phenomenological effective media equations which can, in many cases, describe the conductivity over the entire composition range of a binary mixture, whose morphology lies between the symmetric and asymmetric extremes. The type of distribution or morphology is described in terms of a single parameter D.

Earlier, less general versions of these equations were published by McLachlan [1]. However, in this form they are only suitable for dealing with the conducting side of the case where the conductor surrounds or tends to surround the insulator or the insulating side of the case where the insulator tends to surround the conductor. The present equations allow one to treat the cases when an imperfect insulator is mixed with an imperfect conductor and either conductor or insulator tends to be the host.

Although this paper concentrates on the electrical conductivity (resistivity), the equations presented here should also apply to the dielectric constant, the magnetic susceptibility and thermal conductivity of and gaseous diffusion in inhomogeneous macroscopic media.

II THEORY

For a more complete review of Bruggeman's theories the reader is referred to a previous paper by Landauer[2] and the references therein. Bruggeman's symmetric theory [3], [4], [5], [6], [7] treats the two (or more) constituents, with conductivities σ_1 and σ_2 and volume fractions f_1 and f_2 on a completely symmetrical basis. The theory is for a random mixture of particles of the two constituents, in the correct volume ratio, which together completely fill the media. The cases for 1, 2 and 3 dimensions (d) can be combined in the equation.

$$f_1(\sigma_1 - \sigma_m)/(\sigma_1+(d-1)\sigma_m) + f_2(\sigma_2 - \sigma_m)/(\sigma_2 +(d-1)\sigma_m) = 0 \qquad (1)$$

If $f_1 = f$ and the conductivity $\sigma_1 = 0$ and of $\sigma_2 = \sigma_h$, eq.1 becomes

$$\sigma_m = (1 - (d/d-1)f)\sigma_h. \qquad (1a)$$

Note the metal-insulator transition when the insulator volume fraction is 2/3 (insulator area fraction = ½).

Equation 1 can also be written as

$$f_1(\rho_m - \rho_1)/(\rho_m + (d-1)\rho_1) + f_2(\rho_m - \rho_2)/(\rho_m + (d-1)\rho_2) = 0 \quad (1')$$

here $\rho_x = 1/\sigma_x$. If $f = f_1$, $\rho_1 = 0$ and $\rho_2 = \rho_h$ Eq.1' becomes

$$\rho_m = (1-df)\rho_h. \quad (1'a)$$

Note the finite to zero resistance transition or short circuit at $f = 1/3$ in 3d.

Bruggeman's asymmetric theory [3], [4], [5], [6], [7] in 3d considers spheres (discs in 2d), with a conductivity σ_d, dispersed in a host constituent of conductivity σ_h. The volume fraction of spheres (f) can range from 0 to infinitely close to one. This is achieved by having an effectively infinite range of sphere sizes each of which remains coated with the host constituent at all volume fractions, i.e., if the host component is an insulator, the medium will always be an insulator. The asymmetric equation can be written as

$$(\sigma_m - \sigma_d)^d/\sigma_m = (1-f)^d(\sigma_h - \sigma_d)^d/\sigma_h \quad (2)$$

for $d = 2$ and 3.

When $\sigma_d = 0$ eq.2 becomes

$$\sigma_m = (1-f)^{(d/(d-1))}\sigma_h. \quad (2a)$$

Note this is the case where the high conductivity component surrounds the low conductivity or insulating dispersion. For a detailed discussion of the asymmetric theory and the case of ellopsoidal particles, where values for the exponent of other than d in eq.2 can be obtained, the reader is referred to the review by Meredith and Tobias [8].

For the case where the high conductivity dispersion is surrounded by the low conductivity component ($\rho_d = 0$), the equation

$$\rho_m = (1-f)^{d}\rho_h \tag{2'a}$$

applies. Here $\rho_x = 1/\sigma_x$ and <u>f is the volume fraction of the dispersed component</u>. Equations 1a, 1'a, 2a and 2'a are plotted in Landauer[2] and McLachlan [1].

The problem now is to determine the conductivity of a system, which lies between the symmetric and asymmetric methods of distributing the two constituents. The starting assumption, made by McLachlan[1] is that in a very large metal-insulator system there will be a critical fraction $f = f_c$ such that for $f < f_c$, $\sigma_m \neq 0$ and for $f > f_c$, $\sigma_m = 0$ or for a conductor-perfect conductor system when $f < f'_c$, $\rho_m > 0$ and for $f > f'_c$, $\rho_m = 0$. The ansatz was then made that these various systems can be characterised by one parameter D. The defining relationships between f_c or f'_c and D are

$$f_c = (d-1)/D \tag{3}$$
$$f'_c = 1/D \tag{3'}$$

depending on whether the high conductivity component tends to surround the low conductivity component (f_c) or vice versa (f'_c). In future all primed f_c's and equation numbers will refer to cases where a low conductivity host tends to surround a high conductivity dispersion.

The actual equations given in McLachlan[1] are

$$\sigma_m = (1 - Df/(d-1))^{d/D}\sigma_h \tag{4}$$
$$\rho_m = (1 - Df)^{d/D}\rho_h. \tag{4'}$$

These equations interpolate between eqs. 1a amd 1'a (D = d) and eqs. 2a (D = d - 1 ; conductor host) and 2'a (D = 1 ; insulator host). Plots of these equations, for these and intermediate values of D, are given in McLachlan [1].

As these equations are very limited in their applicability, it is highly desirable to have equations which predict (even phenomenologically) the conductivity of a mixture, in which both components can have conductivities between 0 and ∞, in terms of these conductivities and a single morphology parameter.

The following extension to eq.4 is proposed:

$$\frac{f(\Sigma_d - \Sigma_m)}{\Sigma_d + ((d-1)/(D+1-d))\Sigma_m} + \frac{(1-f)(\Sigma_h - \Sigma_m)}{\Sigma_m + ((d-1)/(D+1-d))\Sigma_m} = 0 \qquad (5)$$

where $\Sigma_d = \sigma_d^{D/d}$, $\Sigma_h = \sigma_h^{D/d}$ and $\Sigma_m = \sigma_m^{D/d}$.

f is as usual the volume fraction of the dispersion. This equation obviously reduces to eq.1 for D = d and it can be shown to be equivalent to eq.4 when D = d - 1 + ε(an infinitesimal) and σ_d = 0. This equation should allow one to treat systems where $\sigma_h \gg \sigma_d$.

The similar extension for eq.4' is

$$\frac{f(P_m - P_d)}{P_m + (D-1)P_d} + \frac{(1-f)(P_m - P_h)}{P_m + (D-1)P_h} = 0 \qquad (5')$$

where $P_d = \rho_d^{D/d}$, $P_h = \rho_h^{D/d}$ and $P_m = \rho_m^{D/d}$.

f is again the volume fraction of the dispersion. This equation can be shown to reduce to eq.1 for D = d and is equivalent to eq.4' when D = 1 and ρ_d = 0.

Equations 3 and 3' can be used to rewrite eqs. 5 and 5' with the parameter f_c or f_c' instead of D. When comparing eqs. 5 and 5' with experimental work both D and f_c or f_c' will be given.

Figure 1a shows 3d curves for eqs.1 (curve A), 2 (curve B) and 5 (with D = 3 (curve A) and 2.0001 (curve C)) for σ_h = 1 and σ_d = 0.01. Note the near perfect agreement between eq.2 and eq.5, with D = 2 + ε, and σ_d = 0 has been lost. However, as curve C lies between the Bruggeman asymmetric equation and the Clausius-Mossotti relationship (curve D) it should still be a useful relationship.

Figure 1b shows 3d curves for eqs. 1' (curve A), 2' (curve B) and 5' (with D = 3 (curve A) and 1 (curve C)) for ρ_d = 0.01. Note that the agreement between eq. 2' and eq. 5' for D = 1 is poor but, because it lies close to the Clausius-Mossotti relationship for small f and between eq. 2' and the Clausius-Mossotti relationship for larger f, it should still be a valid equation to describe the resistivity of this class of mixture.

Equations 5 and 5' can also be written as quadratic equations and σ_m (ρ_m) solved for in terms of $\sigma_h(\rho_m)$, $\sigma_d(\rho_d)$ and D.

As shown by McLachlan [1], near a metal insulator transition, eq.4 can be written as

$$\sigma_m = [Dx/(d-1)]^{d/D} \sigma_h \qquad (6)$$

where $x = f_c - f$ (f is insulator fraction). Defining x by $x = f - f_c'$ and for ρ_h large and x small, one can obtain from eq. 5'

$$\Delta\sigma_m = \frac{1}{\rho_d}\left(\frac{x}{2(D-1)}\right)^{d/D} \qquad (6')$$

This behaviour, although mathematically similar to the predictions of percolation theory [9], is physically very different, in that the threshold f_c is related to the value of the exponent (d/D). This contradicts the belief that the value of the critical exponent β should be universal while f_c depends on other factors such as the type of lattice. In spite of this obvious difficulty the next section will show that good agreement, with at least some experimental data, is obtained using eqs. 5 and 5'. The equations might be found to fit some data better if D/d and (d - 1)/(D + 1 - d) in eq. 5 and D/d and (D - 1) in eq. 5' were independent parameters. This could allow for percolation type behaviour near f_c but assure, in a phenomenological way, the correct limit when the media consist entirely of the conducting component. The next section of this paper will concentrate on how well experimental data can be fitted using eqs. 5 and 5' and one single morphology parameter.

III EXPERIMENTAL EVIDENCE

As the equations presented in the previous section are, except at certain extremes, phenomenological, they are justified by comparison with experiment in this section. The data are digitized from enlargements of the graphical data presented in the literature. Using a sophisticated fitting programme most of the data are fitted to the relevant equation using the appropriate number of parameters; usually the host resistivity (conductivity), the dispersed component resistivity (conductivity) and D. If the situation and the data warrents it more complicated expressions for the resistivity can be used. The computer uses the experimental resistivity (conductivity) and starting parameters to calculate an area or volume fraction called Fit from eq. 5 or 5'. Fit is then compared with the given (or calculated) area or volume fraction (Frn) and the quantity $\chi^2 = \Sigma_N ((Fit-Frn)/0.01)^2$ is minimised by the programme by varying the non-fixed parameters. If $\sigma = \sqrt{\chi^2/(N-P)} = 1$, where N is the number of experimental points and P is the number of variable parameters, it is claimed that data has been fitted to an accuracy of 0.01 in the area or volume fraction. σ and the errors given with the fitted parameters in the text are calculated by the programme.

Figure 2 shows the results for the two-dimensional speckle pattern metal films of Smith and Lobb[10]). The full circles are for an isotropic film and the curve (eq.2 : conductor host) through the results is for D = 1.667 (Metal area fraction 0.4).

The experiments of Laibowitz et al[11]) on the resistivity as a function of area fraction for gold films, provide an example for the use of eq. 5' (insulator host) in two dimensions. The results and the theoretical best fit are shown in Figure 3 where ρ(substrate) = 1.8 ± .1 x $10^7 \Omega$, ρ(Au) = 8.3 ± .1Ω, D = 1.31 ± .001 (f_c = 0.76) and σ = 1.8.

In three dimensions the first case to consider is where the good conductor is host. The data selected is that of Smith and Anderson[12]) for a thick arc-sprayed film of Ni (good conductor) and V_2O_5 (poor conductor). An advantage of this data is that the resistivity for pure Ni and pure V_2O_5 films is given so that there is in effect only one parameter D. The data and the results of this fit are shown in Figure 4 where $\sigma(V_2O_5) = 0.485 \Omega\text{-cm}^{-1}$ (given), σ(Ni) = 5565Ω-cm^{-1} (given), D = 2.42 ± .02(f_c = .413) and σ = 2.103. If all three parameters are allowed to vary $\sigma(V_2O_5)$ = 0.492 -cm , σ(Ni) = 5526Ω-cm , D = 2.42 ± .02 (f_c = .413) and σ=2.090. These values of σ are well within the digitizing error. The fact that this is a conductor host situation shows that the more malleable nickel is tending to wrap around the V_2O_5 It is interesting to note that the point at f(Ni) = 0.134 also lies to the left of theoretical expressions shown by Smith and Anderson[13]) and that eq. 5 lies between the theoretical expressions used by Smith and

Anderson[13]). The Smith and Anderson theory[13]) is an effective media theory with a variable number (Z) of contacts at each node and a variable contact conductance distribution i.e., a two parameter theory. Both curves are for Z = 12. The dashed curve is for fixed contact conductance while the dotted curve has a contact conductance distribution.

When considering the three dimensional insulator host situation the cermet systems (very well reviewed by Abeles[14]) are the obvious ones to examine. Unfortunately, only a limited number of the systems discussed by Abeles can be easily fitted using eq. 5', as both the oxides and the metallic components resistivity may vary with composition. Shown in Figure 5 are the data of Abeles and Hanak[15]) on the Al-SiO$_2$ system at 4.2K(□) and 300K(Δ). The parameters for the theoretical line through the 4.2K points (A) are $\rho(SiO_2) = 1.0 \pm 0.9 \times 10^8 \Omega cm$, $\rho(A\ell) = 4.00 \pm .16 \times 10^{-5} \Omega cm$, $D = 1.75 \pm .01$ and $\sigma = 0.8$, while at 300K (B) they are $\sigma(SiO_2) = 4.2 \pm 8.6 \times 10^7 \Omega$-cm, $\rho(A\ell) = 6.4 \pm .25 \times 10^{-5} \Omega$-cm, $D = 1.84 \pm 0.2$ and $\sigma = 1.3$.

Figure 6 shows experimental data, for an emulsion in which the resistivity ratio ρ_h/ρ_d is 15.7, taken by Meredith and Tobias[16]). The curves A, B and C are eq. 5 with D = 3, 1.9 and 1 respectively and are one parameter (D) fits done directly on the plotter. The curve with D = 1.9 would appear to fit the data well. The medium was created by continually circulating the two fluids which were emulsified by the pump blades. The fit by eq. 5' seems too good to be coincidental and the D value of 1.9 and not 1.0 may be due to the fact that the emulsions are in a dynamic equilibrium and/or that the dispersion has adopted some non-random structure.

Equations 4 and 5' have also been compared with systems where one component is fluid and is squeezed out of the media, changing the relative volume fractions and the resistivity.

In the first of these a random packing of rubber sections, (0.8mm < d < 10mm), immersed in a brine solution, is placed in a square piston and cylinder and the axial and tranvserse resistances are measured separately, using suitable placed electrodes, as the piston is moved down step by step. By measuring the resistance, the piston displacement and the weight of brine extruded at each point, the resistivity and relative volume fractions can be calculated. The experiment [17] is a model system for the formation of sandstone and the data follows the curves given by eq. 4 reasonably well for D values between 2.05 and 2.15. The D value increases with increasing size range of rubber sections. McLachlan [18] has also proposed that eq. 4 can be used as a phenomenological equation for the conductivity of water containing geological formations in terms of the ground water conductivity and a single morphology parameter.

The resistivity of calcined coal powder (0.2 to 0.8mm) has been measured as a function of volume fraction between two conducting pistons in an insulating cylinder [19]. This data agrees reasonably well with eq. 5'. A σ of 1.62 is obtained for the volume fraction of coal changing from 0.4 to 0.9 and the resistivity from over 300 to 5mΩ-cm. In the fitting programme the resistivity for the host (air) was fixed so that it was 10^6 times larger than that of the coal. (Using factors of 10^5 or 10^7 made very little difference). The fact that data could be fitted reasonably well by a single D value (2.65 ± 0.02) is surprising as compaction is occurring via elastic and plastic deformations, slipping and crushing. Equation 5' could be used to determine the bulk resistivity of other substances obtainable only in the form of powders or small crystals.

The experimental data used to justify eq.5 in this paper was for isotropic or nearly isotropic media. For anisotropic media it would now appear that two independant parameters are necessary. The first is d/D which should be linked with the geometric shape of the dispersion particle. The second is either $(d-1)/(D+1-d) \equiv f_c/(1-f_c)$ or $(D-1) \equiv (1-f_c')/f_c'$ which determines the critical composition. For instance, the resistivity of graphite flakes has been measured as a function of the volume fraction of graphite. The results fitted the modified eq.5 with a resistivity of 2.02 mΩ-cm, a d/D of 3.92, a f_c of .208 (vol fraction of graphite) and a σ of 0.49. The resistivity of air was taken as 10^6 times the resistivity of the graphite. A d/D of 3.92 according to asymmetric media theory (see for instance Meredith and Tobias[8]) corresponds to oblate ellipsoids.

IV DISCUSSION

The above fits of the equations to various experiments should convince the reader that eqs. 5 and 5', if not exact, are valuable phenomenological effective media formula for a wide variety of conducting mixtures. They can also easily be used in computer programs to fit data in terms of a limited number of parameters.

The notion that there are conducting mixtures, whose morphology lies between the symmetric and asymmetric extremes, which may be characterised by a single parameter, is new and a tremendous simplification of a very complex situation.

Equations 5 and 5' should also hold for the dielectric constant, magnetic susceptibility and thermal conductivity as well as electrical conductivity of and diffusion in hetrogeneous mixtures. The idea of a continual change in morphology, characterisable by one (or more) parameter, may be valid in other instances where effective media theories can be used. The procedure would be to derive rigorous (if possible) theories for the symmetric and antisymmetric situations and look for a suitable continuous interpolation between these, using one (or more) geometrical parameter.

Equations 5 and 5' are claimed to hold for macroscopic media. What this means is that the mean free paths of the electrons in the two components should be small compared with the grain sizes or determined by scattering from the grain boundaries. If calculations show that the "mean free path" of a microscopic mixture is of the atomic dimensions or less, these equations should also hold for such media.

References

1) McLachlan DS, J. Phys. C $\underline{18}$ 1891 (1985)
2) Landauer R, Electrical Transport and Opitical Properties of

Inhomogeneous Media eds. JC Garland and DB Tanner (American Institute of Physics Conference Proceedings 40) p2 1978

3) Bruggeman DAG, Ann. Physik. (Leipz.) 24 636 (1935)
4) Bruggeman DAG, Ann. Physik. (Leipz.) 24 665 (1935)
5) Bruggeman DAG, Ann. Physik. (Leipz.) 25 645 (1936)
6) Bruggeman DAG, Phys. Z. 37 906 (1936)
7) Bruggeman DAG, Ann. Physik. (Leipz) 24 160 (1937)
8) Meredith RE and Tobias CW, Advances in Electrochemistry and Electrochemical Engineering 2 ed. CW Tobias (Interscience, New York) p 15 (1962)
9) Kirkpatrick S, Rev. Mod. Phys. 45 574 (1973)
10) Smith LN and Lobb CJ, Phys. Rev. B20 3653 (1979)
11) Laibowitz RB, Voss RF and Allessandrini EI, Percolation, localization and superconductivity (NATO ASI Series B, Physics V109) eds. AM Goldman and SA Wolf (New York : Plenum Press) p 145 (1983)
12) Smith DPH and Anderson JC, Phil. Mag. B43 811 (1981)
13) Smith DPH and Anderson JC, Phil. Mag. B43 797 (1981)
14) Ables B, Applied Solid State Science vol 6; ed. Wolfe R (New York : Academic Press) ch.1 (1976)
15) Ables B and Hanak JJ, Phys. Lett. A34 165 (1971)
16) Meredith RE and Tobias CW, J. Electrochem Soc. 108 286 (1961)
17) McLachlan DS, Button MB, Adams SR, Gorringe VM, Kneen JD, Muoe J, and Wedepohl E (submitted for publication)
18) McLachlan DS (submitted for publication)
19) McLachlan DS, Falcon R and Chandler HD (to be published)

Figure Captions

Fig. 1   Figure 1a shows normalized conductivity curves in 3d for conducting components, where $\sigma_h = 1$ and $\sigma_d = 0.01$, as a function of the poor conductor (dispersion) volume fraction.   Curve A is eq. 1 and eq. 5 with D = 3, curve B is eq. 2, curve C is eq. 5 with D = 2.0001, curve D is the Clausius-Mossotti relationship and curve E is the average for layers parallel to the current.

Figure 1b shows normalized resistivity curves in 3d for resistive components, where $\rho_h = 1$ and $\rho_d = 0.01$, as a function of the low resistivity (dispersion) volume fraction.   Curve A is eq. 1' and eq. 5' with D = 3, curve B is eq. 2', curve C is eq. 5' with D = 1, curve D is Clausius-Mossotti relationship and curve E is the average for layers perpendicular to the current.

Fig. 2   The relative conductivity of a speckle pattern metal film as a function of the metal area fraction (1-f). The curve shown is for D = 1.667.   For details see Smith and Lobb[10].

Fig. 3   The conductivity of a two dimensional discontinuous gold film as a function of the metal area fraction[11]. The theoretical curve is eq. 5' with ρ(substrate) = 1.8×10^7, ρ(Au) = 8.3, D = 1.31 and an error $\sigma = 1.8$.

Fig. 4   The resistivity of a thick plasma arc sprayed film of nickel and vanadium oxide against the volume fraction of nickel[12].   The solid theoretical curve is eq. 5 with σ(Nickel) = $5565(\Omega\text{-cm})^{-1}$, $\sigma(V_2O) = 0.485(\Omega\text{-cm})^{-1}$, D = 2.42 and the error $\sigma = 2.1$ The dotted and dashed curves are obtained from the effective media theory of Smith and Anderson[13].

Fig. 5　　　The resistivity [15] of thick sputtered $A\ell$-SiO_2 as a function of the volume fraction of SiO_2 at 4.2K (■) and 300K (∆). The theoretical lines A(4.2K) and B(300K) are eq. 5' and the values of the parameters are given in the text.

Fig. 6　　　The resistivity of an emulsion [16], in which the resistivity ratio ρ_h/ρ_d is 15.7, as a function of the dispersion volume fraction. A possible theoretical fit (curve B) is eq. 5' with D = 1.9. Curve A is for D = 3 and curve C for D = 1.

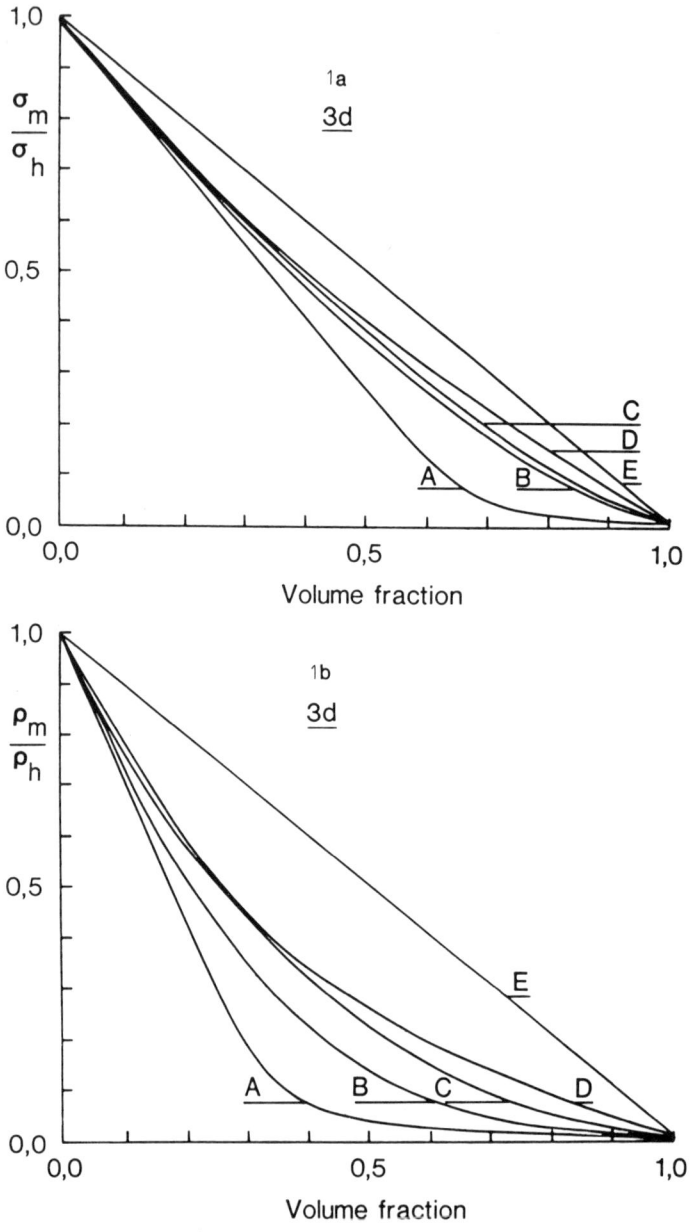

Fig. 1

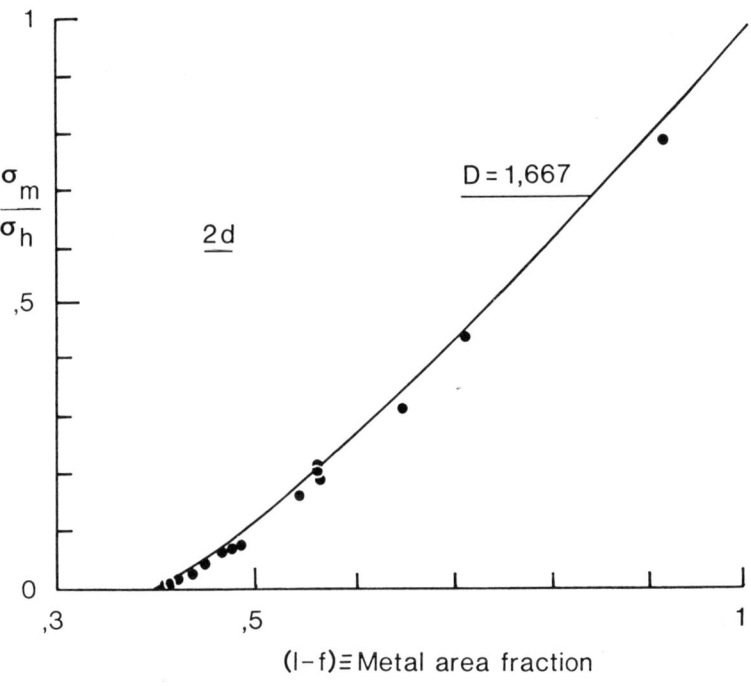

Fig. 2

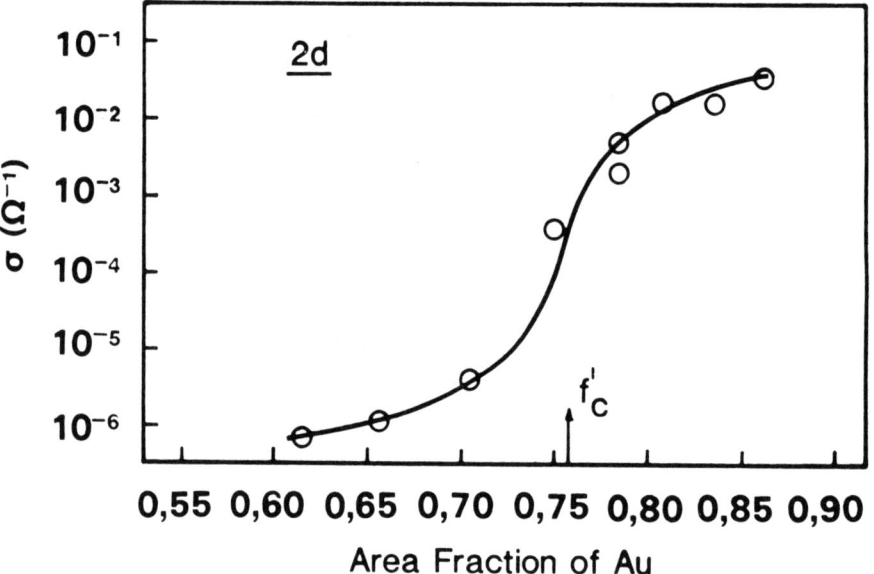

Fig. 3

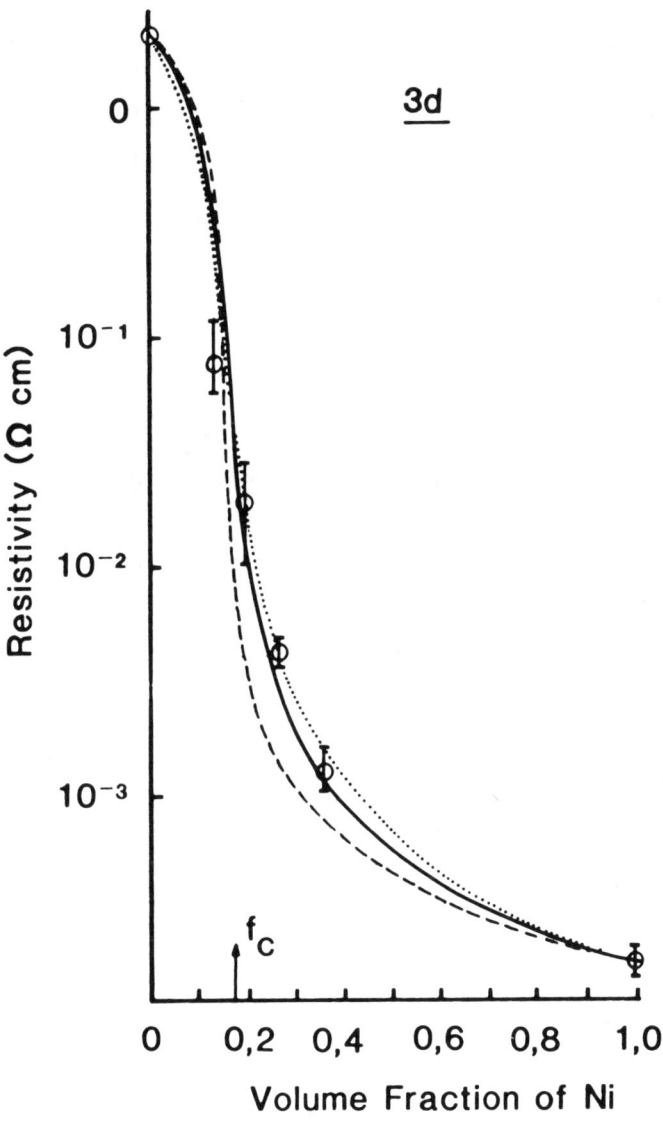

Fig. 4

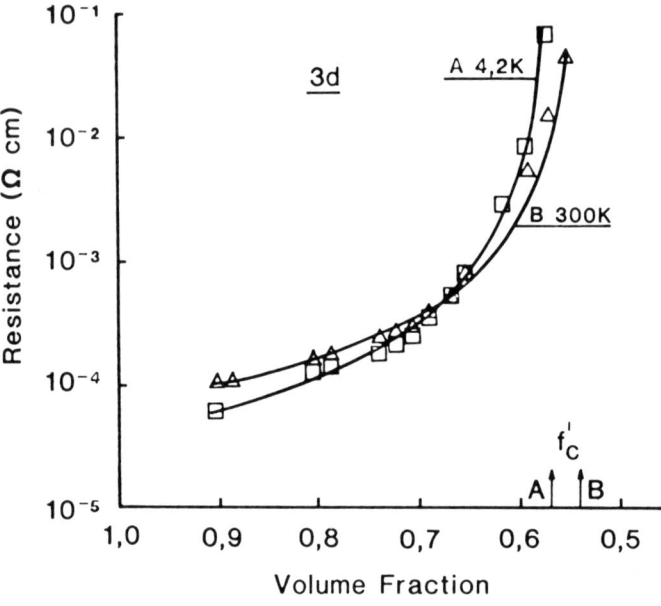

Fig. 5

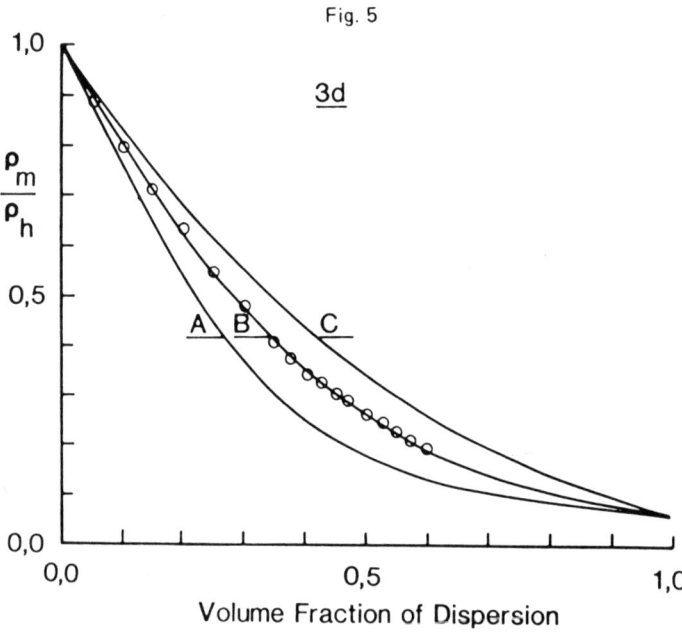

Fig. 6

NON-MARKOVIAN DYNAMICS OF
ONE-DIMENSIONAL TRAPPING PROBLEMS[*]

J. Masoliver[**]
Institute for Nonlinear Sciences
University of California, San Diego
La Jolla, CA 92093 U.S.A.

Bruce J. West
La Jolla Institute
Center for Studies of Nonlinear Dynamics[***]

Katja Lindenberg
Department of Chemistry, B-014
and
Institute for Nonlinear Sciences
University of California, San Diego
La Jolla, CA 92093

ABSTRACT

We calculate the trapping times and survival probabilities for an excitation in a one-dimension system with a random trap distribution. The excitation dynamics are generated by fluctuations that may have a finite correlation time, leading to non-Markovian dynamics of the excitation.

---

[*]Research supported in part by NSF grant DMR-8315950 and DARPA grant DAAG-29-85-K-0246.
[**]Fulbright Fellow. Permanent address: Department de Matematiques, E.T.S. Enginyers Telecomunicacio, Universitat Politecnica de Catalunya, Jordi Girona Salgado s/n, 08034 Barcelona (Spain).
[***]Affiliated with the University of California at San Diego. 3252 Holiday Ct., Suite 208 La Jolla, CA 92038 U.S.A.

1. Introduction

The trapping of excitations at defect sites has traditionally been treated as a diffusive process or as its discrete equivalent, a random walk process.[1-4] The underlying assumptions in these approaches are that the dynamics are Markovian and that they take place in a sequence of small steps. Recently investigators have examined processes in which the step lengths can be arbitrarily long but the Markov property is preserved.[5,6] The obverse problem of a non-Markov process continuous in space has recently also attracted considerable attention.[7-9] It is this last problem that we address herein in the excitation trapping context. We restrict the present analysis to one dimensional continuous processes.

Consider a classical one-dimensional system with randomly located trapping points. The dynamical process on this system that we will consider is described by the differential equation

$$\dot{X} = f(t) \tag{1}$$

where $X(t)$ is the position of the excitation at time t and $f(t)$ is a fluctuating function whose statistics are to be specified. The usual case is that in which $f(t)$ is Gaussian and delta-correlated,

$$< f(t)f(t') > \; = 2D\delta(t-t') \quad , \tag{2}$$

with zero average value. The phase space description of the process (1) is then given by the diffusion equation[10]

$$\frac{\partial}{\partial t} P(x,t|x_o) = D \frac{\partial^2}{\partial x^2} P(x,t|x_o) \tag{3}$$

where $P(x,t|x_o)dx$ is the probability that the dynamic variable $X(t)$ is in the interval $(x,x+dx)$ given that initially $X(0) = x_o$. The discrete analog of (3) is a nearest neighbor random walk.[2,4] In the presence of traps one generally wishes to know the "survival probability" $F(t)$ that an excitation has not been trapped at or before time t.[11] This survival probability can be determined from the solution of the diffusion equation (3) subject to absorbing boundary conditions at the two trapping points sandwiching x_o.[6]

When $f(t)$ is no longer a Gaussian process but one that is still stable, then the phase space description is no longer of the form (3) but is in general an integro-differential equation.[5,6] If $f(t)$ has delta-correlated moments, then this integro-differential equation is still local in time, i.e. it is then a master equation. The discrete analog of this process is described by a fractal random walk.[2,4] Because of the spatial nonlocality which allows the process to discontinuously jump over traps, the trapping problem can no longer be posed as a boundary value problem and has only been solved approximately.[5,6]

When the moments of $f(t)$ are no longer delta-correlated, the process ceases to be Markovian and even the master equation no longer gives a proper description of the process. In this case not only is the phase space equation for $P(x,t|x_0)$ not obvious, but the relation of the survival probability $F(t)$ to $P(x,t|x_0)$ is also obscure.[7-9] Such non-Markovian effects arise, for example, when there is a preference to continue moving in a given direction rather than reversing direction. Inertial effects would illustrate such an asymmetry.

We have recently developed a new approach[9] to the trapping problem for a process described by Eq. (1) in which the fluctuations $f(t)$ are dichotomous and arbitrarily correlated, with

$$< f(t)f(t') > = \phi(t-t') \, . \tag{4}$$

A subclass of these processes is that for which $\phi(t-t')$ is exponential, thereby making the process $f(t)$ [but not $X(t)$] Markovian.[7,8] For this subclass one can write a master equation in an extended phase space and the trapping problem can then be cast in the traditional form.[8] However, when $f(t)$ is not a Markov process such an approach is not feasible. Rather than dealing with the phase space equation, our approach deals directly with trajectories to construct the appropriate phase space.

2. Trajectories

Let $f(t)$ alternate between the values a and $-a$ and let the random times that $f(t)$ retains a given value be governed by the distribution $\psi(t)$. Consider one particular realization of $f(t)$, with $f(0) = a$ (cf. Fig. 1). The process $X(t)$ for this realization is obtained by integrating (1):

$$\begin{aligned}
X(t) &= x_0 + at \, , & 0 &\leqslant t \leqslant t_1 \\
&= x_0 + at_1 - a(t-t_1) \, , & t_1 &\leqslant t \leqslant t_1 + t_2 \\
&= x_0 + at_1 - at_2 + a(t-t_1-t_2) \, , & t_1 + t_2 &\leqslant t \leqslant t_1 + t_2 + t_3 \\
&\vdots & &\vdots
\end{aligned} \tag{5}$$

and $X(0) = x_0$. The t_n are the time intervals during which $f(t)$ retains a given value. A typical trajectory is shown in Figure 2.

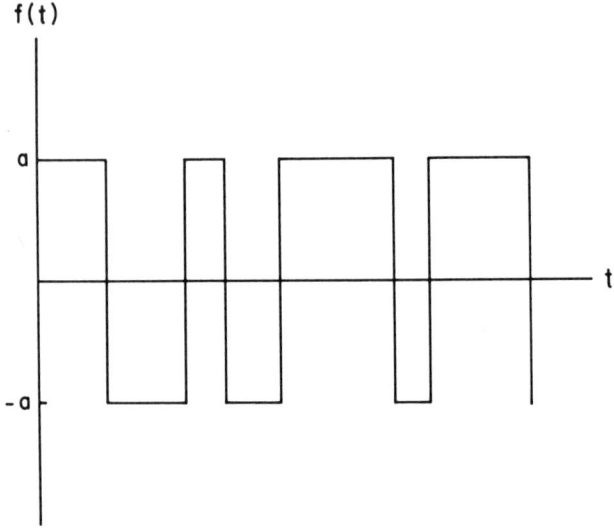

Fig. 1: A typical realization of a symmetric dichotomous process.

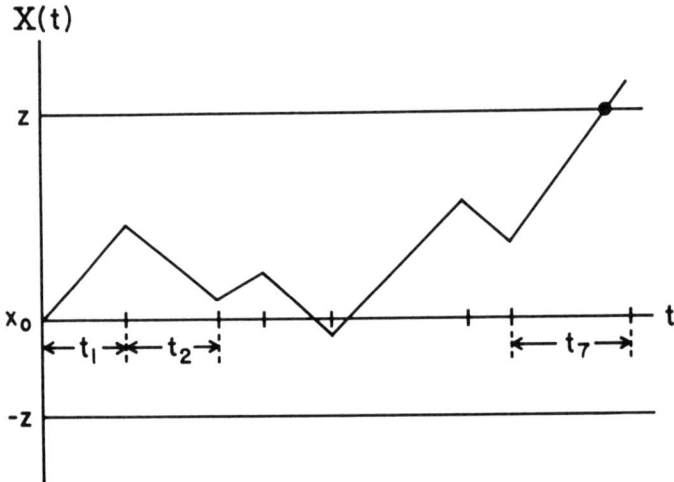

Fig. 2: A typical trajectory for a dynamic process $X(t)$ starting at x_0. This trajectory is driven by the dichotomous process of Fig. 1.

Let us now suppose that there are trapping points located at $x = z$ and $x = -z$, and that we wish to know the survival probability of the process in this range. Towards this end we define a conditional first passage time density as follows: $p(t;x_0,z)dt$ = probability that the process $X(t')$ crosses z or $-z$ in the time range $t \leqslant t' \leqslant t+dt$ without ever having crossed either of these levels during the time span $0 \leqslant t' < t$. To calculate $p(t;x_0,z)$ it is useful to denote each time range t_n between switches as an "interval" and to define the auxiliary probability $p_n(t;x_0,z)dt$ = probability that the first crossing of z or $-z$ occurs during the nth interval and in the time range $(t,t+dt)$. Clearly

$$p(t;x_0,z) = \sum_{n=1}^{\infty} p_n(t;x_0,z) \quad . \tag{6}$$

To construct $p_n(t;x_0,z)$, consider a trajectory such as that of Eq. (5). We wish to insure that no crossing of $\pm z$ occurred in the first $(n-1)$ intervals and that a crossing does occur during the nth interval. The level z is not crossed during the first interval if $X(t_1) < z$, i.e. if $t_1 < (z-x_0)/a$. The probability that this is true is

$$\text{Prob }(t_1 < (z-x_0)/a) = \int_0^{(z-x_0)/a} \psi(t_1)dt_1 \quad , \tag{7}$$

where we have assumed that $f(t)$ is an ordinary renewal process,[12] i.e. that a switch occurred at $t = 0$. A crossing of $-z$ does not occur during the second interval if $t_2 < (z+x_0+at_1)/a$. The probability of this being true is

$$\text{Prob }(t_2 < (z+x_0+at_1)/a) = \int_0^{(z+x_0+at_1)/a} \psi(t_2)dt_2 \quad . \tag{8}$$

We continue in this fashion until the $(n-1)^{st}$ interval. During the nth interval a crossing is insured if t_n is sufficiently long. Thus, if n is even then a crossing of $-z$ (which is the only possible one) occurs if $t_n < (z+x_0+\cdots+at_{n-1})/a$. The probability for this crossing during the nth interval then is

$$\text{Prob }(t_n > (z+x_0+\cdots+at_{n-1})/a) = \int_{(z+x_0+\cdots+at_{n-1})/a}^{\infty} \psi(t_n)dt_n \tag{9}$$

(a similar expression is obtained for crossing at z when n is odd). Finally, the crossing will occur at time t if $X(t) = -z$. The probability density for this crossing event is the delta function

$$p(t:X(t) = -z) = a\delta(z+x_0+at_1-at_2+\cdots+at_{n-1}-a\Delta_n) \tag{10}$$

where $\Delta_n = t - (t_1+t_2+\cdots+t_{n-1})$. Collecting these results gives the following integral form for the first passage time density:

$$p_{2m}^{(a)}(t;x_0,z) = \int_0^{\tau_1} dt_1\, \psi(t_1) \int_0^{\tau_2} dt_2\, \psi(t_2) \cdots \int_0^{\tau_{2m-1}} dt_{2m-1}\, \psi(t_{2m-1})$$

$$\times \int_{\tau_{2m}}^{\infty} dt_{2m}\, \psi(t_{2m})\, \delta(\tau_{2m}-\Delta_{2m}) \qquad (11)$$

where $m \geq 1$,

$$\tau_{2m-1} \equiv (z-x_0-at_1+at_2+\cdots+at_{2m-2})/a \;, \qquad (12a)$$

$$\tau_{2m} \equiv (z+x_0+at_1-at_2+\cdots+at_{2m-1})/a \;, \qquad (12b)$$

and where we have explicitly indicated the initial value $f(0) = a$. Similar expressions are valid for $p_{2m+1}^{(a)}(t;x_0,z)$ and also for the corresponding probability densities with $f(0) = -a$. We have shown elsewhere[9] that performing the summation indicated in Eq. (6) leads to the integral equation

$$\tilde{p}^{(a)}(s;x_0,z) = \tilde{p}_1^{(a)}(s;x_0,z) + \tilde{p}_2^{(a)}(s;x_0,z)$$

$$+ \int_0^{\tau_1} dt_1 \int_0^{\tau_2} dt_2\, \psi(t_1)\psi(t_2)\, e^{-s(t_1+t_2)}\, \tilde{p}^{(a)}(s;x_0+at_1-at_2,z) \qquad (13)$$

with a similar expression for $\tilde{p}^{(-a)}$, where $\tilde{g}(s)$ denotes the Laplace transform of $g(t)$.

The survival probability of interest in condensed matter is an average of $p(t;x_0,z)$ over both the initial condition x_0 and the distribution of interval lengths $2z$. The initial location of the excitation is generally assumed to be uniformly distributed over non-trapping points. If we let $q(z)$ be the normalized distribution of interval lengths $2z$ between trapping points, then the survival probability is

$$F(t) = c\int_0^{\infty} dz\, q(z)\, F(t,z) \qquad (14a)$$

where

$$F(t,z) = \frac{1}{2} \int_t^{\infty} dt' \int_{-z}^{z} dx_0 \left[p^{(a)}(t';x_0,z) + p^{(-a)}(t';x_0,z) \right] \qquad (14b)$$

and where c is the concentration of traps. It is usually impossible to evaluate

$F(t,z)$ analytically but one can often find an asymptotic (long-time) expression for it. In all the cases to be considered here we have

$$F(t,z) \sim 2z \exp[-t/T_1(z)] \tag{15}$$

where $T_1(z)$ is the mean first passage time to z or $-z$ averaged over the initial position x_0.[6]

3. Results

It is interesting to compare the survival probabilities for different time distributions $\psi(t)$. For completeness we also present the results for a diffusive process. These latter results can be obtained via the traditional boundary value methods.[6]

i) Gaussian, delta-correlated (diffusive process)

For a specified initial position x_0 the mean first passage time to $\pm z$ is given by

$$T_1(z,x_0) = \frac{z^2 - x_0^2}{2D} \tag{16}$$

where D is the diffusion coefficient defined in Eq. (2). This expression is plotted in Fig. 3 as a function of x_0 for $z = 1$ and $D = 1/2$. The average of (16) over a uniform initial distribution then is

$$T_1(z) = \frac{1}{2z} \int_{-z}^{z} dx_0 \, T_1(z,x_0) \tag{17a}$$

$$= \frac{z^2}{3D} \tag{17b}$$

and is shown in Fig. 4 for $D = 1/2$. The survival probability (14a) for various concentrations is shown in Fig. 5 for a random distribution of traps, i.e. with

$$q(z) = 2ce^{-2cz} \quad . \tag{18}$$

The integral in (14a) can only be evaluated analytically in the asymptotic regime where Eq. (15) is known to be valid. Thus, although the numerical results shown in Fig. 5 start from $t = 0$, they are only reliable as t becomes large. In this asymptotic regime one obtains the well-known stretched exponential form[11]

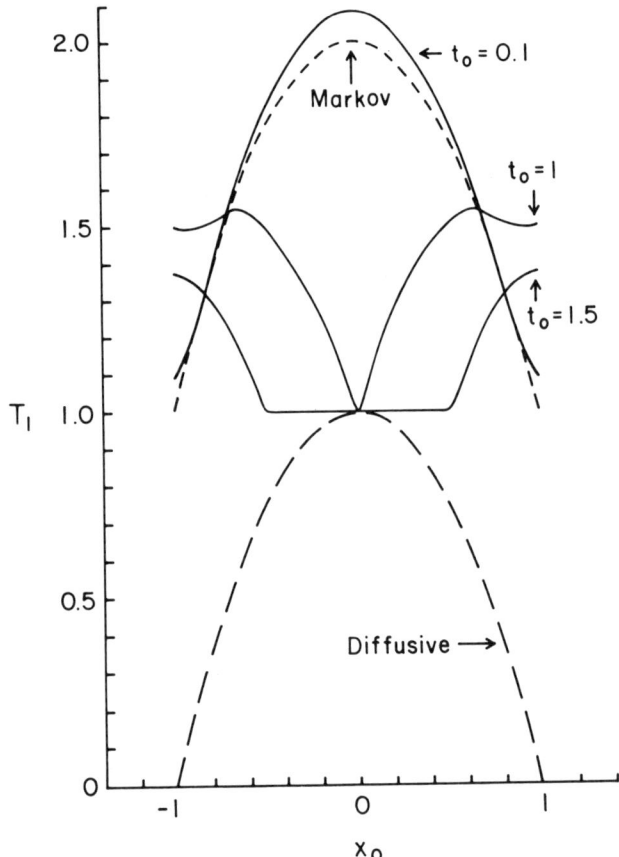

Fig. 3: Mean first passage time $T_1(z,x_o)$ as a function of the initial location x_o with parameter values $\lambda = a = 1$ and $z = 1$. The processes depicted are diffusive (— — —), dichotomous Markov (– – –), and dichotomous inertial (———) for three values of the inertial parameter.

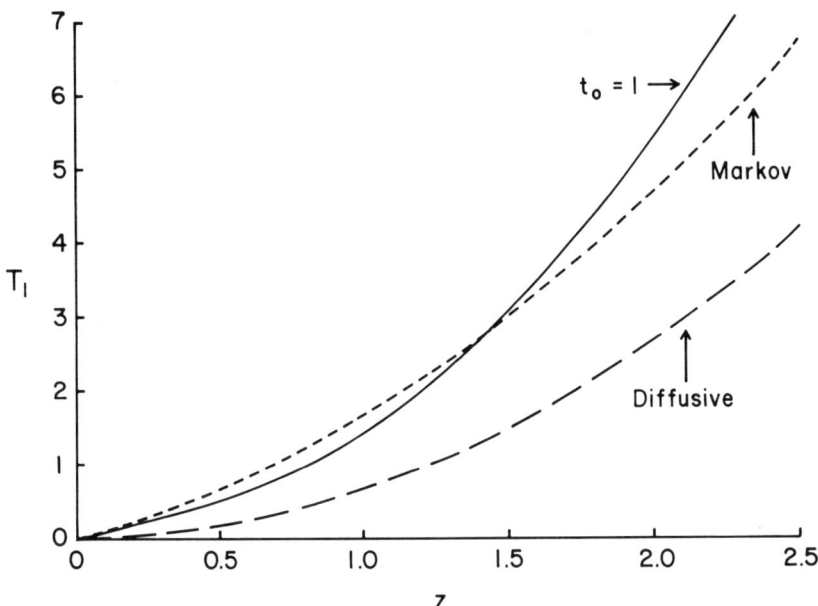

Fig. 4: Mean first passage time $T_1(z)$ as a function of trap separation for diffusive (— — —), dichotomous Markov (– – –) and dichotomous inertial (———) processes. Parameter values are $a=\lambda=1$.

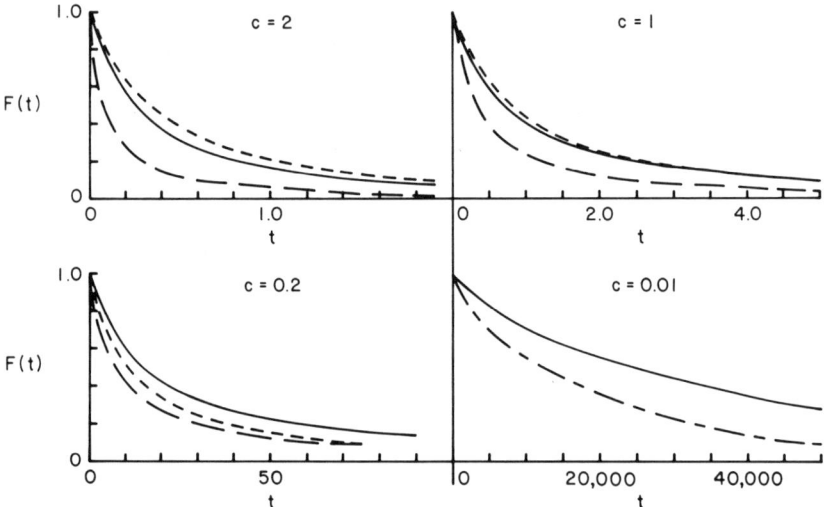

Fig. 5: Survival probabilities F(t) for four different values of the concentration of traps. Note the change in time scale for the different concentration values. Curves are given for the diffusive process (— — —), the dichotomous Markov process (– – –), and the dichotomous inertial process with $t_o = 1$ (———). The curve (__ _ __ _ __) denotes overlapping diffusive/Markov results. The parameter values are $\lambda = a = 1$.

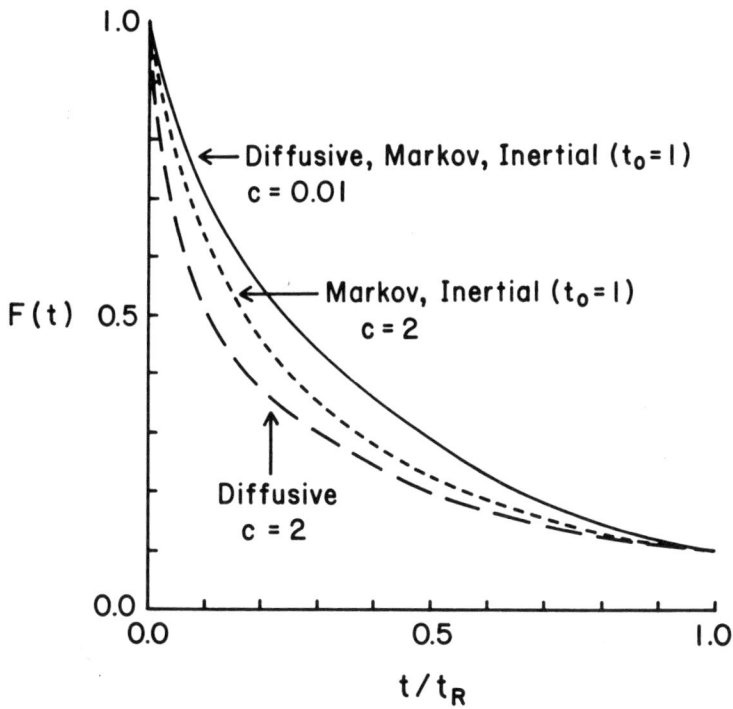

Fig. 6: Survivial probabilities F(t) vs. scaled time variable t/t_R, where $F(t_R) = 0.1$ for each process. The diffusive result with $c = 2$ is represented by the (___ ___ ___) curve. For $c = 2$, the Markov and inertial ($t_o = 1$) results coincide and are shown in the (– – –) curve. For $c = 0.01$ all three results coincide are are shown in the (_____) curve. Parameter values are $a = \lambda = 1$.

$$F(t) \sim t^{1/2} \exp\left[-\frac{3}{2}(c\pi)^{\frac{2}{3}} t^{\frac{1}{3}}\right] . \tag{19}$$

In Fig. 6 the survival probability $F(t)$ is plotted against the scaled time variable t/t_R where t_R is determined by the condition $F(t_R) = 0.1$.

The diffusive results given in Figs. 3-6 are of course well known and are given to provide a basis for comparison for the more complex processes considered subsequently.

ii) Dichotomous Markov process

A dichotomous Markov process $f(t)$ is characterized by the exponential interval distribution function

$$\psi(t) = \lambda e^{-\lambda t} \tag{20}$$

where λ^{-1} is the mean interval length. In this case the integral equations (13) for $\tilde{p}^{(a)}$ and the corresponding one for $\tilde{p}^{(-a)}$ can be converted to the differential equations[9]

$$\frac{d^2}{dx_0^2}\tilde{p}^{(\pm a)}(s;x_0,z) - \frac{s(s+2\lambda)}{a^2}\tilde{p}^{(\pm a)}(s;x_0,z) = 0 \tag{21}$$

together with the boundary conditions

$$\tilde{p}^{(\pm a)}(s;\pm z,z) = 1 \tag{22a}$$

and

$$\frac{d}{dx_0}\tilde{p}^{(\pm a)}(s;x_0,z)\bigg|_{x_0=\mp z} = \mp\frac{\lambda}{a} \pm \frac{(s+\lambda)}{a}\tilde{p}^{(\pm a)}(s;\mp z,z) . \tag{22b}$$

The solution of (21) with (22) is straightforwardly determined to be

$$\tilde{p}^{(\pm a)}(s;x_0,z) = \left\{\frac{1}{(\alpha+r)e^{2rz}-(\alpha-r)^{-2rz}}\right\}$$

$$\times\left\{(\alpha+r)e^{r(z\pm x_0)} - (\alpha-r)e^{-r(z\pm x_0)} + \frac{\lambda}{a}e^{r(z\mp x_0)} - \frac{\lambda}{a}e^{-r(z\mp x_0)}\right\}$$

$$\tag{23}$$

where

$$\alpha = \frac{(\lambda+s)}{a} \tag{24a}$$

and

$$r = \left[\frac{s(s+2\lambda)}{a^2}\right]^{1/2} . \tag{24b}$$

The time-dependence of the survival probability can be obtained from a Laplace inversion of (23). As a practical matter, the analytic inversion of this expression is unattainable except of long times. A small-s expansion of (23) followed by an integration over x_0 and over time as indicated in (14b) yields the exponential form (15). The conditional mean first passage time $T_1^{(\pm a)}(z,x_0)$ from a given initial location x_0 to $\pm z$ with $f(0) = \pm a$ satisfies a differential equation together with boundary conditions obtained directly from Eqs. (21) and (22):[9]

$$\frac{d^2}{dx_0^2} T_1^{(\pm a)}(z,x_0) = -2\frac{\lambda}{a^2} , \tag{25}$$

$$T_1^{(\pm a)}(z,\pm z) = 0 , \tag{26a}$$

$$\frac{d}{dx_0} T_1^{(\pm a)}(z,x_0)\bigg|_{x_0=\mp z} = \pm \frac{\lambda T_1^{(\pm a)}(z,\mp z) - 1}{a} . \tag{26b}$$

The solution of this equation is[8,9]

$$T_1^{(\pm a)}(z,x_0) = \frac{z^2-x_0^2}{2D} + \frac{(z\mp x_0)}{a} \tag{27}$$

where $D = a^2/2\lambda$. The time $T_1(z)$ that enters the survival probability (19) is given by (17a), with

$$T_1(z,x_0) = \frac{1}{2}\left[T_1^{(a)}(z,x_0) + T_1^{(-a)}(z,x_0)\right] \tag{28}$$

whence we obtain

$$T_1(z) = \frac{2}{3}\frac{\lambda}{a^2}z^2 + \frac{z}{a} . \tag{29}$$

In Fig. 3 we show $T_1(z,x_0)$ as a function of x_0 for $z = 1$ and the parameter values $\lambda = a = 1$. We note that the mean first passage time for the non-diffusive process is larger than the diffusive one for all x_0. We also note that $T_1(z,\pm z) \neq 0$ in the non-diffusive case.[8,9] Both of these results are a consequence of the possibility that an excitation in the neighborhood of the boundary has a finite probability of re-entering the interval. In Fig. 4 we show the averaged mean first passage time $T_1(z)$ of Eq. (29) with $\lambda = a = 1$. Again, the diffusive result lies below the Markov one. At large values of $z (z > > 1)$ the two eventually merge [cf. Eqs. (29) and (17b)].

Using these results we have evaluated the survival probability $F(t)$ numerically with the trap distribution (18) and show the results in Fig. 5 for various trap concentrations. For large concentrations the survival probability for the dichotomous Markov process is enhanced over that of the diffusive process. As c decreases, the two results converge because in this case the large z behavior of $T_1(z)$ dominates the results. An analysis of our numerical results indicates that the survival probability for the dichotomous Markov process is also a stretched exponential such as (19) but with different numerical factors. The difference between these processes can be de-emphasized by plotting $F(t)$ as a function of the scaled time variable t/t_R where $F(t_R) = 0.1$. The scale factor t_R is of course different for each process. This plot is shown in Fig. 6. Clearly the scaled processes are qualitatively similar. This similarity suggests that the survival probability for the dichotomous Markov process has a stretching exponent 1/3 as in Eq. (19) but that the prefactor of the exponential may be a different function of time than $t^{1/3}$.

iii) Dichotomous inertial process

We call a dichotomous process $f(t)$ "inertial" if $f(t)$ is forced to retain its value for *at least* a time t_o before switching to the other value. The interval distribution function that we choose to represent this process is

$$\psi(t) = 0 \quad , \quad t \leqslant t_o$$

$$= \lambda e^{-\lambda(t-t_o)} \quad , \quad t > t_o \quad . \tag{30}$$

The integral equation (13) for $\tilde{p}^{(a)}$ and the corresponding one for $\tilde{p}^{(-a)}$ can again be converted to differential equations,

$$\frac{d^2}{dx_0^2}\tilde{p}^{(\pm a)}(s;x_0,z) - \frac{(\lambda+s)^2}{a^2}\tilde{p}^{(\pm a)}(s;x_0,z)$$

$$+ \frac{\lambda^2}{a^2} e^{-2st_0}\tilde{p}^{(\pm a)}(s;x_0,z) = 0 \qquad (31)$$

together with the boundary conditions

$$\tilde{p}^{(\pm a)}(x;z\mp at_0,z) = e^{-st_0} \qquad (32a)$$

and

$$\frac{d}{dx_0}\tilde{p}^{(\pm a)}(s;x_0,z)\bigg|_{x_0=\mp z} = \mp\frac{\lambda}{a}e^{-2st_0} \pm \frac{(s+\lambda)}{a}\tilde{p}^{(\pm a)}(s;\mp z,z) \quad . \qquad (32b)$$

For $\tilde{p}^{(a)}(s;x_0,z)$ these equations are valid if $x_0 < z-at_0$. For $x_0 > z-at_0$, trapping necessarily occurs during the first interval, whence

$$p^{(a)}(t;x_0,z) = a\delta(at-z+x_0) \quad . \qquad (33a)$$

For $p^{(-a)}(s;x_0,z)$ Eqs. (31) and (32) are valid if $x_0 > at_0-z$. For $x_0 < at_0-z$, trapping again occurs during the first interval so that

$$p^{(-a)}(t;x_0,z) = a\delta(at-z+x_0) \quad . \qquad (33b)$$

The survival probability $F(t,z)$, which is again given by Eq. (15), requires knowledge of the mean first passage time. The mean first passage time $T_1^{(\pm a)}(z,x_0)$ again satisfies a differential equation obtainable directly from (31):

$$\frac{d^2}{dx_0^2}T_1^{(\pm a)}(z,x_0) = -\frac{2\lambda}{a^2}(1+\lambda t_0) \qquad (34)$$

together with the boundary conditions

$$T_1^{(\pm a)}(z,\pm z\mp at_0) = t_0 \qquad (35a)$$

and

$$\frac{d}{dx_0}T_1^{(\pm a)}(z,x_0)\bigg|_{x_0=\pm z} = -\frac{\lambda T_1^{(\pm a)}(z,\mp z) - (1+2\lambda t_0)}{a} \quad . \qquad (35b)$$

For $T_1^{(a)}(z,x_0)$ these equations hold if $x_0 < z-at_0$. For $x_0 > z-at_0$,

$$T_1^{(a)}(z,x_0) = (z-x_0)/a \quad . \qquad (36a)$$

Similarly, these equations are valid for $T_1^{(-a)}(z,x_0)$ provided that $x_0 > at_0-z$. Otherwise

$$T_1^{(-a)}(z,x_0) = (z+x_0)/a \quad . \tag{36b}$$

The solutions of (34) and (35) are

$$T^{(\pm a)}(x,x_0) = \frac{\lambda}{a^2}(1+\lambda t_0)(z^2-x_0^2) - \lambda t_0^2$$

$$+ \frac{1}{a}(1+\lambda t_0)[(1-\lambda t_0)z \mp (1+\lambda t_0)x_0] \quad . \tag{37}$$

In Fig. 3 we show the mean first passage time $T_1(z,x_0)$ given by (36) and (37) in (28) for three values of t_0 with $a = \lambda = 1$. For small t_0 the behavior essentially coincides with that of the dichotomous Markov process. As t_0 increases, the boundaries z and $-z$ are reached more rapidly when the excitation begins near the center of the interval because of those realizations that arrive at the boundary during the first interval. The effect becomes more pronounced as t_0 increases. The flat portion of the curve for $t_0 = 1.5$ represents realizations that reach a boundary during the first interval regardless of the initial direction of motion. This occurs in the interval $z - at_0 < x_0 < at_0 - z$. In this range $T_1(z,x_0) = (1/2)[(z+x_0)+(z-x_0)]/a = z/a$ [cf. Eq. (36)]. These results emphasize the fact that the most efficient trapping does not always occur when an excitation begins its migration near a trap. However, the inertial near first passage time may exceed the Markov one for all values of x_0 if t_0 is sufficiently small compared to z/a and λ^{-1}.

The mean first passage time averaged over x_0 is given by $T_1(z) = z/a$ for $z < at_0/2$, and for $z > at_0/2$

$$T_1(z) = \frac{2\lambda}{3a^2}(1+\lambda t_0)z^2 + \frac{1}{a}(1-\lambda^2 t_0^2)z - \frac{\lambda t_0^2}{2}(1-\lambda t_0)$$

$$+ \frac{a\lambda t_0^3}{6z}\left(1 - \frac{1}{2}\lambda t_0\right) \tag{38}$$

and is shown in Fig. 4 for $t_0 = 1$. For small z this average reflects the behavior shown in Fig. 3 resulting in an inertial $T_1(z)$ that lies between the diffusive and the Markov values. for large z (to held fixed) the inertial result crosses above the Markov one, in analogy with the $t_0 = 0.1$ results in Fig. 3. For large z the ratio of the inertial to the Markov and diffusive results approaches the value with the trap distribution (18) and exhibit the results in Fig. 5 for $t_o = 1$ and for the same set of concentrations considered previously. For $c = 2$ the inertial survival probability lies between the Markov and diffusive ones. This behavior reflects the range of z values most heavily weighted in the integral (14a). For $c = 2$ this is the range $z \lesssim 1$. In this range Fig. 4 shows the same relative ordering of the $T_1(z)$. When c decreases to unity the z-regime most heavily weighted increases to around $z \sim 1.5$, where the inertial and Markov first passage times are equal. This equality is apparent in the overlap of the survival probability curves. As the concentration of traps is further decreased, the inertial curve separates from the Markov and diffusive ones, which move closer together and eventually merge. This is a direct consequence of the large z behavior of $T_1(z)$.

The similarities among the various processes are emphasized in the scaled graph of Fig. 6. At the lowest concentrations the three scaled processes are indistinguishable.

4. Conclusions

Herein we have presented a procedure to calculate the trapping dynamics for a class of processes with memory. We have calculated mean first passage times and survival probabilities for several processes using this approach, and have compared the results obtained for three models: diffusive, "dichotomous Markov", and "dichotomous inertial". The results indicate that many of the qualitative features of the trapping dynamics are very sensitive to the correlation properties of the model. We reiterate the more interesting results of our analysis.

The mean first passage time $T_1(z,x_o)$ from an initial location x_o to absorbing traps at $\pm z$ is, of those considered here, the most sensitive measure of the process. The most dramatic effect introduced by a memory is the overall increase of $T_1(z,x_o)$ above the diffusive result. This increase arises from the capacity of the process to re-enter the non-trapping interval even if the excitation is arbitrarily close to an absorbing point. Another unfamiliar consequence of this capacity is the fact that the mean first passage time to trapping of an excitation that starts arbitrarily close to a trapping point does not vanish, i.e. $T_1(z,\pm z) \neq 0$.[8,9] The details of the behavior of $T_1(z,x_o)$ as a function of x_o are process-dependent. We have illustrated this dependence by analyzing inertial processes that emphasize the retention of a given direction of motion by an excitation. The initial condition leading to the most (least) effective trapping rate is determined by the inertial parameter t_o.

The mean first passage time $T_1(z)$ averaged over a uniform initial distribution smooths out much of the structure in $T_1(z,x_0)$. For large z the dichotomous Markov result approaches the diffusive one from above, whereas the inertial $T_1(z)$ lies above both of them. The ratio of the two results approaches $(1+\lambda t_0)$. As z decreases two situations may arise. If t_0 is sufficiently small, then the inertial mean first passage time exceeds the Markov result for all z. For larger values of t_0, the two curves cross, resulting in the inertial time lying between the Markov and diffusive ones below the crossing point [cf. Figure (4)]. [cf. Figure (5)]

The survival probability $F(t)$ reflects the behavior of $T_1(z)$. Thus, for large concentrations of traps the small z-behavior of $T_1(z)$ is weighted most heavily, resulting in the relative ordering of survival probabilities being that of the first passage times for small z. As the concentration of traps decreases the crossing and merging of the $T_1(z)$ curves is directly mimicked by the survival probability curves.

When the survival probabilities of the three processes are each scaled in time as we did in Fig. 6, one observes a general scaling phenomenon that would indicate a common stretched exponential behavior with different numerical coefficients and prefactors for all the processes considered. We stress that these latter results are contingent on the particular choice (18) for the distribution of traps. Whether in fact these are general effects which persist independently of the choice of distribution remains an open question.

6. References

1. *Random Walks and Their Applications in the Physical and Biological Sciences*, AIP Conference Proceedings Vol. 109 (AIP, New York, 1982), eds. M. F. Shlesinger and B. J. West; J. Stat. Phys. 30 (1983), eds. G. H. Weiss and R. J. Rubin.
2. E. W. Montroll and B. J. West, in *Fluctuation Phenomena* (North-Holland, Amsterdam, 1979), eds. E. W. Montroll and J. L. Lebowitz.
3. G. H. Weiss and R. J. Rubin, in *Advances in Chemical Physics* Vol. 52 (Wiley, New York, 1983), eds. I. Prigogine and S. A. Rice.
4. E. W. Montroll and M. F. Shlesinger, in *Nonequilibrium Phenomena II: From Stochastics to Hydrodynamics* (North-Holland, Amsterdam, 1984), eds. J. L. Lebowitz and E. W. Montroll.
5. V. Seshadri and B. J. West, Proc. Natl. Acad. Sci. USA 79, 4501 (1982).
6. K. Lindenberg and B. J. West, J. Stat. Phys. 42, 201 (1986).
7. G. H. Weiss, J. Stat. Phys. 24, 587 (1981); *ibid.* 37, 325 (1984).
8. P. Hänggi and P. Talkner, Zeitr. für Physik B45, 79 (1981); *ibid.*, Phys. Rev. Lett. 51, 2242 (1983); *ibid.*, Phys. Rev. A32, 1934 (1985).
9. J. Masoliver, K. Lindenberg and B. J. West, Phys. Rev. A (Rapid Communication), to appear; *ibid.*, Phys. Rev. A, submitted.

10. M. C. Wang and G. E. Uhlenbeck, Rev. Mod. Phys. **17** 323 (1945).
11. See e.g. J. K. Anlauf, Phys. Rev. Lett. **52**, 1945 (1984).
12. D. R. Cox, *Renewal Theory* (Wiley, New York, 1962).

TRAPPING OF PARTICLES AND DEPOLARIZATION OF SPINS BY RANDOM WALKS ON LATTICES

K.W. Kehr*, J.K. Anlauf, and R. Czech
Institut für Festkörperforschung
der Kernforschungsanlage Jülich
D-5170 Jülich
FEDERAL REPUBLIC OF GERMANY

This contribution describes two relaxation phenomena that are governed by diffusion of particles in random situations. The randomness is introduced through random distributions of trapping centers or of spin rotation frequencies. The particles perform simple random walks as in ordered crystals. These simple models already yield unusual behavior of the survival probability or of the spin polarization.

I. The trapping problem

Consider a lattice with randomly distributed traps of concentration c. A particle is put randomly on this lattice, where it performs random walk until it encounters a trap; it is then regarded as lost. The quantity of interest is the survival probability Ψ_n of the particles in an ensemble of such systems after n steps, if discrete random walk is performed, or the survival probability $\Psi(t)$ after time t in the case of continuous-time random walk. In the discrete case the survival probability is given by

$$\Psi_n = \langle (1-c)^{S_n} \rangle \qquad (1)$$

where S_n is the number of distinct sites visited by a walk of n steps. The average over the random trap configurations has already been performed in (1). Since 1 - c is the probability that a particular site is not a trap, expression (1) is obvious, and the remaining average is over the different random walks. An equivalent expression for the survival probability is

$$\Psi_n = \sum_S p_n(S) \exp(-\lambda S) \tag{2}$$

with $p_n(S)$ the probability distribution of the number of distinct sites visited by an n-step walk and $\lambda = -\ln(1-c)$.

It is tempting to replace in (1) S_n by $\langle S_n \rangle$, the mean number of distinct sites visited with n steps, since this quantity is well-known for large n from random walk theory[1]. This leads to the Rosenstock approximation[2] for the survival probability

$$\Psi_n^R \sim \begin{cases} \exp\left[-\lambda(8n/\pi)^{1/2}\right] & d = 1 \\ \exp\left[-(\lambda\pi n)/\ln(8n)\right] & d = 2 \\ \exp\left[-\lambda(1-p_r)n\right] & d \geq 3 \end{cases} \tag{3}$$

where p_r is the probability of return of a particle to the origin by random walk in 3 and more dimensions. E.g., $p_r = 0.341$ for simple-cubic lattices.

The results of this approximation in continuous time are obtained by substituting n by t/τ where τ is the mean residence time of the particle at a site. In d = 3, the survival probability is then $\Psi(t) = \exp(-\Gamma t)$ where the capture rate $\Gamma = \lambda(1-p_r)/\tau$. This is the rate for diffusion-limited trapping on a lattice. Hence the expression (3) and its variants represent very familiar behavior in d = 3. Unfortunately (3) is not correct asymptotically, the correct asymptotic results in d = 1, 2, 3 were first obtained by Balagurov and Vaks[3] and later proven rigorously by Donsker and Varadhan[4].

First a heuristic argument will be given that yields the correct leading asymptotic behavior. Consider a d-dimensional simple-cubic lattice and a trap-free region of S sites in it. The region is assumed to be of cubic shape with linear dimension $L = S^{1/d}$. The mean number of steps necessary to traverse this region is $\langle n \rangle \simeq S^{2/d}$. The survival probability of a particle put into the trap-free region is asymptotically an exponential function of the step number, $\Psi_n \sim \exp(-n/\langle n \rangle)$. This survival probability must be averaged with the probability $\exp(-\lambda S)$ of occurrence of a region with S free sites,

$$\Psi_n \sim \sum_S \exp(-\lambda S) \exp(-n/S^{2/d}) \qquad (4)$$

The sum is converted into an integral and evaluated by the saddle-point method. The saddle point is at $S^* = (\frac{2n}{d\lambda})^{d/(d+2)}$ and the survival probability is

$$\Psi_n \sim \exp\left[-(\frac{d+2}{d})(\frac{d\lambda}{2})^{2/(2+d)} n^{d/(2+d)}\right] \qquad (5)$$

This result agrees with the general result of Donsker and Varadhan[4] for the leading asymptotic behavior, apart from a numerical factor.

The considerations above show that the largest contributions to the survival probability arise from regions which are relatively trap-free (factor $\exp(-\lambda S)$) and from those random walks that are more 'compact' than the average ones (wing of $p_n(S)$). It is seen immediately that the asymptotic behavior (5) is in contradiction to the results of the approximation (3).

The survival probability can be derived in greater detail for random walks on linear chains. An approximate asymptotic theory is obtained[5] by using published asymptotic expressions for the span distribution $p_n(S)$ in $d = 1$[6]. The appropriate form is the one which converges better at values

$S < S_m$ where S_m is the maximum of $p_n(S)$, $S_m \propto n^{1/2}$. It is sufficient to take only the first term of the second series given in [6].

$$p_n(S) \simeq \frac{8n}{S^3} \left(\frac{\pi^2 n}{S^2} - 1\right) \exp\left(-\frac{\pi^2 n}{2S^2}\right) \tag{6}$$

The scaling variable $x = (\pi\lambda)^{2/3} n^{1/3}$ will be introduced. As a result of a systematic treatment of (5) where (6) is used we obtain

$$\Psi(x) = 8\left(\frac{2}{3}\right)^{1/2} \left(\frac{x}{\pi}\right)^{3/2} \exp\left[-\frac{3x}{2} + \frac{a_1}{x} + \frac{a_2}{x^2} + \frac{a_3}{x^3} + \frac{a_4}{x^4} + ..\right]$$
$$+ O\left[x^{3/2} \exp\left(-\frac{27}{2}x\right)\right] \tag{7}$$

$$a_1 = \frac{17}{18} \quad a_2 = -\frac{7}{54} \quad a_3 = -\frac{26}{243} \quad a_4 = \frac{167}{972} \tag{8}$$

In this approximation $\Psi(x)$ depends on the scaling variable x only.

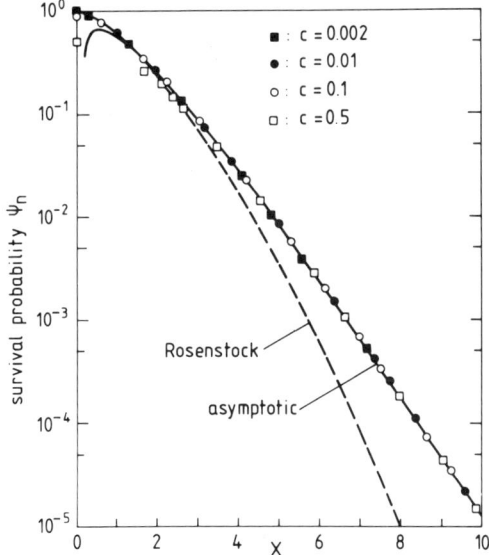

Fig. 1: Plot of survival probability Ψ_n against $x = (\pi\lambda)^{2/3} n^{1/3}$ for different trap concentrations c. Solid line, asymptotic form Eq. (7) including third order correction a_3. Broken line, Rosenstock approximation Eq. (3). Dots, Monte Carlo data.

The survival probability was estimated by numerical simulations of (2). Fig. 1 shows very good agreement between (7) and the simulations at various concentrations $c \leq 0.5$. It is also seen that the Rosenstock approximation deviates significantly from the correct values of Ψ_n when $\Psi_n < 10^{-2}$.

The derivation reviewed above used the asymptotic span distribution $p_n(S)$ which is valid for large n. If the trap concentration is large, a particle is trapped after a few steps. Hence the exact instead of the asymptotic form of the span distribution should be used when Ψ_n is calculated for large c. An exact asymptotic theory can be derived from exact expressions for the span distribution. The result has a similar form to (7). However, there appears an additional prefactor

$$C(c) = c^2/[(1-c)\lambda^2] \qquad (9)$$

and also the coefficients $a_1, \ldots, a_4$ are modified. The expansion of the prefactor is $C(c) = 1 + c^2/12 + \ldots$, hence the modification of the result is less than 3 % for $c \leq 0.5$. However, it becomes important at higher trap concentrations. The modified coefficients a_1 and a_2 will be given below in the context of correlated random walk. It should be noted that the modifications are no more of scaling form, i.e., the coefficients $a_1, \ldots, a_4$ depend now explicitly on λ. Since the correction terms are proportional to x^{-k}, they vanish in the limit $n \to \infty$. High-precision numerical evaluations of Ψ_n using transfer matrix techniques verify the validity of the results.

Corrections to scaling are also obtained from an extension of the model to correlated walks of the particle, in the presence of traps. In the correlated random-walk model in $d = 1$, the particle has a different probability p_f of making a step in the same direction as the previous one than performing a step in the reverse direction. For instance, forward correlations are characterized by $p_f > 1/2$; they lead to a decreased survival probability. Asymptotically, correlated walk is a random walk, with a mean-square displacement $\propto fn$ where f is the correlation factor and $f = p_f/(1 - p_f)$ in $d = 1$. A simple estimate of the effect of correlations on Ψ_n is obtained by rescaling the step number n with the correlation factor, $n \to fn$. The correct asymptotic theory of the survival probability uses exact expressions for the span distribution $p_n^c(S)$ for correlated walks. The leading term of the result is given, to a good approximation, by a rescaling of the span distribution of uncorrelated random walk, together with a shift of the argument,

$$p_n^c(S) \simeq p_{fn}(S + f - 1) \tag{10}$$

One consequence of this form is that the mean number of distinct sites visited by correlated walks in $d=1$ is given by

$$\langle S \rangle_n^c \simeq \langle S \rangle_{fn} - f + 1 \tag{11}$$

This relation was derived recently[7]. When (10) is employed in (2) one obtains

$$\Psi_n^c \simeq (1 - c)^{1-f} \Psi_{fn} \tag{12}$$

Fig. 2 shows the result of simulations of the survival probability for correlated walks at $c = 0.5$, multiplied with $(1 - c)^{f-1}$, for various values of p_f. Asymptotically all points lie on the same curve. The theoretical curves were obtained by evaluating the approximate asymptotic result (7) with the scaling variable $x = (\pi \lambda)^{2/3} (fn)^{1/3}$ and multiplying with $(1 - c)^{f-1}$. The deviations of the data especially for $p_f = 0.9$, from a single curve at small x indicate corrections to scaling. There is also a slight deviation of the data points from the (approximate) theory. This deviation disappears when the exact asymptotic theory of the survival probability for correlated walks is used.

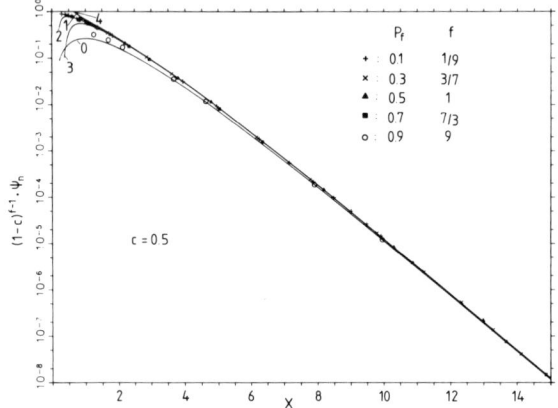

Fig. 2: Plot of $(1-c)^{f-1} \Psi_n$ against x for different correlation factors f. Solid lines, approximation Eq. (7) multiplied by $(1-c)^{f-1}$ and with $x = (\pi\lambda)^{2/3} (fn)^{1/3}$ in different orders (line denoted by i includes all terms up to a_i)

The result of the exact asymptotic theory for correlated walks in d = 1 is[8]

$$\Psi_n = \frac{c^2}{(1-c)^f \lambda^2} 8(\frac{2}{3})^{1/2} (\frac{x}{\pi})^{3/2}$$

$$\exp\left[-\frac{3x}{2} + \frac{a_1}{x} + \frac{a_2}{x^2} + \ldots\right]$$

$$a_1 = \frac{17}{18} - \frac{1}{24}(3f^2 - 1)\Lambda^2$$

$$a_2 = -\frac{7}{54} + \frac{1}{36}(3f^2 - 7)\Lambda^2 + \frac{1}{6}f(f^2 - 1)\frac{\Lambda^3}{\pi} \quad (13)$$

where $\Lambda = \pi\lambda$. The next two terms a_3 and a_4 have also been calculated and will be published elsewhere[8]. It is interesting that a_4 contains a contribution which is different for even and odd step numbers.

The trapping problem is much less amenable to analytical methods in higher dimensions. The leading term in the exponent of the survival probability is given by the result of Donsker and Varadhan. One desires to know the corrections to the leading behavior in order to estimate at which step number the asymptotic behavior according to (5) becomes important. As (2) indicates the quantity required is the distribution $p_n(S)$ of distinct sites visited; properties of $p_n(S)$ are known near the maximum S_m of $p_n(S)$ but not near the saddle point S^* of the product (6) where $S^* \ll S_m$ for large n. Results on the behavior of Ψ_n were obtained by Lubensky from field theory[9]. He found that the survival probability behaves asymptotically as

$$\Psi_n \sim n^\zeta \exp[-\sigma(n)] \tag{14}$$

where

$$\sigma(n) = \frac{1}{2}(d+2) A_1 \left(\frac{2n}{dA_1}\right)^{d/(d+2)}$$
$$- \frac{1}{2}\frac{d^2+d+4}{d+2} A_2 \left(\frac{2n}{dA_1}\right)^{(d-1)/(d+2)} + O\left[n^{(d-2)/(d+2)}\right]$$

The coefficient $A_1 \propto \lambda$, and $\zeta = [4(\delta-1)-d]/[2(d+2)]$ where δ is an exponent characterizing the band-edge singularity of the corresponding electronic problem. The leading term of $\sigma(n)$ agrees with the result of Donsker and Varadhan. The correction terms are of interest here. The coefficient A_2 was derived only approximately in[9].

In an attempt to estimate these corrections in $d=2$ we made numerical simulations of (2) at $c=0.5$. We divided out the leading behavior according to Donsker and Varadhan, and considered $F_n = \Psi_n/\Psi_n^{DV}$. Fig. 3 gives $\ln F_n$ in a doubly-logarithmic representation. The first correction term should appear as a straight line for large argument, with slope 1/4. The results show that a slope of 1/4 is consistent with the behavior for large n, hence the existence of the correction term seems to be verified. To determine the coefficient

A_2 we plotted $(\lambda n)^{-1/4} \ln F_n$ versus $(\lambda n)^{-1/4}$. The ordinate of the graph at abscissa zero should give the required coefficient, however, the extrapolation of the data points appears to be rather abiguous, and we cannot estimate A_2 with reasonable accuracy.

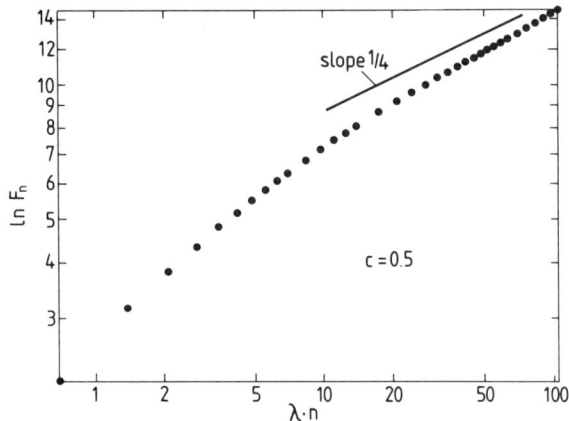

Fig. 3: Plot of $\ln F_n = \ln (\Psi_n/\Psi_n^{DV})$ against λn with $c = 0.5$ for simple random walk from Monte Carlo data. The solid line has a slope of 1/4.

The pyhsical origin of the difficulties is the small probability of finding large trap-free regions. When the survival probability is less than 10^{-12}, the limiting value determined, the relevant trap-free regions occur so rarely that they cannot be found during a simulation. Thus a better numerical estimate of Ψ_n for large n and small c seems to be extremely difficult unless other sampling methods are developed. On the theoretical side the coefficients of all correction terms up to the logarithmic corrections should be determined. Only then the practical relevance of these results can be assessed.

Other numerical evaluations of the survival probability in $d = 2$ and 3 were made by Havlin et al.[10] especially at higher trap concentrations. They showed that the asymptotic behavior according to Donsker and Varadhan is only seen for $\Psi_n < 10^{-13}$, further they examined the dependence on the scaling variables appearing in $d = 2$ and 3.

II) The depolarization problem

This is another problem where the distribution of random walks with different lengths is important. At time t=0 a spin with a definite polarization, say in +x direction, is put randomly on a lattice. At site i the spin performs a rotation with frequency ω_i in the x,y plane. The frequency ω_i is a 'quenched' random variable taken from a probability distribution $p(\omega_i)$; usually a Gaussian distribution is assumed. A physical realization of the model is provided by polarized muons implanted into a metal and exposed to a field transverse to the polarization direction. The random rotation frequencies are due to the dipolar interactions with the host nuclei. The rotation frequency of the spin is a function of time when the spin makes transitions to neighbor sites with different rotation frequencies. The polarization at time t is given by

$$P(t) = \langle\langle \exp[-i \int_0^t dt' \omega(t')] \rangle_{RW} \rangle_{\{\omega_i\}} \quad (15)$$

where one average extends over different realizations of the random walk of the spin for fixed configurations of the $\{\omega_i\}$, and the other one over different configurations of the random frequencies.

In the standard stochastic theory of depolarization, developed by Anderson[11] and Kubo and Tomita[12] it is assumed that $\omega(t)$ represents a Gaussian stochastic process. If the further assumption is made that $\omega(t)$ is also a Markov process, the frequency autocorrelation function must have the form $\langle \omega(t)\omega(0)\rangle = \sigma^2 \exp(-t/\tau_c)$ where τ_c is the correlation time. These assumptions allow an evaluation of (15) in closed form. Here only the limiting cases will be given,

$$P(t) \sim \begin{cases} \exp(-\sigma^2 t^2/2) & \sigma\tau_c \gg 1 \\ \exp(-\sigma^2 \tau_c t) & \sigma\tau_c \ll 1 \text{ and } t \gg \tau_c \end{cases} \quad (16)$$

The first case is the Gaussian decay of the polarization for nearly immobile spins, the second case describes motional narrowing of the lineshape, the Fourier transform of (16).

The problem considered here is the decay of polarization in the case that the stochastic process $\omega(t)$ is induced by the random walk of the spin on the lattice. There is a tendency for returns to the sites already visited and thus a slow decay of frequency correlations, which is especially pronounced in low dimensions. On the linear chain frequency correlations decay according to $<\omega(t)\,\omega(0)> \propto \sigma^2 (t/\tau_c)^{-1/2}$. When it is still assumed that $\omega(t)$ is a Gaussian stochastic process, and the decay $\propto t^{-1/2}$ of the frequency autocorrelation function according to one-dimensional random walk used, a depolarization

$$P(t) \sim \exp(-\text{const } \sigma^2 \tau_c^{1/2} t^{3/2}) \tag{17}$$

is obtained[13]. This result is physically appealing since it represents an intermediate behavior between Gaussian decay for immobile spins and exponential decay for mobile spins that do never return to the same frequencies. There are experiments[14] that do support the result (17). However, the validity of (17) is questionable in view of the following inconsistency. The diffusion process of the spin on a lattice is a Markov but not a Gauss process, and the algebraic decay of the frequency correlations is not compatible with the simultaneous assumption of a Gauss and Markov process.

For simplicity, discrete random walk of the spin will be considered in the further derivations. The depolarization after n steps is then given by

$$P_n = <<\exp(-i \sum_{j=1}^{s} \omega_j h_j)>> \tag{18}$$

The sum extends over all sites visited and h_j is the number of visits to site j. The time interval between two steps is set unity, $\tau = 1$, and ω_j should be considered as the phase angle acquired during one visit to site j. If independent Gaussian distributions are assumed for the rotation angles ω_j, one of the averages in (18) can be performed with the result

$$P_n = \langle \exp(-\tfrac{1}{2}\sigma^2 \sum_{j=1}^{S} h_j^2) \rangle \qquad (19)$$

The remaining average extends over different random walks.

An exact evaluation of (19) is not known yet. A crude approximation is obtained by replacing the number S of distinct sites visited by its mean number $\langle S \rangle$, in $d = 1$ $\langle S \rangle \simeq (8n/\pi)^{1/2}$. It is further assumed that each site within $\langle S \rangle$ is equally often visited, $h_j = n/\langle S \rangle$, or $h_j = (\pi n/8)^{1/2}$. From this approximation a depolarization follows

$$P_n \sim \exp\left[-(\pi/32)^{1/2} \sigma^2 n^{3/2}\right] \qquad (20)$$

This is the discrete analogue of the decay $P(t) \sim \exp(-\text{const} \cdot t^{3/2})$.

A better approximation, which takes the distribution of the number of distinct sites visited into account, is obtained by classifying the different random walks according to their spans S. It will be still assumed that each site, within the span S, is equally often visited, $h_j = n/S$. With this approximation the depolarization is given by

$$P_n \sim \sum_S p_n(S) \exp\left(-\frac{\sigma^2 n^2}{2S}\right) \qquad (21)$$

Note that extended random walks with large S give large remaining contributions to the polarization. Since the large S values dominate P_n, the asymptotic series for $p_n(S)$ will be used that converges right of the maximum S_m of $p_n(S)$, $S > S_m$. Here the first term of the first series given in[6] is sufficient,

$$p_n(S) \simeq 8(2\pi n)^{-1/2} \exp(-S^2/2n) \tag{22}$$

The saddle point S^* of the expression (21) is at $S^* \simeq 2^{-1/3} \sigma^{2/3} n$, and (21) evaluated at this saddle-point yields

$$P_n \simeq 8 \times 3^{-1/2} \exp(-3 \times 2^{-5/3} \sigma^{4/3} n) \tag{23}$$

which is valid for values of σ not exceeding unity. According to this approximation, the polarization decays as a simple exponential function of the number of steps n, not as suggested by (20). Eq. (23) gives in fact an upper bound on the polarization when a Gaussian frequency distribution is assumed. A lower bound on the polarization can also be given[15] which is also an exponential function of the number of steps n. Hence it is proven that the polarization decays only exponentially, with an argument $\propto n$, when the stochastic process is induced by random walks on the linear chain and the ω_j taken from a Gaussian distribution. The physical reason for the slow polarization decay, compared to the form (20), is the existence of extended walks, which lead to exponential decay.

In the numerical simulations it is advantageous to estimate directly (19) rather than to perform the double average (18). Fig. 4 shows that the polarization decay approaches indeed the simple-exponential decay $\propto \exp(-\gamma n)$. The behavior of the decay constant γ is shown in Fig. 5 (d = 1); a behavior $\gamma \propto \sigma^\beta$ with $\beta = 1.30 \pm 0.02$ can be extacted. It is not known whether the deviation from $\beta = 4/3$ is significant or not.

One open question is whether the results given so far are typical for different static frequency distributions $p(\omega)$ or specific for the Gauss distribution. Several other distributions were considered[16] by us, namely a symmetric

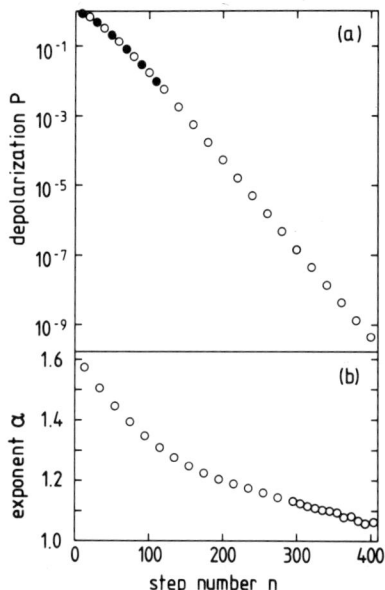

Fig. 4: (a) Depolarization of spin rotation as a function of the number of steps of a one-dimensional random walk. The width σ was 0.1. Open circles, simulation of Eq. (19); closed points, simulation of Eq. 18. (b) Apparent exponent of decay law $\ln P_n \propto n^\alpha$.

exponential distribution with variance σ^2, the rectangular and the dichotomic distributions. Already the decay in the immobile case is different for all three distributions, e.g., the exponential distribution yields $P(t) = (1 + \sigma^2 t^2/2)^{-1}$. The short-time expansion of this decay law is identical with the short-time expansion of the Gaussian case, see (16); this is a general phenomenon. In the mobile case $\sigma\tau \ll 1$ the true asymptotic behavior of $P(t)$ is specific for each distribution as well. However, there is an intermediate time range, depending on $\sigma\tau$, where simple-exponential decay pertains. Let us return to the discrete random walk and consider the polarization decay in the approximation that each site within the span S is equally often visited. The generalization of (21) is then

$$P_n = \sum_S p_n(S) \, C(S) \qquad (24)$$

where C(S) is specific for each distribution. It is important whether the maximum of the product in (24) appears to the right or to the left of the maximum S_m of $p_n(S)$. If the maximum is to the right, a similar result as above is obtained, i.e., $P_n \sim \exp(-\text{const } \sigma^{4/3} n)$. This is due to the fact that C(S) can be approximated by $\exp[-(\sigma^2 n^2)/(2S)]$ as long as $\sigma n/S \ll 1$. This inequality is fulfilled when σ is small and the step number not too large. For example the exponential distribution gives a maximum to the right of S_m when roughly $n < n^* \simeq 20/\sigma^2$. If n becomes large, different results emerge for each distribution. E.g., the exponential distribution yields

$$P_n \sim \exp[-\text{const} \times n^{1/3} \times \ln^{2/3}(\sigma^{3/2} n)] \qquad (25)$$

If σ is small, then P_{n^*} at the change-over point to this behavior is already so small that (25) is practically unobservable. The parameter σ can be made arbitrarily small by regarding continuous-time random walk, where σ is replaced by $\sigma \tau_c$, and selecting short τ_c. Experimentally this may be achieved by adjusting the temperature. The numerical simulations support the picture given here.

Finally, also the polarization decay due to random walks in higher-dimensional lattices is of interest. Since the polarization decays exponentially with step number or time already in the case of random walk in d = 1, one expects simple exponential decay also for random walks in higher dimensions, where revisitation effects are less important. It should be noted that already the Gaussian assumption for $\omega(t)$, combined with the appropriate decay of the frequency

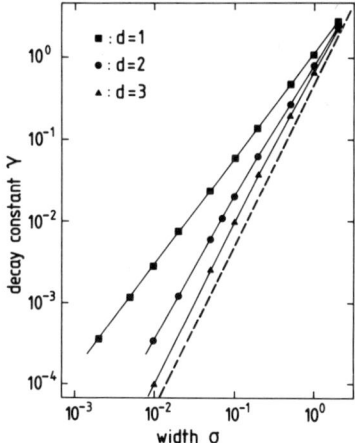

Fig. 5: Dependence of the decay constant in $P_n \sim \exp(-\gamma n)$ on the width σ of the Gaussian frequency distribution; results of simulations for random walk on a linear chain (d=1), on a square (d=2) and on a simple cubic lattice (d=3). [The decay constant $\gamma = \sigma^2/2$ of the simple model where with each step a new frequency is visited has been included for comparison, dashed line.]

correlations, leads to a simple exponential in $d \geq 3$. This is a rigorous lower bound for the actual polarization decay as can be shown with the help of Jensen's inequality. The interesting point is the dependence of the decay constant γ on σ, in analogy to $d = 1$, where a dependence on σ with a broken power appeared. Numerical simulations in $d = 2$ and 3 confirm the expected exponential decay of P_n. In $d = 2$ the dependence on σ, for the smaller values of σ, could be described by $\gamma \propto \sigma^{1.77}$, while in d=3 and $\sigma \ll 1$ the behavior is $\gamma \propto \sigma^2$, cf. Fig. 5. As a detailed analysis of the data in $d = 3$ show, the decay constant approaches the value one obtains from the Gaussian assumption; for the discrete walk it is given by $\gamma = (a - 1/2) \sigma^2$ where $a = 1.516 \ldots$ for the sc lattice. Thus one can conjecture that for random walks in $d \geq 3$ the Gaussian treatment yields an appropriate description in the mobile case.

In conclusion, both the trapping problem and the depolarization problem show that the distribution of different random walks in random media must be taken into account in order to obtain the correct behavior of the quantities of interest. It should be noted that the exponential dependence of the survival probability with a broken power of the step number, or the dependence of the decay constant γ on a broken power of σ result from this average over the distribution of different walks, in the random environment. In the language of Ref.[17] the models discussed here are models of 'parallel relaxation' where the interesting behavior follows from correctly performing the parallel superposition of the individual results. These models of parallel relaxation appear to provide well-understood examples of anomalous behavior. It is desirable to devise equally well-defined models with series relaxation that yield anomalous behavior.

References

1) E.W. Montroll, G.H. Weiss, J. Math. Phys. $\underline{6}$, 167 (1965)
2) H.B. Rosenstock, J. Math. Phys. $\underline{11}$, 487 (1970)
3) R.Ya. Balagurov, V.G. Vaks, Zh. Eksp. Teor. Fiz. $\underline{66}$, 1135 (1974) [Sov. Phys. JETP $\underline{38}$, 968 (1974)]
4) M.D. Donsker, S.R.S. Varadhan, Commun. Pure Appl. Math. $\underline{28}$, 525 (1975), and $\underline{32}$, 721 (1979)
5) J.K. Anlauf, Phys. Rev. Lett. $\underline{52}$, 1845 (1984)
6) G.H. Weiss, R.J. Rubin, Adv. Chem. Phys. $\underline{52}$, 363 (1983)
7) K.W. Kehr, P. Argyrakis, to be published
8) J.K. Anlauf, K.W. Kehr, S.M. Reulein, to be published
9) T. Lubensky, Phys. Rev. A $\underline{30}$, 2657 (1984)
10) S. Havlin, M. Dishon, J.E. Kiefer, G.H. Weiss, Phys. Rev. Lett. $\underline{53}$, 407 (1984)
11) P.W. Anderson, J. Phys. Soc. Jpn. $\underline{9}$, 316 (1954)
12) R. Kubo, K. Tomita, J. Phys. Soc. Jpn. $\underline{9}$, 888 (1954)
13) D. Hone, K.G. Petzinger, Phys. Rev. B $\underline{6}$, 245 (1972)
14) R.E. Dietz, F.R. Merritt, R. Dingle, D. Hone, B.G. Silvernagel, P.M. Richards, Phys. Rev. Lett. $\underline{26}$, 1186 (1971)
15) R. Czech, K.W. Kehr, Phys. Rev. Lett. $\underline{53}$, 1783 (1984)
16) R. Czech, K.W. Kehr, to be published
17) R.G. Palmer, D.L. Stein, E. Abrahams, P.W. Anderson, Phys. Rev. Lett. $\underline{53}$, 958 (1984)

* Present temporary address: Department of Applied Physics, Tokyo Institute of Technology, Tokyo 152, Japan

On a Generalized Transport Equation for Chromatographic Systems

George H. Weiss
National Institutes of Health, Bethesda, MD. 20894

Abstract

The chromatography model first analyzed by Giddings and Eyring[1] is one of the earliest models for transport in a random medium. It is also a forerunner of trapping models discussed by many authors in the context of solid state physics. In this paper I show that the analysis of the Giddings-Eyring model leads to a generalized diffusion equation. Sufficient conditions are given to insure that propagating peaks are asymmetric in an infinitely long column. These are of two types. The first requires that the second moment of the sojourn time of molecules that are in a stationary phase be infinite. The second allows for all transitions between mobile and stationary phases to be first order with random rate constants. Distributions that give a considerable probability for low rate constants for molecules to exit the stationary phase are shown to imply peak asymmetry. Finally I derive a generalized equation governing first passage times for the multistate Giddings-Eyring model.

Introduction

Chromatographic processes have been used in chemistry since the late 19'th century for the separation of molecular species by a differential response to a physical force that depends on the particular technique being used. Typical examples of such processes include electrophoresis, in which an electric field causes the separation, and ultracentrifugation, in which centrifugal force is used for the purpose. With the development of increasingly sophisticated techniques for both the separation and chemical analysis of compounds by chromatography there arose a need for a theory to better design experiments and to interpret experimental results. The earliest of such models were essentially diffusion equations of the form

$$\frac{\partial c}{\partial t} = D \frac{\partial^2 c}{\partial x^2} - v \frac{\partial c}{\partial x} \qquad (1)$$

where D is the diffusion constant and v the (constant) velocity that takes into account the force field[1]. Variations of Eq. (1) exist for more complicated geometries or forces[2], but the basic assumptions behind all of these are ones common to the theory of diffusion in a continuum.

The first attempt to produce a microscopic model for chromatographic kinetics was that of Giddings and Eyring[3]. They viewed the motion of a molecule through a chromatographic column as a series of sojourns in one of two phases which they called mobile and stationary. In their model transitions between these two states occur by a first order process. The molecules are assumed to travel at constant speed in the mobile state and are taken to be trapped and motionless in the stationary state. A similar model was discussed by van Holde[4]. The first passage time to reach the end of the column was chosen as the random variable of interest. It will be recognized that this is a simple version of the hopping transport model as described, say, in the review by Pfister and Scher[5], the major difference being that the chromatography model focusses on the first passage time problem while the usual hopping transport model describes transport and spreading in an infinite medium.

The solution to Eq. (1) in an infinite column and an initial condition of the form $c(x,0) = \delta(x)$ is easily seen to be a spreading Gaussian peak that travels at speed v down the column. If one calculates an elution profile it is sometimes observed that the measured peak is markedly asymmetric, in contradiction to the Gaussian result expected from the solution to Eq. (1). In this paper we examine the question of the conditions under which asymmetry can arise from the kinetics of the interchange between mobile and stationary states. For this purpose we derive and analyze a generalized transport equation. A further result will be an analysis of an associated first passage time problem, equivalent to the solution of a first passage time problem for the continuous time random walk (CTRW). The results will be described in terms of a generalized diffusion equation and its adjoint.

At the outset it should be stated that the kinetics of the two state picture, as initially discussed by Giddings and Eyring[1] and Sen Sarma, Anders, and Miller[6], must always lead to asymmetric peaks since chromatography columns are necessarily finite. If one considers a model in which there are two stationary states possibly corresponding to two kinds of trapping sites, then by making the rate constants (in a model with first order kinetics for the interchange between mobile and stationary states) very different one can induce considerable peak asymmetry as demonstrated by Keller and Giddings[7]. I will be interested in what I might term "intrinsic asymmetry" which refers to peaks that are asymptotically asymmetric in an infinite column. Any first order model which includes a fixed number of rate constants will be intrinsically symmetric in an infinite column. Therefore one wants to consider models that either differ from first order models or have an infinite number of rate constants. Both of these can be shown to lead to intrinsic asymmetry[8].

Analysis of Asymmetry

Let us start by considering an infinite one dimensional column in which an injected molecule can be either in the mobile state or it can be in any one of a possibly infinite number of stationary states. Let $c(x,t)$ be the concentration of mobile molecules at x, and let $c_i(x,t)$ be the concentration of molecules trapped in stationary state i at x. The total concentration at x will then be

$$\rho(x,t) = c(x,t) + \sum_i c_i(x,t) \tag{2}$$

For simplicity I consider the case of a uniform column only, characterized by a diffusion constant independent of time and space, and a uniform convective force. Under these assumptions the transport of matter can be affected only by molecules in the mobile state, hence we can write

$$\frac{\partial \rho}{\partial t} = D \frac{\partial^2 c}{\partial x^2} - v \frac{\partial c}{\partial x} \tag{3}$$

Finally, we must account for interchange between the mobile and stationary states. Since we will be interested in kinetic schemes more general than those embodied in first order, it is necessary to consider the adsorption and desorption steps from the i'th stationary phase in somewhat greater detail. For simplicity we will assume that the transition $M \rightarrow S_i$ is indeed first order, characterized by the rate constant k_i. The desorption event $S_i \rightarrow M$ will be assumed to be a more general process, the time for which is a random variable characterized by the probability density $\psi_i(t)$. This implies that the rate of change of molecules in S_i at x is governed by the equation

$$\frac{\partial c_i(x,t)}{\partial t} = k_i c(x,t) - k_i \int_0^t c(x,\tau) \psi_i(t - \tau) d\tau \tag{4}$$

The last term represents molecules that entered S_i at time τ and remained for in that state for exactly $t-\tau$ units of time.

Although the component concentrations c and c_i are required to specify the model at the microscopic level, only $\rho(x,t)$ would ordinarily be observable, hence one wants to derive an equation in terms of this variable alone. This is easily done through the introduction of Laplace transforms with respect to time. Let, for example, $\hat{\rho}(x,s) = \mathcal{L}\{\rho(x,t)\}$ with a similar notation for the remaining functions. If we assume that initially all molecules are in the mobile state then the Laplace transform of Eq. (4) is

$$\hat{c}_i(x,s) = \frac{k_i}{s}(1 - \hat{\psi}_i(s))\hat{c}(x,s) \tag{5}$$

which, together with Eq. (2), allows us to write

$$\hat{c}(x,s) = \hat{Q}(s)\hat{\rho}(x,s) \tag{6}$$

in which

$$\hat{Q}(s) = [1 + \frac{1}{s}\sum_i k_i(1 - \hat{\psi}_i(s))]^{-1} \tag{7}$$

Equation (6) is equivalent in the time domain to

$$c(x,t) = \int_0^t Q(t - \tau)\rho(x,\tau)d\tau \tag{8}$$

so that, according to Eq. (3), we can write as the transport equation

$$\frac{\partial \rho(x,t)}{\partial t} = \int_0^t Q(t - \tau)[D\frac{\partial^2 \rho(x,\tau)}{\partial x^2} - v\frac{\partial \rho(x,\tau)}{\partial x}]d\tau \tag{9}$$

The classical transport equation is obtained from this through the ansatz $Q(t) = \delta(t)$ so that there is no microscopic structure.

The Laplace transform of Eq. (9) allows us to answer questions related to the symmetry of a concentration peak moving down an infinite column. So far we have used the terminology of concentration, but one can also think about chromatographic processess by assuming that the initial concentration is unity, i.e., the normalized function $\rho(x,t)/\int\rho(x,t)dx$ can be regarded as a probability density. Henceforth I will regard $\rho(x,t)$ as such. Since the fluctuating state of a migrating molecular species insures peak spreading, one can obtain qualitatively similar effects to the complete transport equation by assuming that diffusion is negligible, which is equivalent to setting D=0 in Eq. (9). On the assumption that $\rho(x,0) = \delta(x)$, or pulse loading, one finds that the Laplace transform of Eq. (9) is

$$v\hat{Q}(s) \frac{\partial \hat{\rho}}{\partial x} + s\hat{\rho} = \delta(x) \qquad (10)$$

with the solution

$$\hat{\rho}(x,s) = \frac{1}{v\hat{Q}(s)} \exp\left(-\frac{sx}{v\hat{Q}(s)}\right) \qquad (11)$$

While this equation is difficult to invert in general it leads to a simple expression for the Laplace transform of the spatial moments

$$\mu_n(t) = \int_0^\infty x^n \rho(x,t)dx / \int_0^\infty \rho(s,t)dx \qquad (12)$$

It is easy to see that

$$\mu_n(s) = \frac{n!}{s^{n+1}} (v\hat{Q}(s))^n \qquad (13)$$

In order to determine whether the peak is asymptotically asymmetric we need a measure of this property. One measure of asymmetry using in statistics is the normalized third cumulant[9]

$$\gamma(t) = (\mu_3(t) - 3\mu_2(t)\mu_1(t) + 2\mu_1^3(t))/\sigma^3(t) \qquad (14)$$

where $\sigma^2(t) = \mu_2(t) - \mu_1^2(t)$. A symmetric peak is characterized by $\gamma=0$. We can determine asymptotic behavior of $\gamma(t)$ through an examination of the behavior of $\hat{\gamma}(s)$ as $s \to 0$. If the $\psi_i(t)$ have moments up to order 2 then, as $s \to 0$, $\hat{Q}(s)$ can be expanded in a Taylor series where the coefficients can be derived from Eq. (7);

$$\hat{Q}(s) = Q_0 + Q_1 s + Q_2 s^2 + \ldots \qquad (15)$$

In this case the combination of Eqs. (13)-(15) can be used to show that

$$\gamma(t) \approx \frac{3}{\sqrt{2}} \frac{(Q_1^2 + Q_0 Q_2)}{(Q_0 Q_1^3)^{1/2} t^{1/2}} \quad (16)$$

which tends to 0 as $t \to \infty$. Hence in this case the peak is asymptotically symmetric.

It is intuitively clear that the phenomenon of tailing requires that some molecules must remain in stationary states for a long time. One might think that this would require the average sojourn times in at least some stationary states to be infinite. However, this is not necessary. Even if the first moments are finite one can have asymptotic asymmetry provided that the second moments of sojourn times are infinite. This can be guessed from Eq. (16) in which the term Q_2 appears in the numerator. If we specifically assume that

$$\psi_i(s) = 1 - sT_i + a_i s^\alpha + \ldots \quad (17)$$

as $s \to 0$, where $1 < \alpha < 2$ then it follows from Eq. (7), that

$$Q(s) \approx Q_0 + Q_1 s^{\alpha-1} + \ldots \quad (18)$$

where now

$$Q_0 = (1 + \sum_i k_i T_i)^{-1}, \qquad Q_1 = Q_0^2 \sum_i k_i a_i \quad (19)$$

It is fairly straightforward to show using Eq. (13), that

$$\gamma(t) \approx \frac{3}{2^{3/2}} \left(\frac{Q_0}{Q_1}\right)^{1/2} \frac{(\alpha - 2)}{(\alpha - 1)^{1/2}} \Gamma^{1/2}(5 - \alpha) t^{\frac{\alpha-1}{2}} \quad (20)$$

as $t \to \infty$, so that the asymmetry increases with time.

These results are valid even when there is only a single state in the stationary phase, provided only that $Q(s)$ satisfies Eq. (7).

It is also possible to define a first order model in which the adsorption and desorption steps are first order and the peak exhibits asymptotic asymmetry. However, for this to happen one needs to assume that the rate constants are random variables. First order kinetics implies that $\hat{\psi}(s)=k/(k+s)$ where k is the rate constant appearing in the desorption step. Now a molecule trapped in a stationary state at x must be labelled with the appropriate adsorption and desorption rate constants, respectively. Hence instead of $c_i(x,t)$ we will have $c(k,k'; x,t)$. Let the joint probability density of (k,k') be denoted by $p(k,k')$. Then since Eq. (4) remains valid one finds that the Laplace transform of the concentration of stationary molecules is related to that of the mobile molecules by

$$<\hat{c}_{st}(k,k';x',s)> = c_m(x,s) \int_0^\infty\!\!\int \frac{kp(k,k')}{s+k'} dkdk' \qquad (21)$$

Consequently the transform $\hat{Q}(s)$ is

$$\hat{Q}(s) = (1 + \hat{U}(s))^{-1} \qquad (22)$$

in which

$$\hat{U}(s) = \int_0^\infty\!\!\int \frac{kp(k,k')}{s+k'} dkdk'$$
$$= \int_0^\infty \frac{f(k')}{s+k'} dk' \qquad (23)$$

where $f(k')=\int_0^\infty kp(k,k')dk$. I will suppose that $U(0)$ is finite, and investigate the conditions sufficient to insure that $\hat{Q}(s)$ can be expanded in the form of the series in Eq. (17). We write

$$\hat{U}(s) - \hat{U}(0) = -s\int_0^\infty \frac{f(k')dk'}{k'(s+k')} = - \int_0^\infty \frac{f(us)du}{u(u+1)} \qquad (24)$$

In order to insure that $U(s)-U(0) \sim As^\beta$ for $s \to 0$ where A is a constant and $0<\beta<1$ we need only postulate that $f(v) \sim Bv^\beta$ so that

$$\hat{U}(s) - \hat{U}(0) \approx -Bs^\beta \int_0^\infty \frac{u^{\beta-1}}{u+1}\,du = -\frac{\pi Bs^\beta}{\sin(\pi\beta)} \qquad (25)$$

This implies that the properties of low desorption rates are the main determinants of asymptotic asymmetry. This is intuitively reasonable since the very low desorption rates imply long residence times in a stationary state. Equation (25) gives a condition necessary to insure asymptotic peak asymmetry; undoubtedly there are others leading to the same conclusion.

First Passage Times

Having examined some properties of the generalized diffusion equation in Eq. (9) in an infinite medium we next turn our attention to deriving an equation for the first passage time to a point L for a molecule initially at x=0. In the case of an ordinary diffusion equation ($Q(t)=\delta(t)$ in Eq. (9)) it is well-known that the equation for the first passage time for a particle to diffuse from x to an absorbing point L, is the adjoint of the diffusion equation[10]. A parallel theory can be developed in the present case.

In order to see this let us reconsider the derivation of the generalized diffusion equation in Eq. (9) by passing to the appropriate limit starting from the equations governing a nearest neighbor continuous time random walk (CTRW) on a lattice[11]. We will deal only with the case of a homogeneous medium, although the more general inhomogeneous case can be dealt with by the same methods. One can observe that our basic model is a two state model in which the stationary phase is further divided into substates. Instead of considering the model in great detail, let us instead lump all of the stationary states into a single on with a sojourntime density $\psi(t)$ that is a composite of the $\psi_i(t)$. Let $a_r(t)dt$ ($b_r(t)dt$) be the joint probability of waiting a time between t and t+dt then stepping r→r+1 (r→r-1). We consider the simple case in which

$$a_r(t) = p\psi(t), \quad b_r(t) = q\psi(t)$$
$$p + q = 1 \tag{26}$$

where p and q are constants. Let $\omega_r(t)dt$ be the probability that a random walker coming from either direction reaches r at a time between t and t+dt. The $\omega_r(t)$ satisfy

$$\omega_r(t) = p \int_0^t \omega_{r-1}(\tau)\psi(t-\tau)d\tau + q \int_0^t \omega_{r+1}(\tau)\psi(t-\tau)d\tau \tag{27}$$

or, in Laplace transformed variables

$$\hat{\omega}_r(s) = \hat{\psi}(s)[p\hat{\omega}_{r-1}(s) + q\hat{\omega}_{r+1}(s)] \tag{28}$$

The next step in the analysis is to scale the space variable by letting x=rΔL where ΔL will tend to 0, and also make the assumption that p and q are scaled so that p-q=aΔL where <u>a</u> is a constant with the dimensions of inverse length. An expansion of Eq. (28) up to second order in $(\Delta L)^2$ allows us to write

$$\hat{\omega}(x,s) = \hat{\psi}(s)[\hat{\omega}(x,s) - a(\Delta L)^2 \frac{\partial \hat{\omega}}{\partial x} + (\Delta L)^2 \frac{\partial^2 \hat{\omega}}{\partial x^2}] \tag{29}$$

or

$$s\hat{\omega}(x,s) = \frac{s\hat{\psi}(s)}{1-\hat{\psi}(s)}[(\Delta L)^2 \frac{\partial^2 \hat{\omega}}{\partial x^2} - a(\Delta L)^2 \frac{\partial \hat{\omega}}{\partial x}] \tag{30}$$

This is equivalent to Eq. (9) provided we set sΔt=ρ, where ρ is now a dimensionless Laplace transform variable, and D, v, are defined by

$$D = \lim_{\Delta L, \Delta t \to 0} \frac{(\Delta L)^2}{\Delta t}, \quad aD = v \tag{31}$$

and Q(t) is found from

$$\mathcal{L}_\rho\{Q(t)\} = \frac{\rho\psi(\rho/\Delta t)}{1 - \psi(\rho/\Delta t)} \tag{32}$$

Thus $\psi(t)$ changes over a time scale that is proportional to Δt. For example when $\psi(t)=(1/T)\exp(-t/T)$,

$$Q(t) = (\Delta t/T)\delta(t) \tag{33}$$

and for this to lead to a physically sensible model we must have $T=O(\Delta t)$. Similarly when $\psi(t)$ corresponds to a stable law $\hat{\psi}(s)$ has the form, for $s \to 0$

$$\hat{\psi}(s) \approx \exp(-(sT)^\alpha) \tag{34}$$

and again the requirement $T=O(\Delta t)$ is necessary to produce a sensible diffusion approximation. This means that the development in time must take place on a sufficiently short time scale to produce a consistent continuum equation.

We next consider the first passage time problem for the process just described, by letting $U_j(t)$ be the probability density for the random walk to go from point j to an absorbing point N in time t provided that the random walk has just arrived at j (so that $\psi(t)$ is the appropriate sojourn density). It has been shown[11] that the $U_j(t)$ satisfy

$$U_j(t) = \int_0^t a_j(\tau) U_{j+1}(t-\tau) d\tau + \int_0^t b_j(\tau) U_{j-1}(t-\tau) d\tau \tag{35}$$

One can perform the same passage to the limits as led from Eq. (27) to Eq. (32), to find that $U(x,t)$ satisfies

$$\frac{\partial U(x,t)}{\partial t} = \int_0^t Q(t-\tau)[D \frac{\partial^2 U(x,\tau)}{\partial x^2} + v \frac{\partial U(x,\tau)}{\partial x}] d\tau \tag{36}$$

A generalized form of this equation has been derived by Hänggi and Talkner[12]. Equation (36) is to be solved subject to the boundary conditions $U(L,t)=\delta(t)$ at the absorbing boundary and $\partial U(x,t)/\partial x=0$ at a reflecting boundary. As an example we can consider the case in which $D=0$ so that the only cause of boundary spreading is due to the random amounts of time spent in the stationary phase. The Laplace transform $\hat{U}(x,s)$ satisfies

$$s\hat{U}(x,s) = v\hat{Q}(s) \frac{\partial \hat{U}}{\partial x} \qquad (37)$$

and is subject to the boundary condition $U(L,s)=1$. The general solution to this last equation at $x = 0$ is

$$\hat{U}(0,s) = \exp\left(- \frac{sL}{v\hat{Q}(s)}\right) \qquad (38)$$

and moments of the first passage time are generated by

$$T_n = (-1)^n \frac{d^n}{ds^n} \hat{U}(0,s)\Big|_{s=0} \qquad (39)$$

Thus, if $\hat{Q}(s)$ has the expansion given in Eq. (15) one finds that

$$T_1 = \frac{L}{vQ_0}, \quad \sigma^2 = \frac{2L}{v}\left(\frac{Q_1}{Q_0^2}\right) \qquad (40)$$

since there are no fractional powers of s with an exponent $\alpha<1$. When $\hat{Q}(s)$ has the expansion in Eq.(18) the mean first passage time is finite, but the variance of the first passage time is infinite. It is also possible to generate non integer moments of sufficiently low index from $U(0,s)$ but the use of such moments has not generally been explored.

The generalization of these analyses to deal with spatial dependence of D and v is not difficult and leads to the expected result that the functional appearing in brackets in Eq. (36) is the adjoint of the forward operator. The introduction of time-varying coefficients does appear to lead to more serious problems.

References

1. J.C. Giddings, Dynamics of Chromatography. Part I Principles and Theory (Marcel Dekker, Inc., New York) 1965.
2. H. Fujita, Foundations of Ultracentrifugal Analysis (John Wiley and Sons, New York) 1975.
3. J.C. Giddings, H. Eyring, J. Phys. Chem. 59, 416 (1955).
4. K.E. van Holde, J. Chem. Phys. 37, 1922 (1962).
5. G. Pfister, H. Scher, Adv. Phys. 27, 747 (1978).
6. R.N. Sen Sarma, E. Anders, J.M. Miller, J. Phys. Chem. 63, 559 (1959).
7. R.A. Keller, J.C. Giddings, J. Chromatogr. 3, 205 (1960).
8. G.H. Weiss, Sep. Sci. and Techn. 17, 1609 (1982-83).
9. M.G. Kendall, A. Stuart, The Advanced Theory of Statistics, Vol. 1. (Griffin, London) 1958.
10. G.C. Gardiner, A Handbook of Stochastic Processes (Springer-Verlag, New York) 1983.
11. G.H. Weiss, J. Stat. Phys. 24, 581 (1981).
12. P. Hanggi, P. Talkner, Zeit. Phys. B45, 79 (1981).

DIFFUSION, ANOMALOUS DIFFUSION, AND 1/f-NOISE IN CHAOTIC SYSTEMS

Theo Geisel

Institute for Theoretical Physics, University of Regensburg,
D-8400 Regensburg, Federal Republic of Germany

ABSTRACT

Dynamical systems with periodic symmetry can exhibit a transition from a localized chaotic motion to an unbounded chaotic motion. The latter may be characterized as a deterministic "random walk" occurring in the absence of external random forces. Examples of such systems are a driven damped particle in a periodic potential, driven Josephson junctions, and pinned charge-density-waves (CDW) under periodic excitation. Besides, these systems offer a rich variety of other dynamical phenomena, e.g. drifting periodic and chaotic motions.

The onset of diffusion shows analogies to a phase transition with universal scaling properties, where the diffusion coefficient plays the role of an order parameter. Here the mean-square displacement grows linearly in time. There are also special cases, in which the diffusion becomes anomalous.[1] Depending on universality classes the mean-square displacement grows like t^ν $(0 < \nu < 1)$, $t/\ln t$, or t. More recently other cases were found[2] where the growth is faster than linearly, i.e. like $t \ln t$, t^μ $(1 < \mu < 2)$, or t^2.

Experimentally these phenomena manifest themselves through typical power spectra. Their asymptotic form is given for the above cases and compared with experimental results on Josephson junctions, in particular with a recent observation of 1/f-noise.[3] This observation can be explained by an accelerated chaotic diffusion process.[2]

[1] T. Geisel, S. Thomae, Phys. Rev. Lett. 52, 1936 (1984)
[2] T. Geisel, J. Nierwetberg, A. Zacherl, Phys. Rev. Lett. 54, 616 (1985)
[3] R.F. Miracky, M. Devoret, J. Clarke, Phys. Rev. A 31, 2509 (1985)

LEVY WALKS IN CHAOS AND IN TURBULENCE

Michael F. Shlesinger
Office of Naval Research
Physics Division
800 North Quincy Street
Arlington, Virginia 22217

and

Joseph Klafter
Corporate Research Science Laboratory
Exxon Research & Engineering Company
Annandale, New Jersey 08801

ABSTRACT

We introduce a stochastic process called Levy walk which describes correlated motion in space and in time and has inherent scaling properties. For different scaling assumptions this process provides a statistical description of accelerated diffusion in Josephson junctions and diffusion in a turbulent fluid. In the latter case, the process relates Kolmogorov's scaling to Richardson's result in a natural fashion.

I. INTRODUCTION

The discovery of <u>dynamical</u> chaos has led some to predict the demise, or at least the downgrading, of <u>statistical</u> mechanics and <u>stochastic</u> processes as viable or useful descriptions of nature. We, on the other hand, feel that the complexities arising from dynamical equations of motion present new challenges to statistical physics. Some obvious examples are calculating diffusion constants for phase space motion in Hamiltonian systems which possess intrinsic stochasticity, predicting the asymptotic behavior of dissipative systems with fractal basin boundaries, and providing a proper statistical description of motion governed by strange attractors especially in high dimensions. From our perspective the interplay and mixing between nonlinear dynamics and stochastic processes will be very fruitful and exciting area of theoretical study.

In this manuscript we will discuss statistical representations of two nonlinear processes, accelerated phase diffusion in Josephson junctions and turbulent diffusion in fluids. Section II reviews the work of Geisel, Nierwetberg, and Zacherl[1] on accelerated diffusion in chaotic systems. Section III presents a stochastic process called a Lévy walk[2,3] with the same statistical properties of accelerated diffusion found in Section II. A change of scaling in the Lévy walk is shown in Section IV to reproduce the even more rapid diffusion described by Richardson's equation for turbulent diffusion[4-6]. This rapid diffusion as described by the Lévy walk relates Richardson's result to Kolmogorov's scaling[7].

II. ACCELERATED DIFFUSION IN CHAOTIC SYSTEMS

It was generally thought that diffusion, described by

$$\langle x^2(t)\rangle = Dt \tag{1}$$

was a property only of random systems. It is now known that such behavior can result for motion induced by deterministic mappings. For example, the map[8]

$$x_{t+1} = x_t - \mu \sin(2\pi x_t), \tag{2}$$

produces diffusive motion, $\langle(x_t-x_0)^2\rangle \sim Dt$ (the angular brackets indicate an average over initial conditions) when $\mu > \mu_{critical} \sim 0.732$. Mappings[9] can also produce anomalous diffusion where the diffusion constant is rigorously zero, i.e. $\langle(x_t-x_0)^2\rangle \sim t^\alpha$, $\alpha < 1$. This occurs if the map $f(x_t)$ closely approaches the 45° line in the plot of x_{t+1} vs. x_t, so the trajectory gets stuck for many iterations in the same region for a long time $T(x_0)$. This idea was introduced by Pomeau and Manneville[10] in their dynamical theory of intermittency, and used by Geisel and Thomae[9] to find the universal z behavior

$$\langle(x_t-x_0)^2\rangle \sim \begin{cases} t^{\frac{1}{(z-1)}} & , z > 2 \\ t & , 1 < z < 2 \end{cases} \quad (3)$$

for maps with the following behavior near the origin,

$$x_{t+1} = x_t + ax_t^z \quad \text{for } x_t \to o^+ \quad (4)$$

with $a > o$ and $z > 1$. The slower than diffusive behavior results because some initial conditions lead to trajectories which spend extremely long times in the region near the origin. Geisel, Nierwetberg, and Zacherl[1] have extended this analysis to mappings which have a near tangent to the 45° line, but in a neighboring unit cell at its boundary, (see Fig. 1). When the trajectory leaves the unit cell this represents a motion to a neighboring cell. Periodic boundary conditions bring the new initial position back to the original unit cell. In the previous analysis the time $T(x_0)$ spent near the origin now becomes the time spent in correlated (constant velocity) motion moving between unit cells. The mapping[1]

$$x_{t+1} = f(x_t)$$

where

$$f(x) = (1+\varepsilon)(x-m) + a(x-m)^z + m-1 \quad (m \lesssim x < m+1/2) \quad (5)$$

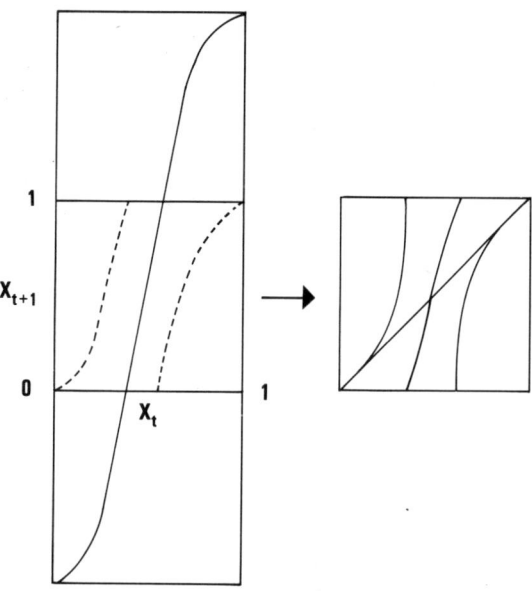

FIG. 1. The mapping of Geisel, Nierwetberg and Zacherl. The trajectory gets trapped for many iterations in the upper right and lower left hand corners and is transferred out of the unit cell successively resulting in a correlated motion that leads to accelerated diffusion for z > 3/2.

for $(x-m) \ll 1$ and $\epsilon \to 0$ when averaged over all initial leads to

$$\langle (x_t - x_0)^2 \rangle \sim \begin{cases} t^2, & z > 2 \\ t^{3 - \frac{1}{(z-1)}}, & \frac{3}{2} < z < 2 \\ t \ln t, & z = 3/2 \\ t, & 1 < z < 3/2 \end{cases} \qquad (6)$$

The cases for $z > 3/2$ are called accelerated diffusion. The mapping Eqn. (5) has been used to approximate the complex voltage phase behavior in resistively shunted Josephson junctions. In the next section we present a stochastic process with precisely the same behavior of Eqn. (6).

III. LÉVY FLIGHTS AND WALKS[2,3,11]

Our main concern here is to present a stochastic process which has the same four types of behavior found for the dynamical system in Eqn. (6). We first analyze a Lévy flight. We use this term to denote a random walker which jumps instantaneously between successively visited sites, however distant. The preface Lévy to the flight implies that the distribution of jump distances has an infinite second moment. The term Lévy walk describes the same random walk but with the walker forced to visit the intervening sites between successively visited sites of the Lévy flight. In this case the jump is not instantaneous, but has a finite velocity. See Figure 2. However, different length jumps may occur at different velocities.

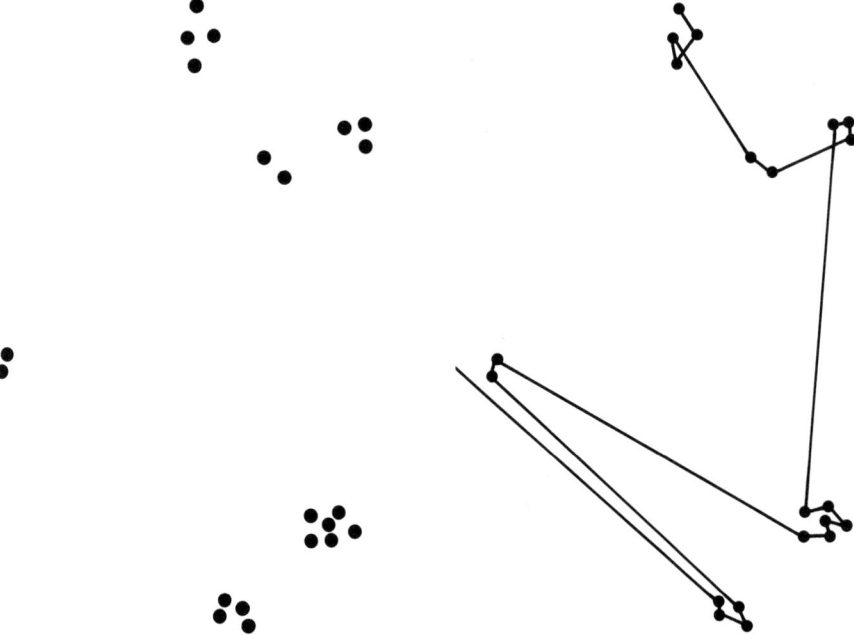

FIG. 2. The set of points on the left are those visited by a random Lévy flight and are characterized by a fractal dimension β. The points on the right are the same as those on the left, except that they are connected by straight lines. The exponent γ determines the velocity with which the straight lines are traversed.

To simplify the analysis of these stochastic processes we limit our discussion to a one dimensional periodic lattice.

Let $p(x)$ be the probability that a random walker, in a single step, has a displacement x. Also denote by $P_n(x)$ the probability that the walker, if originally at the origin, is at site x after n steps. Then

$$P_{n+1}(x) = \sum_{x'} P_n(x-x') \, p(x') . \tag{7}$$

Fourier transforming over the lattice sites, we find using the convolution theorem, that

$$\tilde{P}_n(k) = \sum_x e^{ikx} P_n(x) = [p(k)]^n$$

or equivalently,

$$P_n(x) = \frac{1}{2\pi} \int_0^{2\pi} [p(k)]^n e^{-ikx} dk \tag{8}$$

For jumps restricted to nearest neighbors with equal probability

$$p(x) = \frac{1}{2} [\delta_{x,+1} + \delta_{x,-1}]$$

and

$$p(k) = \cos k \sim 1 - \frac{1}{2} k^2 \text{ as } k \to 0 . \tag{9}$$

For any unbiased jump distribution, with a finite variance $\overline{x^2} \equiv \sum_x x^2 p(x)$, one can always expand $p(k)$ as in Eqn. (9)

$$\begin{aligned} p(k) = \sum_x e^{ikx} p(x) &\sim 1 - \frac{1}{2} \overline{x^2} k^2 \text{ as } k \to 0 \\ &\sim \exp(-\frac{1}{2} \overline{x^2} k^2) . \end{aligned} \tag{10}$$

Substituting this $p(k)$ in Eqn. (8) for $P_n(x)$ yields the standard gaussian behavior,

$$P_n(x) \sim (2\pi n \, \overline{x^2})^{-1/2} \exp(-x^2/2n \, \overline{x^2}) \tag{11}$$

in accord with the Central Limit Theorem.

When the vogue in probability theory was to see what were the weakest conditions under which gaussian behavior would result, P. Lévy sought exceptions to the Central Limit Theorem. He eventually considered random walk processes where $\overline{x^2}$ was infinite, so the trajectory possessed no largest length scale. If for large x,

$$p(x) \sim |x|^{-1-f}, \text{ with } 0<f<2$$

then $\overline{x^2}$ is infinite, and[10]

$$\begin{aligned} p(k) &\sim 1 - \text{const.} |k|^f \text{ as } k \to 0 \\ &\sim \exp(-\text{const.} |k|^f) \end{aligned} \quad (12)$$

and

$$P_n(x) \sim \frac{1}{2\pi} \int_0^2 \exp(-\text{const.} |k|^f) \exp(-ikx) dk. \quad (13)$$

The probability $P_n(x)$ is known as a Levy density and its integral representation can only be calculated analytically for a few select cases, e.g. f = 1 corresponds to

$$P_n(x) = \frac{1}{\pi} \frac{An}{x^2 + (An)^2} \quad (14)$$

where A is a constant. The exponent f can be shown[11] to be the fractal dimension of the set of points visited by the random walker. For the Lévy flight governed by Eqns. (12) and (13) the mean square displacement,

$$\langle x^2(t) \rangle \equiv \sum_x x^2 P(x,t) = \infty \quad (15)$$

since $\overline{x^2}$ is infinite for a single jump. How can Eqn. (15) be transformed to reproduce the mapping results of Eqn. (6)?

A manner in which to obtain a finite time dependent result for $\langle x^2(t) \rangle$ is to account for the time taken to move between the initial and final positions of a jump. We will assume that the random walker moves with a constant velocity, i.e. longer jumps take a longer time. Different size jumps may have different velocities. To properly account for this feature we must generalize Eqn. (7) to

$$Q(x,t) = \sum_{x'} \int_0^t Q(x-x', t-\tau) \Psi(x',\tau) \, d\tau + \delta_{x,0}\delta(t) \qquad (16)$$

where $Q(x,t)$ is the probability for reaching site x <u>exactly</u> at time t for a random walker starting at the origin at $t = 0$. The quantity $\Psi(x,t)$ is the probability density for the random walker to reach site x at time t in a single jump. Thus one can reach site x at time t by first reaching site $x-x'$ at time $t-\tau$ (accounting for all possible pathways) and then in the remaining time τ the random walker makes a jump of length x'. All possible intermediate sites are summed over.

We write the memory function $\Psi(x,t)$ as

$$\Psi(x,t) = \psi(t|x)p(x) \qquad (17)$$

where $p(x)$ is the probability of making a jump of length x, or in a different interpretation making x correlated jumps of unit length all in the same direction.
We further write

$$\psi(t) = \sum_x \Psi(x,t) \qquad (18)$$

where $\psi(t)dt$ is the probability that if a jump ended at time $t = 0$ then the next jump terminates in the time interval $(t, t+dt)$. The probability $P(x,t)$ for being at site x at time t is somewhat different from $Q(x,t)$ because the walker might reach site x at an earlier time $t-\tau$ and then not move for a time τ, i.e.

$$P(x,t) = \int_0^t Q(x, t-\tau) \int_\tau^\infty \psi(t') \, dt' \, d\tau \qquad (19)$$

Both Fourier and Laplace transforming Eqn. (19) we find

$$\sum_x \int e^{-st} e^{ikx} P(x,t) dt - \tilde{P}^*(k,s) = \tilde{Q}^*(k,s) \frac{1-\psi^*(s)}{s} \qquad (20)$$

and from Eqn. (16)

$$\tilde{Q}^*(k,s) = \frac{1}{1-\tilde{\Psi}^*(k,s)} . \qquad (21)$$

For a random walk without a bias

$$\int_0^\infty e^{-st} \langle x^2(t) \rangle dt \equiv \sum_x x^2 P*(x,s)$$

$$= -\frac{\partial^2}{\partial k^2} \tilde{P}*(k=0,s)$$

$$= \frac{1}{s[1-\psi*(s)]} \frac{\partial^2}{\partial k^2} \tilde{\Psi}*(k=0,s) \qquad (22)$$

Let us choose

$$p(x) \sim \frac{1}{|x|^{2+\beta}} \text{ as } x \to \infty \qquad (23a)$$

and

$$\psi(x|t) = \delta(|x|-t) \qquad (23b)$$

Note $\psi*(s) = \sum_x \Psi*(x,s) = \int_0^\infty \exp(-st) p(t) dt$.

For $< \beta \leq 1$

$$\psi*(s) \sim 1 - s\langle t \rangle + \text{const. } s^{1+\beta} \text{ as } s \to 0 \qquad (24)$$

where $\langle t \rangle \equiv \int_0^\infty t \psi(t) dt$ is the average time spent on a jump. For $-1 < \beta < 0$, $\langle t \rangle$ is infinite and

$$\psi*(s) \sim 1 - \text{const. } s^{1+\beta}, \text{ as } s \to 0. \qquad (25)$$

For these two β regimes, $-1 < \beta \leq 1$,

$$\partial^2 \tilde{\Psi}*(k=0,s)/\partial k^2 \sim s^{\beta-1} \qquad (26)$$

Substituting Eqns. (23) - (26) in the expression for the mean square displacement, Eqn. (22), gives

$$\langle x^2(t)\rangle \sim \begin{cases} t^2, & -1 < \beta \leq 0, \langle t\rangle = \infty \\ t^{2-\beta}, & 0 < \beta < 1, \langle t\rangle < \infty, \langle t^2\rangle = \infty \\ t \ln t, & \beta = 1, \langle t^2\rangle \text{ log divergent} \\ t, & \beta > 1, \langle t^2\rangle < \infty \end{cases} \quad (27)$$

This is the equivalent of the dynamical result if we set

$$\beta = \frac{2-z}{z-1} \quad (28)$$

The t^2 behavior originates from the first moment of $\psi(t)$ diverging, the next two cases originate from $\langle t\rangle$ finite, but $\langle t^2\rangle$ infinite, and the standard Brownian motion result corresponds to $\langle t^2\rangle$ finite.

In the Josephson junction mapping if the voltage phase rotates, say, counterclockwise, n times, this corresponds to the Levy walker taking n correlated steps, say to the right, while clockwise rotations would correspond to correlated steps to the left. The t^2 result should not be misinterpreted to mean that wave motion is occurring. In fact, the motion is quite random and involves correlated jumps over an infinite number of scales. The meaning of $\langle t\rangle$ being infinite is that there is no largest characteristic time scale (and in this case also correlated jump distances have no largest scale), as can be seen in the following pedogogical example.
Choose

$$\psi(t) = \frac{1-N}{N} \sum n^j b^j \exp(-tb^j), \quad N,b < 1 \quad (29)$$

where jumps occur on an infinite of time scales b^{-1}, b^{-2}, b^{-3}, ...etc., but an order of magnitude longer time scale in base b^{-1} occurs with an order of magnitude less probability in base N^{-1}. About N^{-1} events occur on a time scale of b^{-1} before an intermittent interval of length b^{-2} occurs. Then about N^{-1} events separated in time by an interval b^{-1} occur again and this happens about N^{-1} times before an intermittent interval of length b^{-3}

occurs, and so on with intermittences of all time scales entering in a self-similar manner. These intermittent intervals can represent the time a random walker spends in a correlated motion, and one expects the set of event times to have a fractal dimension of log N/log b. To see this analytically let us Laplace transform Eqn. (29) to obtain

$$\psi^*(s) = \frac{1-N}{N} \sum_{j=1}^{\infty} \frac{(Nb)^j}{b^j + s} = N\psi^*(s/b) + (1+N)\frac{b}{b+s} \qquad (30)$$

The homogeneous part of this equation,

$$\psi^*(s) = N\psi^*(s/b)$$

has the solution

$$\psi^*(s) = s^f K(s)$$

where

$$f = \ln N / \ln b$$

and $K(s)$ is a function periodic in $\ln s$ with period $\ln b$. It can be shown[10] that f is the fractal dimension of the set of times marking the switching between different rotation states.

IV. RICHARDSON'S LAW

At first glance it appears that $\langle x^2(t) \rangle = t^2$ is the fastest possible motion for the Levy walker because it involves an infinite correlation length. This is true if all the jumps occur at the same constant velocity. If however, a distribution of velocities exists then one can arrive at a more rapid diffusion which is associated with turbulence.

Let us choose in contrast to Eqn. (23b),

$$\psi(x|t) = \delta(|x| - t^\gamma). \qquad (31a)$$

$$\text{with } \gamma = 3/2 \qquad (31b)$$

but keep the same form of p(x) as in Eqn. (23a). This corresponds to

Kolmogorov's scale dependent velocity argument[7] where $v(x) \sim x^{1/3}$. Eq. (31a) can then be expressed as $\delta(x-v(x)t)$.

Analysis similar to the calculations of Section III show that if

$$-1 < \beta < -1/3$$

then

$$\langle x^2(t) \rangle \sim t^3 \qquad (32)$$

Eqn. (32) was given by Richardson in 1926 to describe turbulent diffusion. In our approach we envision a random walker in a distribution of vortices. The distribution of correlation lengths $p(x)$ corresponds to the distribution of vortex sizes. Larger vortices have more energy and a larger velocity so the time t to complete a correlated motion of length x is proportional to $x^{2/3}$ rather than x as in the Josephson junction accelerated diffusion problem.

Richardson suggested that

$$\frac{\partial P(x,t)}{\partial t} = \frac{\partial}{\partial x} K(x) \frac{\partial P(x,t)}{\partial x} \qquad (33)$$

with $K(x) = \text{const.} \; x^{4/3}$. This scaling leads to

$$\frac{d\langle x^2(t) \rangle}{dt} = [\langle x^2(t) \rangle^{1/2}]^{4/3} \qquad (34)$$

Richardson's famous "4/3" law which is equivalent to Eqn. (32). Our random walk picture is equivalent to a non-local in space and time master equation,

$$\frac{dP(x,t)}{dt} = \int_{x'} (P(x-x') \; t-\tau) \; \Phi(x',\tau)$$
$$- P(x,t-\tau) \; \Phi(x',\tau)) d\tau \qquad (35)$$

where[10]

$$\frac{\Phi^*(k,s)}{s + \tilde{\Phi}^*(k=0,s)} = \tilde{\Psi}^*(k,s) \tag{36}$$

Correction to Richardson's law due to the breakdown of the concept of homogeneous turbulence because the turbulence is concentrated on a fractal set d_f has been suggested by Mandelbrot[12]. If D is the euclidean dimension then it has been concluded[13] that the Richardson 4/3 is replaced by $(4+2(D-d_f))/3$. Such corrections can be incorporated into the Lévy walk by choosing $\gamma > 3/2$ in Eqn. (31b). Then if $\gamma(1+\beta) < 2$ the behavior

$$\langle x^2(t) \rangle \sim t^{2\gamma} \tag{37}$$

ensues asymptotically. Several slower behaviors, down to the Brownian motion result, can be obtained for different choices of γ and β, as we saw in the equation for the accelerated diffusion example in Eqn. (27). Much further work remains to be completed for the statistical analysis of a Lévy walk restricted to various fractal sets.

CONCLUSION

The Lévy walk with a constant velocity was able to reproduce the statistical behavior of accelerated diffusion which has also been derived from a deterministic one dimensional map. Lévy walks with a distribution of velocities correlated to vortex sizes can generate the proper mean square displacement of a particle in a turbulent fluid. It would be of interest to explore if a dynamical map could also achieve this last result.

REFERENCES

1. T. Geisel, J. Nierwetberg, and A. Zacherl, Phys. Rev. Lett. **54**, 616 (1985).
2. M. F. Shlesinger and J. Klafter, Phys. Rev. Lett. **54**, 2551 (1985)
3. M. F. Shlesinger and J. Klafter, in Proceedings of the Cargese Summer School "On Growth and Form: A Modern View" eds. H. E. Stanley and N. Ostrowski (in press).

4. L. F. Richardson, Proc. Roy. Soc. London A110, 709 (1926).
5. H. G. E. Hentschel and I. Procaccia, Phys. Rev. A27, 1266 (1983).
6. F. Wegner and S. Grossman, Zeits. für Physik B59, 197 (1985).
7. A. N. Kolmogorov, C. R. (Dokl.) Acad. Sci., USSR 30, 301 (1941).
8. T. Geisel and J. Nierwetberg, Phys. Rev. Lett. 48, 7 (1982).
9. T. Geisel and S. Thomae, Phys. Rev. Lett. 52 1936 (1984).
10. Y. Pomeau and P. Manneville, Comm. Math. Phys. 74 189 (1980).
11. E. W. Montroll and M. F. Shlesinger, in Studies in Statistical Mechanics, Vol. 11, eds. J. Lebowitz and E. W. Montroll, (North-Holland, Amsterdam) 1984.
12. B. B. Mandelbrot on Turbulence and the Navier-Stokes Equation, ed. R. Teman (Springes, Berlin 1976).
13. H. G. E. Hentschel and I. Procaccia, Phys. Rev. Lett. 49, 1158 (1982).